Lean Six Sigma for the Healthcare Enterprise

Sandra Furterer

Lean Six Sigma for the Healthcare Enterprise

Methods, Tools and Applications

Second Edition

Springer

Sandra Furterer
Integrated Systems Engineering
The Ohio State University
Columbus, OH, USA

ISBN 978-1-4471-7616-9 ISBN 978-1-4471-7617-6 (eBook)
https://doi.org/10.1007/978-1-4471-7617-6

This Springer imprint is published by the registered company Springer-Verlag London Ltd., part of Springer Nature.
The registered company address is: The Campus, 4 Crinan Street, London, N1 9XW, United Kingdom

I dedicate this book to my loving family, my husband Dan, and my children Kelly, Erik, and Zachary who provide constant support and the true purpose in my life. My children provide the incentive for me to strive for excellence to be a guiding example for their lives.

I also dedicate this book to my parents, Joan, and the late Mel Brumback, who instilled in me the value of education and a lifelong love of books and learning, and also to my siblings, Dan, Neil, Tim, and Kathy, who encouraged the competitive spirit and being our best in all we do.

Preface

The purpose of this book is to provide an enterprise Lean Six Sigma methodology to enable alignment and optimization of processes across the entire organization. Healthcare has finally embraced Lean Six Sigma tools, even though it has been much later than many industries, there is great value and advantage for applying these tools and methods in healthcare. Healthcare is struggling with shrinking revenues and trying to control costs, what better environment could there be to apply Lean Six Sigma.

In the second edition of this book, we added new case studies in Chaps. 8, 10, 13, and 14. Chapters 13 and 14 provide insights and applications of advanced technologies for improving processes including modeling and simulation and Artificial Intelligence (AI). We also updated the figures throughout the text to enhance visibility and accessibility. In Chap. 1, we provide a description of the author's Strategic Business Process Architecture methodology that can enable enterprise-wide improvement. In Chap. 2, we provide a description of the Lean Six Sigma methodology that can guide our process improvement projects. In Chap. 3, we provide an application of Lean Six Sigma that successfully reduced the patient Length of Stay and Door to Doctor time in a hospital's Emergency Department. In Chaps. 4 and 5, we provide an application of Lean Six Sigma and 5S in an operating room. In Chap. 6, we provide an application of Lean tools applied for preparing patients for surgery. In Chap. 7, we provide an application of how Lean Six Sigma was used to reduce the number of CT scans in an Emergency Department. In Chap. 8, we describe how we applied Lean Six Sigma to optimize light levels for fragile newborns in a NICU. In Chap. 9, we used Lean Six Sigma to design and implement sepsis protocols to reduce the mortality related to sepsis infections. In Chap. 10, we improved satisfaction and timeliness in interventional radiology processes through Lean Six Sigma. In Chap. 11, we provide an innovative Patient Care Operating Model that can reduce delays in the inpatient value chain activities. In Chap. 12, we discuss labor management flaws and best practices in healthcare. In Chap. 13, we applied modeling and simulation to reduce time and cost of implementing process improvements. In Chap 14, we show two applications of incorporating Artificial Intelligence for improving processes in healthcare. In Chap 15, we provide insights into several critical success factors that can enhance your organization's Lean Six Sigma program and a brief view into the future of Lean Six Sigma in healthcare. These case studies are representative of healthcare case studies, but generalized to protect the confidentiality of specific healthcare organizations.

Columbus, USA Sandra Furterer

Acknowledgments My thanks and appreciation goes to all of the Lean Six Sigma team members, project champions, and Lean Six Sigma mentors that worked so diligently and courageously on the Lean Six Sigma projects that enabled us to enhance and improve the patients' experience in the healthcare environments. As with any successful operational excellence program, there are so many wonderful participants that guide, mentor, champion, encourage, and make change happen, I thank you for your dedication to helping make the patients' experience in the hospitals one of healing, compassion, and care.

Special thanks go to several people for their contributions to development and production of the book at Springer UK.

Competing Interests The author has no competing interests to declare that are relevant to the content of this manuscript.

Contributors

Erik Furterer
Emily Kurz
Theodore Allen
Nathalia C. P. Sorgenfrei

Contents

1	**Enabling Enterprise Performance Excellence Through Strategic Business Process Architecture Modelling Techniques**	1
	Introduction	1
	Underlying Concepts Evolving to Business Architecture	2
	Business Architecture	3
	Strategic Business Process Architecture Methodology	4
	Phase I: Enterprise Planning	5
	Phase II Strategic Business Process Architecture Modeling	7
	Lean Six Sigma	9
	Conclusions	10
	References	10
2	**Lean Six Sigma Roadmap**	11
	Lean Six Sigma Overview	11
	Lean Six Sigma Applications in Private Industry	13
	Define Phase	14
	Develop Project Charter	15
	Perform Stakeholder Analysis	17
	Perform Initial Voice of Customer (VOC) and Identify Critical to Satisfaction (CTS)	18
	Select Team and Launch the Project	22
	Create Project Plan	23
	Summary	26
	Measure Phase	27
	Define the Current Process	27
	Define Detailed Voice of Customer (VOC)	29
	Define the Voice of the Process (VOP) and Current Performance	36
	Validate the Measurement System	39
	Define the Cost of Poor Quality (COPQ)	40

Analyze Phase ... 41
Develop Cause-And-Effect Relationships ... 42
Determine and Validate Root Causes ... 44
Develop Process Capability ... 54
Improve Phase ... 57
Identify Improvement Recommendations ... 57
Perform Cost/Benefit Analysis ... 59
Design Future State ... 60
Establish Performance Targets and Project Scorecard ... 60
Gain Approval to Implement, then Implement ... 61
Train and Execute ... 61
Control Phase ... 62
Measure Results and Manage Change ... 62
Report Scorecard Data and Create Process Control Plan ... 63
Apply P-D-C-A Process ... 65
Identify Replication Opportunities ... 65
Develop Future Plans ... 66
Summary ... 66
References ... 66

3 Hospital and Emergency Department Throughput Improvement ... 69
Introduction ... 69
Define Phase ... 70
Develop Project Charter ... 71
Perform Stakeholder Analysis ... 73
Perform Initial Voice of Customer and Identify Critical to Satisfaction ... 73
Select Team and Launch the Project ... 75
Create Project Plan ... 77
Measure Phase ... 77
Define the Current Process ... 77
Define Detailed Voice of Customer ... 78
Define the Voice of Process and Current Performance ... 78
Validate Measurement System ... 83
Analyze Phase ... 84
Develop Cause-And-Effect Relationships and to Determine and Validate Root Causes ... 84
Develop Process Capability ... 92
Improve Phase ... 99
Identify Improvement Opportunities and Plans ... 99
Perform Cost/Benefit Analysis ... 103
Design Future State ... 104

Establish Performance Targets and Project Scorecard 104
Pilot Improvements 105
Train and Execute 107
Control Phase 110
Measure Results and Manage Change 110
Report Scorecard Data and Create Process Control Plan 111
Identify Replication Opportunities 111
Develop Future Plans 112
Conclusions 113
Project Critical Success Factors 113
Reference 113

4 Operating Room Turnaround and On-Time Starts Improvement 115
Lean Six Sigma Project for Operating Room (OR) Turnaround and On-Time Starts of Cases 116
Define Phase 116
Develop Project Charter 116
Perform Stakeholder Analysis 117
Identify Critical to Satisfaction 120
Select Team and Launch the Project 123
Create Project Plan 123
Measure Phase 124
Define the Current Process 125
Define Detailed Voice of Customer 130
Define the Voice of Process and Current Performance 130
Validate Measurement System 136
Analyze Phase 137
Develop Cause-and-Effect Relationships 137
Determine and Validate Root Causes 138
Develop Process Capability 138
Improve Phase 144
Identify Improvement Opportunities and Plans 144
Perform Cost/Benefit Analysis 145
Design Future State 145
Establish Performance Targets and Project Scorecard 145
Pilot Improvements 146
Control Phase 152
Measure Results and Manage Change 152
Report Scorecard Data and Create Process Control Plan 158
Identify Replication Opportunities 158
Develop Future Plans 160
Summary and Project Critical Success Factors 161
Reference 162

5 Applying 5S to Improve OR Organization 163
Introduction 163
Define Phase 164
Project Charter 165
Stakeholder Analysis 167
Suppliers-Inputs-Processes-Outputs-Customers 167
Project Kick off 167
Measure Phase 169
Analyze Phase 173
Improve Phase 175
Control Phase 176
Create 5S Committee 176
Create Audit Control Plan and Schedule 176
Periodic Spring-Cleaning Schedule 177
Assess Improvement 177
Conclusions 179
Reference 188

6 Leaning the Surgical Outpatient Screening Process in an Acute Care Hospital 189
Define Phase 189
Develop Project Charter 190
Perform Stakeholder Analysis 192
Perform Initial Voice of Customer and Identify Critical to Satisfaction 194
Select Team and Launch the Project 194
Create Project Plan 195
Measure Phase 196
Define the Current Process 196
Define Detailed Voice of Customer (VOC) 198
Define the Voice of Process (VOP) and Current Performance 198
Validate Measurement System 200
Analyze Phase 201
Develop Cause and Effect Relationships and Determine and Validate Root Causes 201
Develop Process Capability 205
Improve Phase 205
Identify Improvement Opportunities and Plans and Design Future State 205
Design Future State 221
Establish Performance Targets and Project Scorecard 221
Pilot Improvements and Train and Execute 221
Control Phase 222
Measure Results and Manage Change 222
Report Scorecard Data and Create Process Control Plan 224

Identify Replication Opportunities ... 225
Develop Future Plans ... 225
Project Critical Success Factors ... 225
Summary ... 225
References ... 226

7 **Appropriate Use of CT Scan in the Emergency Department** ... 227
Introduction ... 227
Define Phase ... 229
Develop Project Charter ... 229
Perform Stakeholder Analysis ... 231
Identify Critical to Satisfaction ... 232
Select Team and Launch the Project ... 233
Create Project Plan ... 233
Measure Phase ... 233
Define the Current Process ... 234
Define Detailed Voice of Customer ... 237
Define the Voice of Process and Current Performance ... 238
Analyze Phase ... 241
Develop Cause and Effect Relationships and Determine and Validate Root Causes ... 241
Improve Phase ... 242
Identify Improvement Opportunities and Plans and Design Future State ... 242
Pilot Improvements ... 246
Train and Execute ... 250
Control Phase ... 250
Measure Results and Manage Change ... 250
Report Scorecard Data and Create Process Control Plan ... 250
Identify Replication Opportunities ... 251
Develop Future Plans ... 251
Conclusions and Critical Success Factors ... 252
References ... 252

8 **Applying Lean Six Sigma to Optimize Light Levels in the NICU** ... 255
Introduction ... 255
Methodology ... 256
DMAIC Methodology Applied ... 257
Define Phase ... 257
Measure Phase ... 258
Analyze Phase ... 263
Improve Phase ... 266
Control Phase ... 268
Managerial Implications ... 270

Lessons Learned and Key Challenges ... 270
Conclusions ... 271
Appendix of Results ... 271
Reference ... 274

9 Sepsis Protocol Design Project ... 277
Introduction ... 277
Define Phase ... 278
Develop Project Charter ... 278
Perform Initial Voice of Customer and Identify Critical to Satisfaction ... 281
Select Team and Launch the Project ... 281
Create Project Plan ... 281
Measure Phase ... 281
Define the Current Process ... 282
Define Detailed Voice of Customer ... 285
Define the Voice of Process and Current Performance ... 286
Validate Measurement System ... 286
Analyze Phase ... 287
Develop Cause-and-Effect Relationships and Validate Causes ... 287
Develop Process Capability ... 288
Improve Phase ... 289
Identify Improvement Opportunities and Plans and Design Future State ... 289
Establish Performance Targets and Project Scorecard ... 293
Pilot Improvements, and Train and Execute ... 293
Control Phase ... 294
Measure Results and Manage Change ... 294
Report Scorecard Data and Create Process Control Plan ... 299
Identify Replication Opportunities ... 299
Develop Future Plans ... 299
Summary and Project Critical Success Factors ... 299
References ... 299

10 Improving Satisfaction and Timeliness in Interventional Radiology Processes Through Lean Six Sigma ... 301
Introduction ... 301
Define Phase ... 302
Measure Phase ... 307
Analyze Phase ... 319
Improve Phase ... 325
Control Phase ... 338
Conclusions ... 340

11 Innovating the Patient Care Operating Model: Integrated Value Care 343
Define Phase 344
Develop Project Charter 344
Perform Stakeholder Analysis 346
Perform Initial Voice of Customer and Identify Critical to Satisfaction 349
Measure Phase 350
Define the Current Process 350
Define Detailed Voice of Customer 351
Define the Voice of Process and Current Performance 355
Analyze Phase 357
Develop Cause-and-Effect Relationships and Determine and Validate Root Causes 357
Why are There Delays in the Inpatient Processes? 357
Improve Phase 370
Identify Improvement Opportunities and Plans, and Design Future State 370
Control Phase 383
Measure Results and Manage Change 383
Create Scorecard Data Framework and Create Process Control Plan 384
Conclusions 384
Reference 384

12 Labor Management in Healthcare: Flaws and Best Practices 385
Introduction to Labor Standards 385
Healthcare Development of Labor Standards 387
Standards Based on Organization Structure, Not Work Performed 387
Standard Setting Based on Benchmark Databases Not Based on Actual Work Performed 388
Fixed Departments Expected to Fluctuate with Patient Volume 389
Standard is Developed Based on Aggregate Volume, But Work is Managed Daily 389
Summary of the Fatal Flaws of Healthcare Productivity Setting Practices 390
Recommended Standard Development Methodology 392
Conclusions 393
References 393

13 Applying Modeling and Simulation to Reduce Time and Cost of Implementing Process Improvements ... 395
Introduction ... 395
Modeling and Simulation Background ... 396
Helpful Terms and Definitions ... 396
Modeling and Simulation Approach/Method ... 397
Common Modules for Modeling ... 398
Modeling and Simulation Applied to an Emergency Department Case Study ... 401
Define Phase ... 401
Measure Phase ... 402
Design Phase ... 405
Validate Phase ... 409
Implement Phase ... 411
Conclusions ... 421
Reference ... 421

14 Applications of Artificial Intelligence for Improving Processes in Healthcare ... 423
Introduction ... 423
Applications of AI in Healthcare ... 424
Challenges of AI in Healthcare ... 424
Ethical Challenges with the Use of AI ... 425
Examples of Application of AI Tools for Enhancing Process Improvement in Healthcare ... 426
Application of the Co-Pilot AI Tool with a Value Stream Map (VSM) for the Emergency Department Throughput ... 426
Application of the Co-Pilot AI Tool with a Process Architecture Map (PAM) for the Operating Room (OR) Turnaround and On-Time Starts ... 440
Conclusions ... 451
References ... 452

15 Critical Success Factors and Future of Enterprise Performance Excellence ... 453
Critical Success Factors for Enterprise Performance Excellence ... 453
Introduction ... 453
Key Components ... 454
Value and Strategic Alignment ... 454
Customer Focus ... 457
Culture and Change Management ... 458
Teamwork and Engagement ... 459
Infrastructure and Methodology ... 460
Quality and Lean Tools ... 463
Measurement (Metrics) ... 464

The Future and Challenge of Lean Six Sigma in Healthcare 464
Information Technology that Enables Processes, Services, and Measurement 466
New and Varied Problem-Solving Approaches 467
Process Innovation Through Design for Six Sigma and TRIZ 468
Customer Focus and Engagement: Voice of the Customer and Quality Function Deployment 468
Teamwork and Engagement 468
Quality Management System Incorporating Business Process Management 468
Conclusions 469
References 469

Index 471

About the Author

Dr. Sandra Furterer, Ph.D., MBA, CSSBB, Master Black Belt, CQE, CMQ/OE is a ***Professor of Practice*** at The Ohio State University, in the Department of Integrated Systems Engineering. She is the Curriculum Director for the College of Engineering's Bachelor of Science in Engineering Technology program. She was a tenured Associate Professor in the Engineering Management Systems and Technology Department at the University of Dayton. She teaches online courses as a part-time instructor for Kennesaw State University and California State University Dominguez Hills. Dr. Furterer was an Assistant Professor in the Industrial Distribution and Logistics program at East Carolina University from 2006 to 2007. She was a Visiting Assistant Professor and Assistant Department Chair in the Industrial Engineering and Management Systems Department at the University of Central Florida from 2004 to 2006.

She has applied Lean Six Sigma, Systems Engineering, and Engineering Management tools in healthcare and other service industries. Dr. Furterer designed and deployed the Enterprise Performance Excellence program for a community-based acute care hospital system. Dr. Furterer led the Business and Process Architecture team in the Retail Systems application development area of a major retailer, implementing best practices to achieve operational excellence in information systems application development processes, and helping to enable the business through application of Business Architecture and process improvement. She also helped to develop and deploy a Lean Six Sigma training and mentoring program for the retailer in its corporate headquarters.

Dr. Furterer received her Ph.D. in Industrial Engineering with a specialization in Quality Engineering from the University of Central Florida in 2004. She received an MBA from Xavier University, and a Bachelor and Master of Science in Industrial and Systems Engineering from The Ohio State University. She is an ASQ Certified Six Sigma Black Belt, Certified Manager of Quality/Organizational Excellence, Certified Quality Engineer, an ASQ fellow, and a certified Six Sigma Master Black Belt, a Senior Member in IISE, and recent past president in IISE's Society of Engineering and Management Systems (SEMS).

Dr. Furterer is an author or co-author of over 40 academic journal articles, over 50 conference proceedings, and 11 reference textbooks on Lean Six Sigma, Design for Six Sigma, Lean Systems, and Systems Engineering, including being the Co-Editor for the *ASQ CQIA Handbook* and the *ASQ CMQ/OE Handbook*, and the Editor of the *ASQ CQPA Handbook*. She published *Systems Engineering: Holistic Life Cycle Architecture, Modeling and Design with Real-World Application*, CRC Press, 2022. She also wrote the *Lean Systems* chapter in *Maynard's Industrial and Systems Engineering Handbook, Sixth Edition, 2022.* Her latest textbook publication was published in 2026, *Engineering Economic Analysis: Excellent Decision-Making Techniques (First Edition)*, Cognella Publishing, 2026.

Chapter 1
Enabling Enterprise Performance Excellence Through Strategic Business Process Architecture Modelling Techniques

Abstract We start this book with a description of an Enterprise Performance Excellence (EPE) methodology that links the Lean Six Sigma methodology discussed in Chap. 2 with other systems engineering methodologies that can help to optimize the entire enterprise's processes. EPE helps to understand the organization from a cross-functional enterprise view so that the Lean Six Sigma program can be aligned with the strategic and operational plans as well as focused on key bottleneck processes within the healthcare organization. This methodology starts with understanding the value chains of the organization and functionally decomposing the business functions that provide services to the customers or patients. Following is a description of the EPE methodology.

Keywords Business process · Business strategy · Business process management · Change initiative · Strategic business

Introduction

Six Sigma is both a quality management philosophy and a methodology that focuses on reducing variation, measuring defects, and improving the quality of products, processes, and services.

There are several critical success factors to a successful Lean Six Sigma program, including:

- Basing the Lean Six Sigma project selection on key business priorities (Furterer, 2008).
- Focusing on key business areas (Furterer, 2008).
- Aligning the program with strategic initiatives and prioritize the projects based on business strategy (Furterer, 2008).
- Defining a vision of what the organization will look like after implementing Lean Six Sigma and embracing the change (Sureshchandar et al., 2001).
- Top management support through leadership. This factor includes management's setting of goals and providing leadership and direction (Hoffman & Mehra, 1999).

S. Furterer, *Lean Six Sigma for the Healthcare Enterprise*,
https://doi.org/10.1007/978-1-4471-7617-6_1

- Strategy that is complementary and integrated with existing policies and strategies (Dale, 1994).
- Clear links between strategic goals and the change strategies (Newman, 1994).

These critical success factors point to the need to align an organization's Lean Six Sigma program with the organization's strategic plan, and to the key business processes that enable the enterprise to meet the customers' needs and expectations. It also includes the need to understand the Lean Six Sigma program's vision. Many times, an organization's Lean Six Sigma projects are identified from a bottom-up approach, not one that is top-down and aligned with the enterprise goals and strategic plans. In some organizations, people who want to get certified as a Green or Black Belt will often choose a project themselves based on a perceived need for improving a process. Although these projects often achieve improvement and financial savings, there is no direct visibility or traceability to the organization's strategic goals.

Business architecture is a relatively recent body of knowledge that comes out of the information systems realm (Bieberstein et al., 2008). Business architecture can provide an enterprise-wide understanding of the business. It attempts to connect the business strategies to planned change initiatives focusing on Information Technology (IT) projects that can provide tactics to meet the business strategies. In many organizations, the Business Architecture is documented and developed by the IT organization as a way to understand the business processes. The business processes enable the extraction of key business elements that support required capabilities of the business to meet customers' needs. This provides traceability from the businesses' strategies to the business requirements through to the implemented IT. Demonstrating the alignment between IT initiatives and the business strategies helps to ensure that resources of people, time, and money are applied appropriately.

Business architecture modeling techniques and methods can be used to also provide prioritized alignment with the Lean Six Sigma program goals and the enterprise's Six Sigma projects and improvement initiatives. This can provide alignment between the business strategies and goals and the organization's improvement plans.

We will first describe the concepts that helped to evolve the business architecture methods and tools. We will then discuss the activities that can be used to integrate business architecture methods and models to gain alignment between an enterprise's Six Sigma improvement efforts, key processes, and the business' strategic goals.

Underlying Concepts Evolving to Business Architecture

Business architecture modeling techniques incorporate the concepts of quality management, business process management, and information systems analysis.

Quality management encompasses the evolution from statistical process control, total quality management, business process reengineering (BPR), and Six Sigma bodies of knowledge. Business process management (BPM) encompasses documenting, understanding, managing, and controlling key processes in the business enterprise. The goal of BPM is to improve products and services through a structured approach to performance improvement that centers on systematic design and management of a company's business processes (Chang, 2006). Business processes and workflow are more easily automated through off-the-shelf Business Process Management Systems (BPMS). Business architecture has components of many of these concepts, providing for the synthesis of business processes and information system methodologies.

BPM and BPMS both support Lean Six Sigma process improvement through documentation of processes, and automated collection of process metrics. An enterprise can use business architecture principles and methods as the foundation for identifying the processes that need to be improved to enable the business strategies. It provides the documentation of the business processes and the identification of the optimized business processes and components that can be used across multiple business units, markets, and processes. It enables standardization of processes and alignment of process improvement initiatives with the key processes that enable the business' strategic plan.

Business architecture is helping to provide a context and prioritization of business strategies that can help to focus Lean Six Sigma improvement efforts.

Business Architecture

Business architecture helps us to understand the business processes, and the 3-to-5-year strategies of the business. It helps us to understand the business processes that enable meeting the needs of our customers, and to understand the business capabilities that provide business functionality to meet these needs. Business architecture helps us to define the optimized business components that support the business functionality and the factors that contribute to variability in the business processes (Rosen, 2008).

Business architecture provides models that describe the business entities (business processes and relevant business information), their relationships, dynamics, and rules that govern their interaction to achieve enterprise-wide objectives.

Business Architecture Elements

The elements of the business architecture describe the business enterprise, shown in Fig. 1.1.

The business architecture first includes understanding the customer's needs and expectations. Next, we capture and document the business strategies and goals, as well as the external and internal influencers on the business. The relationship between the business goals and the capabilities that support the goals should be understood. We use the value chain showing a chain of activities that provide value to the customer (Porter, 1985) to understand and decompose our processes. The functional decomposition provides a hierarchical organization of functions and the processes that they include. Each value chain and the subsequent business functions will be used to further decompose the processes. This ensures traceability from the value chains

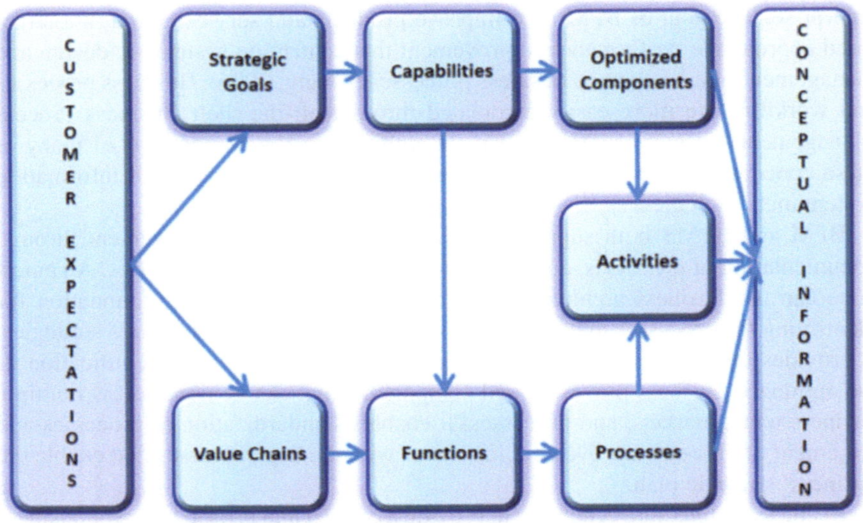

Fig. 1.1 Conceptual business architecture elements

to business processes that provide customer value and provides an organizational inventory of the high-level activities that they perform to meet the customer's needs.

The business capabilities enable the business functions. How the business implements its processes is likely to change frequently, the basic capabilities of a business tend to remain constant. The business processes and their activities describe the sequence of activities that enable the business to meet the customer's expectations and provide value through the value chains.

The business components are identified to optimize the activities that support the business. These components consist of the activities that require similar people, process, and technology. They allow the standardization of the business processes by componentizing the activities that can be used in multiple areas of the business, across many business units and markets (IBM Corporation, 2005). The conceptual information describes the information and roles that exist in the business, which are part of the business processes.

Strategic Business Process Architecture Methodology

The following strategic business process architecture (SBPA) methodology provides a "how to" description for developing and leveraging business architecture that helps to ensure alignment with your Lean Six Sigma initiatives in an organization (Furterer,

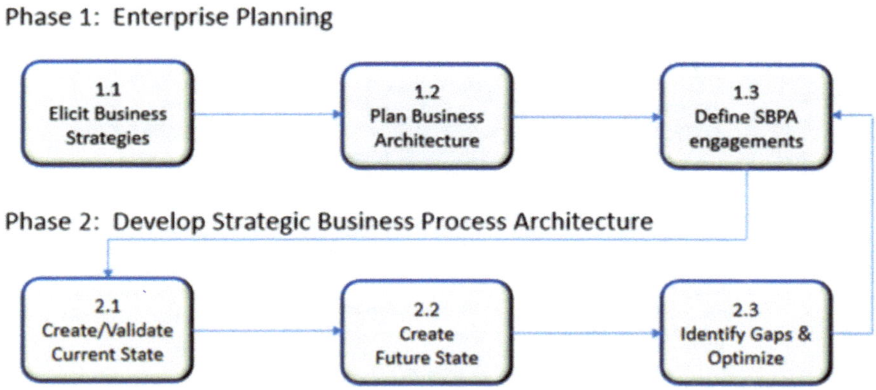

Fig. 1.2 Business architecture development methodology

2009a, 2009b). The SBPA models can be used as the basis for identifying and prioritizing the key processes and extracting improvement projects in an organization's Lean Six Sigma program.

The business architecture methodology, Fig. 1.2, is composed of two main phases:

Phase I Enterprise Planning.

Phase II Strategic Business Process Architecture Modeling.

The SBPA helps to integrate and provide a synthesis across the enterprise and align to customer's needs; to identify the business capabilities that are required to support the entire enterprise and to optimize the business processes. The SBPA provides the inputs to the information system architecture (information, application, and technology) to create the IT roadmaps that support the business strategies. The methodology also provides a roadmap for process optimization and improvement.

Phase I: Enterprise Planning

Phase I, Enterprise Planning, is used to elicit, harvest, understand, and document the business strategies and plan the related business architecture engagements that will enable these strategies.

The objectives, steps, and deliverables of each activity in Phase I, Enterprise Planning, are shown in Fig. 1.3 and described below.

Elicit Business Strategies

Purpose: The purpose of this step is to elicit, validate, and understand the future strategies of the business. This will help to align the business strategies and the future IT and process improvement initiatives.

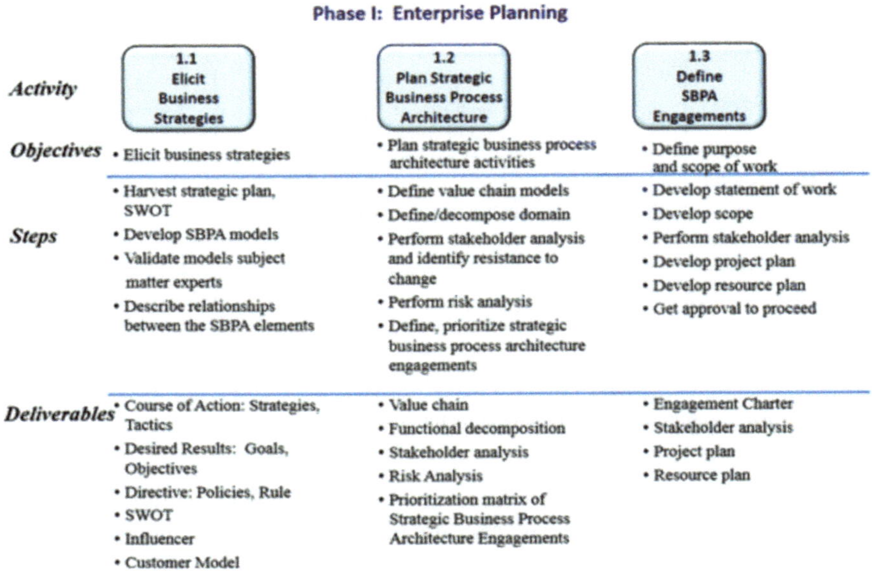

Fig. 1.3 *Phase I*. Enterprise planning

There are several steps for eliciting the business strategies from the enterprise. First the business architect harvests the business strategic plans, and Strengths, Weaknesses, Opportunities, and Threats (SWOT) for each business unit. Next, we develop the SBPA models. Since the models are our translation from the business planning documents to our strategic business models, we must validate them with the business leaders. We then develop the relationships between the elements of the strategic business models. Our last step in eliciting the business strategies is to identify any gaps in the alignment as we developed the traceability between the relationships of the strategic elements of the models.

Plan Strategic Business Process Architecture

Purpose: The purpose of this step is to plan the SBPA engagements based on the business strategies and critical business processes. It is important to understand the critical stakeholders in the enterprise. Since resources are always limited, this step helps us to focus our resources, time, and talents on the most critical business processes and strategies while building the SBPA models.

In the Plan SBPA activities, we first develop the value chains for the enterprise, so that these can be used to prioritize the business processes to model, based on the business strategies. We decompose the value chains into the business functions so that we can create the business process models from the prioritized value chains and subsequent business functions. We define the SBPA engagements that will derive the SBPA models aligned to the prioritized business strategies. We perform a stakeholder analysis to identify the stakeholders and team members that should be involved in the SBPA engagements.

Define Purpose/Scope of Strategic Business Process Architecture Engagements

Purpose: The purpose of this step is to define the scope, purpose, and objectives of the SBPA engagements, as well as to develop the project, risk, and communication plans of the engagements. There are usually not enough resources to create the entire business process architecture at once. So, these activities are designed to break the work into manageable business process architecture engagements, based on the priorities identified in the prior phase.

In these activities, the purpose and scope of the SBPA engagements should be defined. A statement of work or engagement charter can be developed that describes the scope of the engagements. A SIPOC can be used to scope the engagements. A detailed stakeholder analysis for the planned engagements will then be developed.

Phase II Strategic Business Process Architecture Modeling

Phase II, SBPA Modeling, starts the business architecture modeling that is used to document the existing and proposed SBPA for the enterprise. In this phase, we build the existing business architecture models to create a repository that can be used to understand the existing enterprise and drawn upon for process improvement. It is used to help us understand the gaps between the current state and what we desire in the future state. To improve the business processes and related technologies that support them, it is critical to understand the current processes, the existing inefficiencies, waste, and failure points, so that they can be eliminated in the future state processes.

The future state analysis provides the business architecture modeling for the future state within the planning horizon. This provides an understanding of the value chains, business processes, and the people's roles that perform the processes. It provides a view of the handoffs and integration between departments and business units. This business architecture modeling provides the rationalized and optimized business components. The business components are the activities or sets of activities that provide the identified business capabilities that help to realize the business strategies and associated initiatives. These business components represent streamlined elements of the process that we want to incorporate in an optimized process in the future state.

The objectives, steps, and deliverables of each activity in Phase II, Develop SBPA, are shown in Fig. 1.4 and described below.

The purpose of the Develop SBPA phase is to build the existing or "current state" SBPA models, to create a repository that can be used to understand the enterprise as it exists today. Additionally, the future state SBPA is developed that aligns to the business strategies elicited from the business.

The current state analysis phase can be used to build the existing SBPA repository for purposes of business planning or to build a view related to an IT or process

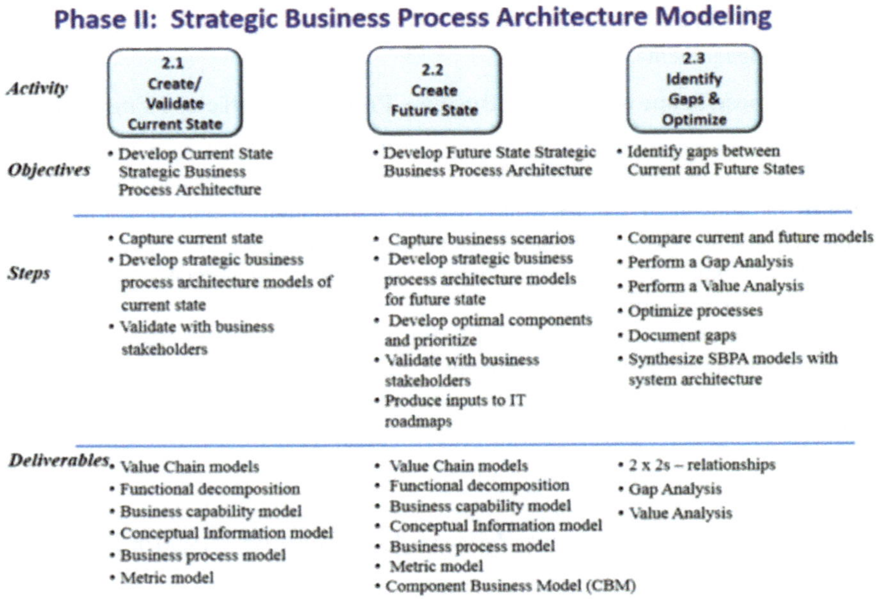

Fig. 1.4 *Phase II*. Strategic business process architecture modeling

improvement change initiative (program or major project). This phase would be performed on an as-needed basis for each function where the current state SBPA models do not currently exist for those functions identified to be part of the scope of the engagement. This phase could also be performed when a view of the current state is needed for business planning activities.

The purpose of the future state analysis is to build the SBPA models based on the defined future state, or where we want to go based on business strategic planning or definition of the change initiative. The Future State SBPA models would be an input to the Information and Application Architectures and the initiative roadmaps for the enterprise. This phase could be performed when a view of the future state is needed for enterprise planning activities or related to an initiative.

Create/Validate Current State

This step is to create the current state SBPA models if they do not exist, or to validate the models if they do exist.

Purpose: The purpose of the current state analysis phase is to build the existing or "current state" SBPA models, to create a repository that can be used to understand the enterprise as it exists today. If the SBPA models do not yet exist for the functions defined as part of the scope of the change initiative, then the information should be elicited from the business, and the business architect should develop the SBPA models. If the business process architecture models exist in the current state repository, then the business architect should validate these with the business to ensure the information still represents the current state of the business.

Frequency: The current state analysis phase can be used to build the existing SBPA repository for purposes of business planning or to build a view related to a change initiative (program or major project). This phase would be performed on an as-needed basis for each program where the current state SBPA models do not currently exist for those functions identified to be part of the scope of the program initiative. This phase could also be performed when a view of the current state is needed for business planning activities.

Create Future State

Purpose: The purpose of the Create Future State is to build the SBPA models based on the defined future state, or where we want to go based on business strategic planning or definition of the change initiative. The Future State SBPA models would be an input to the Information and Application Architectures and the IT or process improvement roadmaps for the enterprise.

Frequency: The future state analysis would be used during enterprise planning as part of the What-If planning and also during the IT initiatives. This phase would be performed on an as-needed basis for each program where the current state SBPA models do not currently exist for those functions identified to be part of the scope of the program initiative. This phase could also be performed when a view of the future state is needed for enterprise planning activities.

Identify Gaps and Optimize

Purpose: The purpose of this step is to identify gaps between the current state and the future state, as well as optimize the processes that are part of the scope of the engagement. The goal is to rationalize the processes and identify optimized components that will optimize the processes in each instance that the components exist.

Lean Six Sigma

Once the business architecture models are derived the next step is to optimize the processes. This is where Lean Six Sigma enters the picture. The Lean Six Sigma projects can be identified and scoped that align to the key business processes and the business strategies. Resources can be applied to these projects, from a central governing steering committee, to ensure alignment between the business strategic initiatives and the Six Sigma projects, and to assess the improvement projects' success and financial savings.

Conclusions

Through the derivation and development of the SBPA, we can optimize the enterprise and enhance operational excellence. SBPA applies the underlying principles of business process management through documenting, controlling, and improving processes that align with the business strategies.

We can leverage the improvement methodologies through the Lean Six Sigma program to improve the processes. The SBPA modeling and data collection helps to provide a prioritization of the key processes that need improvement, which align with the business strategies and goals of the enterprise. The business architecture helps the organization focus on the important Lean Six Sigma improvement projects and the processes that truly matter to the business and enable us to better meet our customer's expectations through streamlined and optimized processes.

References

Bieberstein, N., Laird, R., Jones, K, Mitra, T. (2008). *Executing SOA: A practical guide for the service-oriented architect*. Upper Saddle River, NJ: IBM/Pearson.

Chang, J. F. (2006). *Business process management systems: Strategy and implementation*. Auerbach Publications.

Dale, B. (1994). A framework for quality improvement in public sector organizations: A study in Hong Kong. *Public Money & Management, 14*(2), 31–36.

Furterer, S. L. (2009b). *Lean six sigma in service applications and case studies*. CRC.

Furterer, S. L. (2008). Lean six sigma program success factors in a retail application. In *International Conference on Industry, Engineering, and Management Systems*.

Furterer, S. L. (2009a). *Blazing the trail to operational excellence: Leveraging information systems business architecture methods to enable operational excellence*. IIE Magazine, Norcross.

Hoffman, J., & Mehra, S. (1999). Management leadership and productivity improvement programs. *International Journal of Applied Quality Management, 2*(2), 221–232.

IBM Corporation (2005). *Component Business Models, Making Specialization Real*. IBM Business Consulting Services, IBM Institute for Business Value.

Newman, J. (1994). Beyond the vision: cultural change in the public sector. *Public Money and Management*.

Porter, M. E. (1985). *Competitive advantage: Creating and sustaining superior performance*. Free.

Rosen, M. (2008). The business motivation model: Matching the means to the ends. *Cutter IT Journal, 21*(3), 6–10.

Sureshchandar, G. S., Rajendran, C., & Anantharaman, R. N. (2001). *International Journal of Service Industry Management, 12*(Part 3/4), 378–412.

Chapter 2
Lean Six Sigma Roadmap

Abstract The purpose of this chapter is to provide a roadmap to applying the Lean Six Sigma (LSS) methodology and the key tools that can be applied when implementing LSS in healthcare processes. This chapter was originally published in (Furterer, Lean Six Sigma in service: Applications and case studies, CRC Press, 2009) and adapted for the healthcare enterprise. A little over twenty-one percent (21.3%) of the industrial workforce is engaged in educational services and healthcare, and social assistance occupations (US Census Bureau). Service industries are the fastest growing segment of our economy and now account for almost 70% of economic activity, and over 85 million employees (US Census Bureau North American Product Classification). Healthcare processes include both clinical and nonclinical processes, and both types can be improved using LSS methodologies and tools.

Keywords Control chart · Measure phase · Quality function deployment · Statistical process control · Risk priority number

Lean Six Sigma Overview

Lean Six Sigma (LSS) is an approach focused on improving quality, reducing variation, and eliminating waste in an organization. It is the combining of two improvement programs, Six Sigma and Lean Enterprise. Six Sigma is a quality management philosophy and methodology that focuses on reducing variation, measuring defects (per million output/opportunities), and improving the quality of products, processes, and services. The concept of Six Sigma was developed in the early 1980s at Motorola Corporation. Six Sigma became popularized in the late 1990s by General Electric Corporation and their former CEO Jack Welch. Lean Enterprise is a methodology that focuses on reducing cycle time and waste in processes. Lean Enterprise originated from the Toyota Motor Corporation as the Toyota Production System (TPS) and increased in popularity after the 1973 energy crisis. The term "Lean Thinking" was coined by James P. Womack and Daniel T. Jones in their book, Lean Thinking (Womack & Jones, 1996). The term "Lean Enterprise" is used to broaden the scope of

S. Furterer, *Lean Six Sigma for the Healthcare Enterprise*,
https://doi.org/10.1007/978-1-4471-7617-6_2

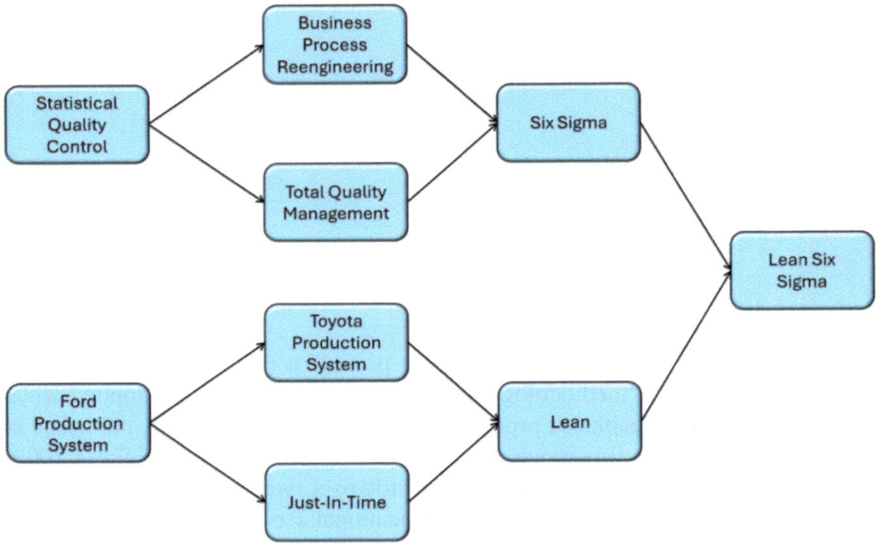

Fig. 2.1 Evolution of quality and productivity to Lean Six Sigma

a Lean program from manufacturing to embrace the enterprise or entire organization (Alukal, 2003).

Figure 2.1 shows the evolution to the combined methods of Lean and Six Sigma.

In the 1920s, at Western Electric, the concepts of control charts and statistical process control (SPC) were developed by Walter Shewhart. Dr. Edward Deming installed SPC in Japanese manufacturing as he assisted Japan in their rebuilding efforts after World War II. Japan's successes in the 1970s have re-popularized SPC in United States businesses. Total Quality Management (TQM) was a natural outgrowth of SPC adding process improvement methodology. In the 1980s, Business Process Reengineering (BPR) and TQM became popular. BPR encouraged completely throwing out the old process and starting over, many times within the context of implementing major information system changes. TQM focused on a less structured approach with the principles of quality and process improvement. These methodologies evolved to Six Sigma.

On the productivity side, the Ford Production System was used to assemble cars, which was the basis for the TPS. Just-in-time production philosophies joined with TPS which evolved into Lean. Now Lean and Six Sigma are merging to capitalize on the best of both improvement philosophies and methodologies.

Six Sigma uses the DMAIC (Define, Measure, Analyze, Improve, and Control) problem-solving approach, and a wide array of quality problem-solving tools. Use of these tools varies based on the type of process studied and the problems that are encountered. There are many powerful tools in the Lean tool set that help to eliminate waste, organize, and simplify the work processes.

Lean Six Sigma Applications in Private Industry

The concept of combining Lean Manufacturing and Six Sigma principles began in the middle to late 1990s and has quickly taken hold as companies recognized the synergies. There are many examples of manufacturing companies implementing a combined effort of Lean and Six Sigma. An early example, starting in 1997, was by an aircraft-engine-controls firm, BAE Systems Controls, in Fort Wayne, Indiana. They blended Lean-manufacturing principles with Six Sigma Quality tools. Their "Lean Sigma" strategy was "designed to increase velocity, eliminate waste, minimize process variation, and secure its future in the evolving aerospace market" (Sheridan, 2000). They started with implementing Lean initiatives and then identified a synergy between Lean and the Six Sigma quality program that had been launched while the company was a part of General Electric. BAE Systems Controls implemented the following lean initiatives: (1) kaizen events, (2) takt-time-driven one-piece-flow product cells, (3) kanban pull system and point-of-use storage bins on the plant floor, (4) lean production cells, (5) mistake proofing, and (6) use of a multi-skilled workforce. As part of the Six Sigma program, they implemented statistical methods and team leadership with the use of Black Belts. In BAE Systems Control's implementation of LSS, they improved productivity by 97% and customer lead time by 90%. Their value-added productivity increased 112% in 5 years, work in process was reduced by 70%, product reliability improved by 300%, and there were zero lost workdays in 1999 (Sheridan, 2000).

Another early innovator combining Lean and Six Sigma was Maytag Corporation. Maytag implemented Lean Sigma in 1999. They designed a new production line using the concepts of Lean and Six Sigma. Maytag reduced utilized floor space to one-third of that used by Maytag's other product lines. Maytag also cut production costs by 55%. Their Lean Sigma effort helped Maytag to achieve savings worth millions of dollars (Dubai Quality Group, 2003).

Lean Six Sigma has been implemented at Northrop–Grumman, an Aerospace Company. They had already begun to implement Lean Thinking when they embarked upon their Six Sigma program. Northrop integrated the Workout™ events (a problem-solving process developed at GE) with the Lean Thinking methods and Kaizen events. They used Six Sigma's strategy and methods within their Product Teams, not as a stand-alone program. Their formal process integrated WorkOut™, Kaizen, and DMAIC into the Six Sigma Breakthrough WorkOut™. Subject matter experts and a Black Belt were used on their project team. They performed a 4- to 5-day Define/Measure phase. They then performed the Measure, Analyze, and Improve phases for roughly 30 days each. The final activities included a post-WorkOut phase as the Control, Integrate, and Realize phase (McIlroy & Silverstein, 2002).

Lockheed Martin Aeronautical Systems reduced costs, improved competitiveness, customer satisfaction, and the first-time quality of all its manufactured goods. They had separate Lean and Six Sigma projects, depending on the objective of the project and the problem that needed to be solved (Kandebo, 1999).

Table 2.1 DMAIC activities

Define	Measure	Analyze	Improve	Control
1. Create project charter, plan and stake-holder analysis 2. Perform initial VOC and identify CTS 3. Select team and launch project	4. Define the current process and VOP performance 5. Define detailed VOC 6. Validate measure system	7. Develop cause-and-effect relationship 8. Determine and validate root causes 9. Develop process capability	10. Design future state with costs and benefits 11. Establish performance targets and project scorecard 12. Gain approval, train and pilot	13. Implement improvement recommenda-tions & manage change 14. Incorporate process control plans and scorecards 15. Implement continuous improvement cycle, plan, do, check, act (PDCA)

The Six Sigma DMAIC problem-solving methodology is used to improve processes. The phases of the DMAIC are well defined and standardized, but the steps performed in each phase can vary based on the reference used. The Define phase is where the scope of the project charter is developed. The goal of the Measure phase is to understand and baseline the current process. In the Analyze phase, we analyze the data collected in the Measure phase to identify the root causes of the problems identified. In the Improve phase, the improvement recommendations are developed and implemented. The goal of the Control phase is to ensure that the improvements had a positive impact and that they will be sustained and controlled. In Table 2.1, a description of the activities that can be performed within each phase of the DMAIC problem-solving methodology (Brassard & Ritter, 2001) is given.

The development of processes in a Women's Center will be used as the case study for describing the DMAIC roadmap in this chapter.

Define Phase

The purpose of the Define phase is to delineate the business problem and scope the project and process to be improved. The following steps can be applied to meet the objectives of the Define phase:

1. Create project charter, plan and stakeholder analysis.
2. Perform initial VOC and identify CTS.
3. Select team and launch project.

Table 2.2 shows the main activities mapped to the tools or deliverables most typically used during each step of the Define phase.

Table 2.2 Define activities and tools/deliverables

#	Define activities	Deliverables
1	Create project charter, plan and stakeholder analysis	• Project charter • SIPOC • Stakeholder analysis • Communication plan
2	Perform initial VOC and identify CTS	• Change plan • CTS summary
3	Select team and launch project	• RACI (Responsibilities Matrix) • IFR (Items for Resolution) • Project plan

Develop Project Charter

The first step in the Define phase is to identify and delineate the problem. The project charter can help to identify the elements that help to scope the project and identify the project goals.

A project charter template is provided in Table 2.3.

Following is a description of the elements of the project charter that helps to scope and define the business problem.

Project name. The name of the project. Describe the process to be improved along with the project goal.

Project overview. Provides a project background and describes basic assumptions related to your project.

Problem statement. A clear description of the business problem. What is the challenge or the problem that the business is facing? The problem statement should consider the process that is impacted. Define the measurable impact of the problem. The team should be specific as to what is happening, when it is occurring, and what the impact or consequences are to the business of the existing problem.

Goal of the project. What is the quantifiable goal of the project? It may be too early in the problem-solving method to identify a clear target, but at least a placeholder should be identified relating to what should be measured and improved.

Scope statement. The scope should clearly identify the process to be improved, and what is included or excluded from the scope for the LSS project. The scope can also address the organizational boundaries to be included and possibly, more importantly, which should be excluded. It can also include a temporal scope of the timing of the process and data collection activities. The deliverable scope includes what specifics should be delivered from the project, such as improvement recommendations, implementation plan, and so forth.

Projected financial and other benefits. Describes potential savings, revenue growth, cost avoidance, cost reduction, Cost of Poor Quality (COPQ), as well as less tangible benefits such as impact to morale, elimination of waste, and inefficiencies.

Table 2.3 Project charter for the women's center

Project Charter

Project Overview, Problem Statement, & Goals

Provides a description of the problem to be solved. Be as descriptive as possible. Include improvement goals that are SMART (Specific, Measurable, Attainable, Realistic, and Time-Based).

Project Name: Women's Center Process Design and Improvement

Project Overview: Many outpatient facilities are focusing on providing comprehensive services to women in a comfortable setting. In a qualitative study of women who had received a mammogram in the prior three years, without a history of cancer, satisfaction was related to the entire experience, not just the actual mammogram procedure. The authors of the described study found seven satisfaction themes from the focus groups: (1) appointment scheduling, (2) facility, (3) general exam, (4) embarrassment, (5) exam discomfort/pain, (6) treatment by the technologist, and (7) reporting results. This supports the focus of designing a seamless experience for women in the Women's Center through applying Systems Engineering Tools.

Problem Statement: The results for patients who visited the former Women's Center were transmitted to the primary care provider (PCP) within an average of 57 hours. There was large variability in how and when the test results were communicated to the patients. The desire is to open the new facility with streamlined processes that align to the patients' needs as defined and collected through the Voice of the Customer. .

Goal(s): To design the processes and metrics which result in optimal flow for the new Women's Center and align to the Voice of the Customer.

Scope Statement: Includes the processes, stakeholders, procedures, resources, materials, equipment, training and required system changes. The scope includes the new Women's Center diagnostic services (Mammography, Ultrasound, Stereotactic Biopsy, and Bone Densitometry), registration, appointment scheduling, and providing spiritual care.

Projected Financial & Other Benefits: Patient satisfaction, increased capacity due to efficient workflow, and resultant revenue are potential benefits of this project.

SIPOC (Suppliers-Input-Process-Output-Customer)

The SIPOC is a useful tool in the Define phase to help scope the project and understand the process. SIPOC shows the interrelationships between the customers and suppliers and how they interact with the process. It also identifies the inputs used in the process steps and the outputs of the process. The process steps transform the inputs into the outputs. The best way to construct the SIPOC is to identify the five to seven high-level process steps that bound the process. For each process step, identify the inputs to the process and who supplies the inputs. Next identify the outputs of each process step and the customer of the output. Table 2.4 is an example of a SIPOC for the Women's Center processes.

Perform Stakeholder Analysis

It is critical to clearly identify the customers and stakeholders that are impacted by the process because the quality of the process is defined by the customers. Quality is measured by first understanding then exceeding the customers' requirements and expectations. There is a high cost of an unhappy customer: ninety-six percent (96%) of unhappy customers never complain; 90% of those who are dissatisfied will not buy again; and each unhappy customer will tell his or her story to as many as 14 other people (Pyzdek, 2003).

Table 2.4 SIPOC

SIPOC				
Suppliers	Inputs	Process	Outputs	Customers
Patient	Patient information Medical need order	Schedule screening	Screening appointment	Patient
Patient	Insurance information Patient information	Authorize screening	Insurance authorization for service	Patient center
Patient	Patient information Order authorization Insurance information Payment	Perform screening	Screening films	Radiologist
Radiologist	Screening films	Read/Send results	Results report screening films	Physician
Physician	Results report screening films	Review Screening results	Results report diagnosis	Patient
Physician center	Result report Diagnosis order(s)	Provide results	Results report	Patient

Customers and stakeholders can be my peers, people who report to me, my boss, other groups within the organization, suppliers, and external customers. The customers can include both internal and external customers of the process. Each process does not always interface directly with an external customer of the company but will have internal customers. The internal customers are people who receive some output from the process, such as information, materials, product, or a service step. It is ultimately the boundary of the process that is being improved that determines who the customer is.

The stakeholder analysis definition identifies the stakeholder groups, their role and how they are impacted, as well as their concerns related to the process. There is an additional column which provides a quick view of whether the impact is positive (+), such as reducing variation, or negative (–), such as resistant to change. This is a high-level view that will be further detailed in the Measure phase. Table 2.5 is an example of a stakeholder analysis.

The next step in the stakeholder analysis is to understand the stakeholders' attitudes toward change, as well as potential reasons for resistance. Additionally, the team should understand the barriers to change as a result of the resistance. Next, activities, plans, and actions should be developed that can help the team to overcome the resistance and barriers to change. A definition of how and when each stakeholder group should participate in the change effort should be developed in the Define phase and then updated throughout the DMAIC project.

The stakeholder commitment scale with respect to stakeholder receptivity shown in Table 2.5 can be used to summarize where the stakeholders are regarding their acceptance or resistance to change. The team should determine, based on initial interviews and prior knowledge of the stakeholder groups, the current level of support or resistance to the project. Strongly supportive indicates that these stakeholders are supportive of advocating for and making change to happen. Moderately supportive indicates that the stakeholders will help, but not strongly. The neutral stakeholders will allow the change and not stand in the way but not go out of their way to advocate for the change. Moderately against, stakeholders do not comply with the change and have some resistance to the project. Strongly against, stakeholders will not comply with the change and will actively and vocally lobby against the change. A strategy to move the stakeholders from their current state to where the team needs them to be by the end of the project should be developed. This change strategy should include how the team will communicate with the stakeholders and any activities in their action plan to gain support and implement change.

Perform Initial Voice of Customer (VOC) and Identify Critical to Satisfaction (CTS)

In the Define phase, the team can perform an initial VOC data collection to understand the CTS criteria. CTSs are the elements of a process that significantly affect the

Table 2.5 Stakeholder analysis

Stakeholder analysis					
Stakeholder	Type	Primary role	Potential impacts/ Concerns	Initial receptivity	Future receptivity
Patient (women)	Primary	Receive services in the center	Customer service, quality, efficiency	Neutral	Neutral
Referring physicians	Primary	Refer patients to the center and communicate results to the patient	Quality of care	Moderate support	Strongly support
Payers	Secondary	Pay-for services, such as managed care, medicare employers, self-pay	Quality of care; cost-effective care	Neutral	Neutral
Donors	Primary	People who donate money to the center	Quality of care; meet patient requirements	Strongly support	Strongly support
Imaging technologists	Primary	Work in the center and perform procedures	Patient satisfaction; improved/ well-designed work environment Associate satisfaction	Moderate support	Strongly support
Radiologists	Primary	Read and provide results of imaging procedures	Reduced volume and revenue Physician satisfaction	Moderate support	Strongly support
Administration	Secondary	Manage the hospital and center	Volume; revenue; patient satisfaction; physician satisfaction Productivity	Strongly support	Strongly support

(continued)

Table 2.5 (continued)

Stakeholder analysis					
Stakeholder	Type	Primary role	Potential impacts/ Concerns	Initial receptivity	Future receptivity
Patient scheduling	Primary	Registration who registers patients and perform insurance authorizations, centralized scheduling who make the patient appointments	Volume Productivity Timeliness of processes	Moderate support	Strongly support
Information technology	Secondary	Provide phone and computer systems	Meet requirements; on time; on budget	Strongly support	Strongly support
Physicians	Primary	Provide women's services	Reduced volume and revenue Physician satisfaction	Moderate support	Strongly support

output of the process. It is critical to focus throughout the phases of the DMAIC problem-solving process and the Six Sigma project on the CTS.

In the Define and Measure phases, the focus is on collecting information from the customer to understand what is important to them regarding the process, product, or service. Identify the Measure phase in the measure of your process directly related to the CTS criteria. Identify the root causes in the Analyze phase related to the CTS. The improvement recommendations implemented are aligned by eliminating the root causes related to the CTS. The variability to be controlled by implementing control mechanisms in the Control phase should reduce the variability related to the CTS.

Some references refer to identifying the CTQ, but the CTS broadens the elements of the CTS by including CTQ, Critical to Delivery (CTD), and Critical to Cost (CTC). There may also be critical elements in the process to measure related not only to quality, delivery, and cost, but to time, as well. For a Six Sigma project, not everything should be a CTS. The CTS should be specific to the scope of the project and the process to be improved. If there are more than a few CTS identified for the project, the scope is probably too large for a reasonable Six Sigma project to be completed in 3–6 months. The CTS should describe the customer need or requirement, not how to solve the problem.

The steps to identify the CTS are (George et al., 2005) as follows:

1. Gather appropriate VOC data from market research, surveys, focus groups, interviews, etc.

2. Extract key verbatims from the VOC data collections, identifying why a customer would do business with your organization.
3. Sort ideas and find themes, develop an Affinity or Tree Diagram.
4. Be specific and follow-up with customers where needed.
5. Extract CTS' and specifications from customer information.
6. Identify where you are missing data and fill in the gaps.

The VOC is a term used to "talk to the customer" to hear their needs and requirements or their "voice." Many different mechanisms can be used to collect VOC, including interviews, focus groups, surveys, customer complaint and warranty data, market research, competitive information, and customer buying patterns. We will further discuss VOC during the Measure phase, where more detailed and extensive VOC can best be performed. The initial VOC is used to identify the CTS'. In the Define phase, the CTS Summary is a listing of the CTS based on knowledge of the process and the customer to this point. The CTSs for the Women's Center is shown in Table 2.6.

Table 2.6 Critical to Satisfaction (CTS)

	Critical to Satisfaction (CTS)	
	CTS' are basic elements that can be used in driving process measurement, improvement, and control. They are elements of a process that significantly affect the output of the process. What are the characteristics of the process critical as perceived by the customer?	
	Title	Description
1	Patient throughput time	The length of stay, or time from when the patient arrives until they are discharged from the Emergency Department.
2	Patient wait time during visit	The time that the patient waits between major steps of the process.
3	Quality of care	The quality of care that the patient receives, especially from a diagnosis perspective
4	Patient satisfaction	The satisfaction of the patient as determined through patient satisfaction surveys.

Select Team and Launch the Project

The Six Sigma project team should be selected based on those team members who have the knowledge of the process and have the commitment to work on the project. The roles and responsibilities of the project team members should be clearly defined.

A team is a group of people working together to achieve a common purpose. Teams need a clearly defined purpose and goals which are provided through the Six Sigma project charter. They also need well-defined roles and responsibilities, which can be provided through developing a responsibilities matrix. The responsibilities matrix, in Table 2.7, identifies the team members, their roles, and high-level responsibilities on the Six Sigma project. Another important component of forming the team is to brainstorm and identify ground rules. The ground rules identify how the team will interact with each other and ensure that the behavioral expectations are clearly defined at the start of the project. The team's common set of values and ethics can be established during the development of the team ground rules.

Sample team ground rules:

- Treat everyone with respect.
- Listen to everyone's ideas.
- When brainstorming, do not evaluate ideas.

Table 2.7 Responsibilities matrix

Role: Responsibility	Team leader	Black belt	Champion	Process owner	Team members
Facilitate meeting	X				
Manage project	X				
Mentor team members	X	X			
Transfer knowledge of Six Sigma tools		X			
Remove roadblocks			X		
Monitor project progress			X		
Approve project			X		
Implement improvements				X	
Subject matter expertise				X	
Apply Six Sigma tools					X
Perform statistical analysis					X
Perform data collection					X

- Contribute fully and actively participate.
- Come to team meetings prepared.
- Make decisions by consensus.
- Identify a back-up resource to complete tasks when not available.

Create Project Plan

The LSS project plan is developed in this last step of the Define phase. The resources, time, and effort for the project are planned. A project plan template is provided in Table 2.8. Additional tasks can be identified within each phase and major activity.

Excel or a project planning software such as Microsoft Project can be used to track tasks completed against the project plan. An important part of the project planning is

Table 2.8 Project plan

Estimate how long each phase will take during the Define phase. The dates do not have to be etched in stone, but can be changed as more information about the project is obtained/progresses

Task #	Task name	Start date	End date	Resources/ Owner
1	*Define phase*			
1.1	Create project charter, plan and stakeholder analysis			
1.2	Perform initial VOC and identify CTS			
1.3	Select team and launch project			
2	*Measure phase*			
2.1	Define the current process and VOP (voice of process)			
2.2	Define detailed VOC (voice of customer)			
2.3	Validate measurement system			
3	*Analyze phase*			
3.1	Develop cause-and-effect relationships			
3.2	Determine and validate root causes			
3.3	Develop process capability			
4	*Improve phase*			
4.1	Design future state with costs and benefits			
4.2	Establish performance targets, project scorecard			
4.3	Gain approval to implement, train, and pilot			
5	*Control phase*			
5.1	Implement improvement recommendations and manage change			
5.2	Incorporate process control plans and scorecards			
5.3	Implement continuous improvement cycle, plan do, check, act (PDCA)			

to perform a risk analysis to identify potential risks that could impact the successful completion of the project.

The team can brainstorm potential risks to the project. They can also assess the probability that each risk would occur on a scale of high, medium, or low occurrence. The impact of the risk should also be assessed, that is, if the risk were to occur, what level of impact would it have on the successful completion of the project (high, medium, or low). It is also important to develop a risk mitigation strategy that identifies if the risk occurs, how will the team mitigate the impact of the risk to reduce or eliminate the impact of the risk. Table 2.9 shows a simple risk matrix.

Another tool that is useful while planning and managing the project is an Item for Resolution (IFR) form. This helps the team to document and track items that need to be resolved. It enables the team to complete the planned agendas in meetings, by allowing a place to "park" items that arise that cannot be resolved in the meeting, due to either time constraints or lack of data or access to appropriate decision-makers. Table 2.10 shows an IFR form. It includes a description of the item to be resolved. A priority (high, medium, low) should be assigned to each item. The status of the item, open (newly opened), closed (resolved), hold (on hold-not being actively worked) should be identified. The owner who is responsible for resolving the issue, as well as the dates that the item were opened and resolved, should be completed on the IFR form. A description of the resolution should also be included. This helps the team to keep track of key decisions and ensures that the items are resolved to the satisfaction of all team members. The log of IFRs can also be used during the lessons learned activity after the project is complete to identify where the problems arose and how they were resolved, to incorporate these items into the risk mitigation strategies for follow-on projects.

Another helpful tool that should be developed in the Define phase but be used throughout the LSS project is a communication plan. The communication plan can be used to identify strategies for how the team will communicate with all key stakeholders. It can be useful to help overcome resistance to change by planning how frequently and in the manner that the team will communicate with the stakeholders.

Table 2.9 Project risk analysis

Project risk analysis			
Brainstorm the potential risks to a successful project, identify the likelihood that each will occur, the potential impact, and develop a mitigation strategy			
Potential risk to successful project	Occurrence of risk	Impact of risk	Risk mitigation strategy
Inability to add clinical resources for results and appointments	Low	High	Gain buy-in from leadership for additional resources, early in the project
Inability to obtain historical process data	High	Moderate	Identify data and metrics needs, incorporate new data collection methods and data fields early in the project

Table 2.10 Item for Resolution (IFR) form

#	Issue	Priority	Status	Owner	Open Date	Resolved Date	Resolution

Each key stakeholder or audience of a communicated message should be identified. The objective or message that will be communicated is then developed. The media or mechanism of how to communicate with the audience is then identified. This can be face-to-face, email, websites, etc. The frequency of the communication is important, especially for those more resistant to change having more frequent communication. The last element of the communication plan is to clearly identify who is responsible for developing and delivering the communication to the audience. The communication plan is shown in Table 2.11.

Team Meeting Management

Some best practices for team meeting management are as follows:

- Respect people and their time.
- Determine critical/required participants for emails, meetings, and decisions.
- Cancel or schedule meetings ahead of time.
- Always create a meeting agenda and send it out in advance of the meeting. The agenda should include required and optional participants.
- Recap action items, meeting minutes.
- Using voting in emails to make easy decisions, or agree upon a meeting time.
- Track meeting attendance and resolve habitual lack of attendance.

The planned meeting agenda should include the following:

1. Date, time, and proposed length of the meeting.
2. Meeting facilitator's name.
3. Meeting location.
4. Required and optional attendees.
5. Purpose of the meeting.
6. Desired outcomes.
7. Topic with time and proposed outcome for each topic.

Some tips that the meeting facilitator can use to keep the meeting productive are as follows:

- Listen and restate what you think you heard.
- Ask for clarification and examples.
- Encourage equal participation and circle the group.

Table 2.11 Stakeholder receptivity communication plan

A communication plan is a road map for getting the message across to the desired audience. Identify how to communicate with stakeholders during the project and help move receptivity in a positive direction

Audience	Objective/Why	Key message	Media	Owner	When/ Frequency
Patient (women)	Understand processes	How process improvement project is developing streamlined processes	Newsletter	Administration	Quarterly
Referring physicians	Changes in the processes	Improvement made and safety of them	Grand rounds	Physician director	Quarterly
Payers	Training for process changes	Process changes	Newsletters	Center director	Monthly
Donors	Status of process improvement project	Status of process changes	Meetings	Center director	Weekly
Imaging technologists	Process changes	Process changes	Emails	Center director	As needed
Radiologists	Process changes	Process changes	Emails	Center director	As needed
Administration	Process changes	Process changes	Grand rounds	Physician director	As needed
Patient scheduling	Improvements	Summary of improvement	Audits	Center director	As needed
Information technology	Improvements	Summary of improvement	Emails	Center director	As needed

- Summarize ideas and discussion.
- Corral digressions and get back to the agenda.
- Close the discussion.

Summary

The Define phase is a critical phase of the project. It is important to spend ample time in the Define phase developing the project charter and getting the buy-in of the project champion, the team members, and all stakeholders. The time spent clearly defining the scope of the project will reap dividends by reducing issues during the remaining phases of the project. A process or a problem poorly defined will require the team to revisit the Define phase when improvement efforts bog down or lose focus in subsequent phases.

Measure Phase

The purpose of the Measure phase is to understand and document the current state of the processes to be improved, collect the detailed voice of the customer information, baseline the current state, and validate the measurement system. The activities performed and tools applied during the Measure phase are discussed below:

1. Define the current process.
2. Define the detailed VOC.
3. Define the Voice of the Process (VOP) and current performance.
4. Validate the measurement system.
5. Define the COPQ.

Table 2.12 shows the main activities mapped to the tools or deliverables most typically used during that step.

Define the Current Process

The first step of the Measure phase is to profile the current state. SIPOC and process mapping are excellent tools to document the current process steps, the information that is used, the people who perform the work, and the internal and external customers of the services. In a process improvement effort, there are typically three levels of

Table 2.12 Measure activities and tools/deliverable

#	Measure activities	Deliverables
4	Define the current process and VOP performance	• Process map, value stream map • Data collection plan with operational definitions • Metrics with baseline • Pareto chart • VOP matrix • Benchmarking • Check sheets • Histograms
5	Define the detailed Voice of the Customer (VOC)	• Surveys, interviews, focus groups, affinity diagrams
6	Validate the measurement system	• Measurement system validation

process maps that are used to help by documenting the current or AS-IS process. Table 2.13 shows the three levels and where they should be applied.

An example of a level 2 process map for the screening mammogram for the Women's Center is shown in Fig. 2.2.

It is also important to identify any process measures and related metrics that are used to measure quality and productivity of the processes. The current profile of the people and cultural state should be understood, including the level of skills and training of the employees and their resistance or acceptance levels to change.

The steps to complete a process map are as follows:

1. Identify level (one, two, or three) to map and document.
2. Define the process boundaries.

Table 2.13 Process levels

Level	Type/Name of tool	Purpose
Level 1	Macro- or high level, process in SIPOC	• Scope the improvement project • Provide project and process boundaries • Provide a high-level view of the process
Level 2	Process map	• Identify process improvement areas • Identify process inefficiencies • Identify waste
Level 3	Process map or flow chart	• Identify improvement areas • Identify value-added and non-value-added activities • Provide detailed how-to, procedural level

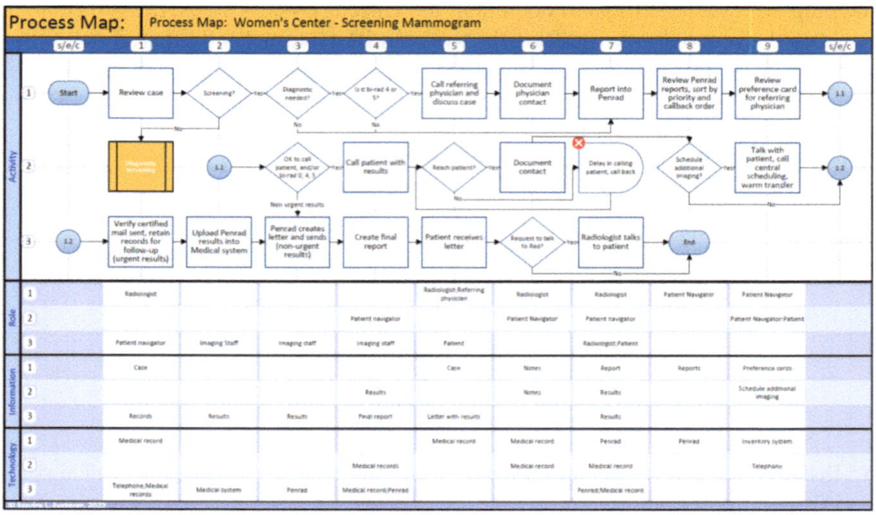

Fig. 2.2 Process map screening mammogram for Women's Center

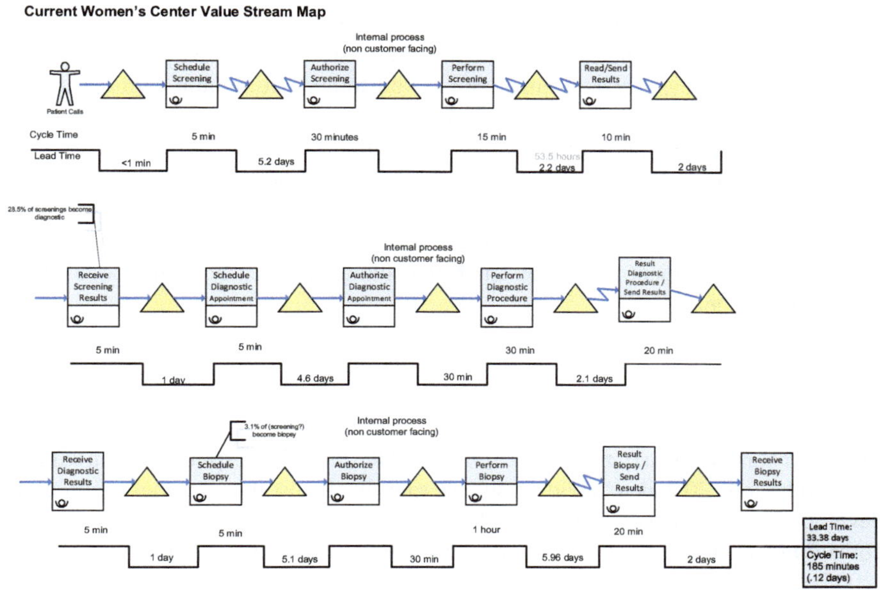

Fig. 2.3 Value stream map for the Women's Center

3. Identify the major activities within the process.
4. Identify the process steps and uncover complexities using brainstorming and storyboarding techniques.
5. Arrange the steps in time sequence and differentiate operations by symbol.
6. Validate the process map by a "walk through" of the actual process and by having other process experts review it for consistency.

Value Stream Map

A Value Stream Map (VSM) identifies the high-level value-added activities that are performed in the process. VSMs are used in Kaizen improvement efforts to quickly improve the process, remove bottlenecks, and reduce or eliminate non-value-added activities. The VSM for the Women's Center is shown in Fig. 2.3.

Define Detailed Voice of Customer (VOC)

In the Measure phase, the voice of the customer information should be collected to define the customers' expectations and requirements with respect to the service delivery process. VOC is an expression for listening to the external customer and understanding their requirements for your product or service. Examples of requirements are their expectations for responsiveness, such as turnaround time on vendor (customer) invoices, or error rates, such as employee (customer) expectations of no errors on their paycheck. The voice of the customer can be captured through interviewing, surveys, focus groups with the customers, complaint cards, warranty information, and competitive shopping. Quality Function Deployment (QFD) can be used to organize the voice of the customer information.

Personal interviews are an effective way to gain the voice of the customer; however, it can be expensive, and training of interviewers is important to avoid interviewer bias. However, additional questioning can occur to eliminate misunderstanding. The objectives of the interview should be clearly defined before the interviews are held.

Customer Surveys

Customer surveys are a typical way to collect VOC data. The response rate on surveys tends to be low, 20% is a "good" response rate. It can also be extremely difficult to develop a survey that avoids and asks the questions that are desired. Customer survey collection can be quite expensive. The steps to create a customer survey are Malone (2005) as follows:

1. *Conceptualization.* Identify the survey objective and develop the concept of the survey, and what questions you are trying to answer from the survey.
2. *Construction.* Develop the survey questions. A focus group can be used to develop and/or test the questions to see if they are easily understood.
3. *Pilot* (*try out*). Pilot the questions by having a focus group of representative people from your population. You would have them to review the questions, identify any unclear or confusing questions, and tell you what they think about the questions asked. You would not use the data collected during the pilot in the actual results of the surveys.
4. *Item analysis.* Item analysis provides a statistical analysis to determine which questions answer the same objectives, as a way to reduce the number of questions. It is important to minimize the number of questions and the total time required to take the survey. Typically, the survey time should be 10 min or less.
5. *Revision.* Revise the survey questions, and roll out the customer survey, or pilot again if necessary.

Focus Groups

Focus groups are an effective way to collect VOC data. A small representative group, typically 7 to 10 people, are brought together and asked to respond to predetermined questions. The focus group objective should be developed, and the questions should support the objective. The participants should be selected by a common set of characteristics. The goal of a focus group is to gather a common set of themes related to the focus group objective. There is no set sample size for focus groups. Multiple

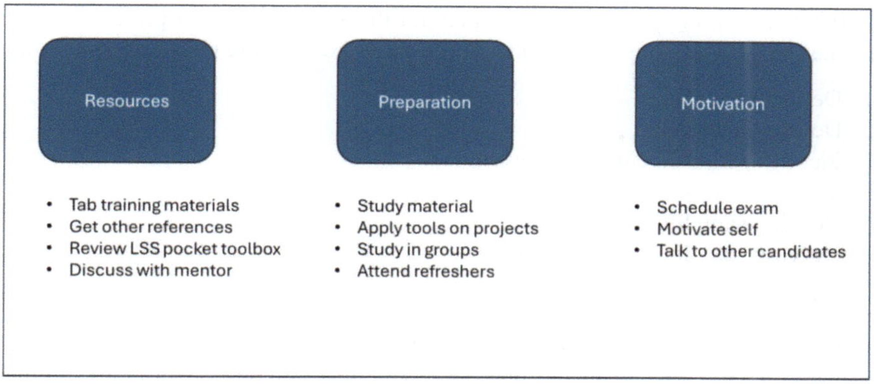

Fig. 2.4 Affinity diagram for Six Sigma Black Belt exam preparation

focus groups are typically run until no additional themes are derived. Advantages of focus groups are as follows:

- They tend to have good face validity, meaning that the responses are in the words of the focus group participants.
- Typically, more comments are derived in an interview with one person at a time.
- The facilitator can probe for additional information and clarification.
- Information is obtained relatively, inexpensively.

Some of the disadvantages of focus groups are as follows:

- The facilitator skills dictate the quality of the responses.
- They can be difficult to schedule.
- It can be difficult to analyze the dialog due to participant interactions.

Affinity diagrams organize interview, survey, and focus group data after collection. The affinity diagram organizes the data into themes or categories. The themes can first be generated, and then the data can be organized into the themes, or the detailed data can be grouped into the themes. An example of a simple affinity diagram for ways to study for a Six Sigma Black Belt exam is shown in Fig. 2.4.

Data Collection Plan

A data collection plan should be developed to identify the data to be collected that are related to the CTS criteria.

The data collection plan ensures.

- Measurement of CTS metrics.
- Identification of the right mechanisms to perform the data collection.
- Collection and analysis of data.
- Definition of how and who is responsible to collect the data.

Table 2.14 shows a data collection plan for the Women's Center project.

The steps for creating a data collection plan in the Measure phase are as follows:

1. Define the CTS.
2. Develop metrics.
3. Identify data collection mechanism (s).
4. Identify analysis mechanism (s).
5. Develop sampling plans.
6. Develop sampling instructions.

A description of each step in the data collection plan development follows:

1. Define the CTS criteria (George et al., 2005):
   CTS is a characteristic of a product or service which fulfills a critical customer requirement or a customer process requirement. CTSs are the basic elements to be used in driving process measurement, improvement, and control.
2. Develop metrics: In this step, metrics are identified that help to measure and assess improvement related to the identified CTS'. Some rules of thumb for selecting metrics are to Evans and Lindsey (2002):

- Consider the vital few versus the trivial many.
- Metrics should focus on the past, the present, and the future.
- Metrics should be linked to meet the needs of shareholders, customers, and employees.

It is vital to develop an operational definition for each metric, so it is clearly understood how the data will be collected by anyone that collects it. The operational definition should include a clear description of a measurement, including the process of collection. Include the purpose and metric measurement. It should identify what to measure, how to measure it, and how the consistency of the measure will be ensured. A summary of an operational definition follows.

Operational Definition

Defining the Measure, Definition.

A clear, concise description of a measurement and the process by which it is to be collected.

Purpose. Provides the meaning of the operational definition, to provide a common understanding of how it will be measured.

Clear way to measure the process.

- Identifies what to measure.
- Identifies how to measure.
- Makes sure the measuring is consistent.

Table 2.14 Data collection plan for Women's Center project

Data collection plan						
Critical to Satisfaction (CTS)	Metric (short title)	Operational definition (metric description)	Data collection source	Analysis mechanism	Sampling plan (size, frequency)	Process to collect and report
Patient throughput time	Total LOS	Time from patient arrives to discharged from ED (physically leaves)	ED bed board system	Statistics	Jan–April	Verify and validate data with team and subject matter experts
Patient throughput time	Triage time	Time from arrival until patient is triaged				
Patient throughput time	EDP time	Time from when the patient is seen by the ED physician to when they complete their disposition				
Patient wait time during visit	Wait time for ED	Time after Triage until the ED physician sees the patient				
Patient wait time during visit	Orders to leave ED	Time from when orders are received to when the patient physically leaves the ED				
Patient satisfaction	% left without being seen	The number of patients left without being seen/total number of patients signed into ED				
Quality of care	Lab time	Time to perform labs: time from ordered by physician to results provided to physician				

(continued)

Table 2.14 (continued)

Data collection plan

Quality of care	Diagnostic time	Time to perform diagnostic tests: time from ordered by physician to results provided to physician				
Quality of care	Register time	Time from arrival in the ED to when the patient is registered				
Quality of care	Admit time	Time from diagnosis to when the patient is admitted to inpatient floor				

3. Identify data collection mechanism(s): Next you can identify how you will collect the data for the metrics. Some data collection mechanisms include customer surveys, observation, work sampling, time studies, customer complaint data, emails, websites, and focus groups.
4. Identify analysis mechanism(s): Before collecting data, consider how you will analyze the data to ensure that you collect the data in a manner that enables the analysis. Analysis mechanisms can include the type of statistical tests or graphical analysis that will be performed. The analysis mechanisms can dictate the factors and levels for which you may collect the data.
5. Develop sampling plans: In this step, you should determine how you will sample the data, and the sample size for your samples. Several types of sampling are as follows:

   - Simple Random Sample: Each unit has an equal chance of being sampled.
   - Stratified Sample: The N (population size) items are divided into subpopulations or strata, and then a simple random sample is taken from each stratum. This is used to decrease the sample size and cost of sampling.
   - Systematic Sample: N (population size) items are placed into k groups. The first item is chosen at random, the rest of the sample selecting every kth item.
   - Cluster Sample: N items are divided into clusters. This is used for wide geographic regions.

6. Develop Sampling Instructions: Clearly identify who will be sampled, where you will sample, and when and how you will take your sample data.

Quality Function Deployment

Quality Function Deployment and the House of Quality is an excellent tool to help translate the customer requirements from the Voice of the Customer into the technical requirements of your product, process, or service. It can also be used to relate the customer requirements to potential improvement recommendations developed during the Improve phase. Figure 2.5 shows the format for the House of Quality.

The steps for creating a House of Quality are Evans and Lindsey (2002) as follows:

1. Define the customer requirements or CTS' characteristics from the Voice of the Customer data. The customer can provide an importance rating for each CTS.
2. Develop the technical requirements with the organization's design team.
3. Perform a competitive analysis, having the customer's rank your product, process, or service against each CTS to each of your competitors.
4. Develop the relationship correlation matrix by identifying the strength of relationship between each CTS and each technical requirement. Typically, a numerical scale of 9 (high strength of relationship), 3 (medium strength of relationship), 1 (low strength of relationship), and blank (no relationship).
5. Develop the trade-offs or relationship between the technical requirements in the roof of the House of Quality. You can identify a positive (+) relationship between the technical requirements, as one requirement increases the other also increases; no relationship (blank) or a negative (–) relationship, there is an inverse relationship between the two technical requirements. An example of a positive relationship can be illustrated in the design of a fishing pole. The line gauge and tensile strength both increase as the other increases. A negative relationship can

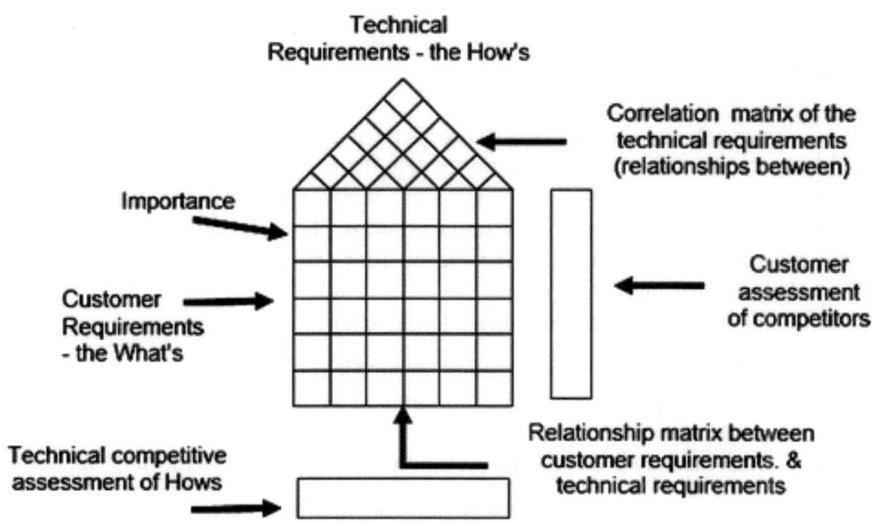

Fig. 2.5 Quality function deployment house of quality

be illustrated by line buoyancy and tensile strength. As tensile strength of the line increases, the buoyancy will be less.

6. The priorities of the technical requirements can be summarized by multiplying the importance weightings of the customer requirements by the strength of the relationships in the correlation matrix. This helps to identify which of the technical requirements should be incorporated into the design of the product, process, or service first.

Define the Voice of the Process (VOP) and Current Performance

There are many tools that can be used to assess the VOP and current performance. We will discuss the VOP matrix, Pareto charts, benchmarking, check sheets, and histograms.

VOP Matrix

A VOP matrix, developed by the author, can be used to achieve integration and synergy between the DMAIC phases and the critical components of the process to enhance problem-solving. The VOP matrix includes the CTS, the related process factors that impact the CTS, the operational definition that describes how the CTS will be measured, the metric, and the target for the metric. An example of a VOP matrix for the Women's Center project is shown in Table 2.15.

Pareto Chart

A Pareto chart helps to identify critical areas causing the majority of the problems. It provides a summary of the vital few rather than the trivial many. It demonstrates the Pareto principle that 80% of the problems are created by 20% of the causes, so that these root causes can be investigated in the Analyze phase. It helps us to arrange the problems in order of importance and focus on eliminating the problems in the order of highest frequency of occurrence.

Following are the steps for creating a Pareto Chart:

Step 1 Define the data categories, defect, or problem types
Step 2 Determine how the relative importance is defined (dollars, number of occurrences)
Step 3 Collect the data and compute the cumulative frequency of the data categories

The Women's Center project team performed a literature search to identify customer requirements that impact satisfaction in a Women's Center. Engelman et al. (2005) performed focus groups to identify the factors and dimensions related to a

Table 2.15 Voice of the Process (VOP) matrix

Voice of the Process (VOP) Matrix				
CTS (Y)	Process factor (X)	Metric	Operational definition	Target (%)
Timeliness of service	Procedure type	Procedure time	Time from when patient goes into procedure room to when they leave	
Timeliness of service	System capacity	Wait time for procedure	Time from when the patient is registered to when they get into the procedure room	
Timeliness of service	Resource availability	Time to register patient	Time from when patient arrives in the facility to when registration is complete	
Quality of service and results	Resource availability	Time to receive results	Time to receive results of exam, from when you finished the procedure	
Customer service	System capacity	Wait time to get an appointment	Time to wait until I get an appointment	
Patient satisfaction	Overall experience	Patient satisfaction results	Patient satisfaction survey results	

patient's experience with a mammography procedure. They identified the number of coded text lines from the focus group transcripts for the factors and dimensions related to the patient experience. We developed a Pareto chart identifying the most important factors and the most important dimensions for each factor. The Pareto chart for the factors is shown in Fig. 2.6.

Benchmarking

Benchmarking is a tool that provides a review of best practices to be potentially applied to improve your processes. In a Six Sigma project, process benchmarking is typically performed. The organization should document the process that they will benchmark, then select who they will benchmark. It is not necessary to benchmark a company in the same industry, but to focus on the process to be benchmarked, and select an organization that is known for having world class or best practice processes. The next step is to work with the organization to collect the data and understand how the data can be used to identify ways to improve your processes and identify potential improvement opportunities to be implemented in the Improve phase. This is similar to Motorola's benchmarking process (Evans & Lindsay, 2002). It is important to be careful when process benchmarking to ensure that you are comparing apples to apples, meaning that the organization's characteristics are similar to your own, so that the benchmarked process applies to your process.

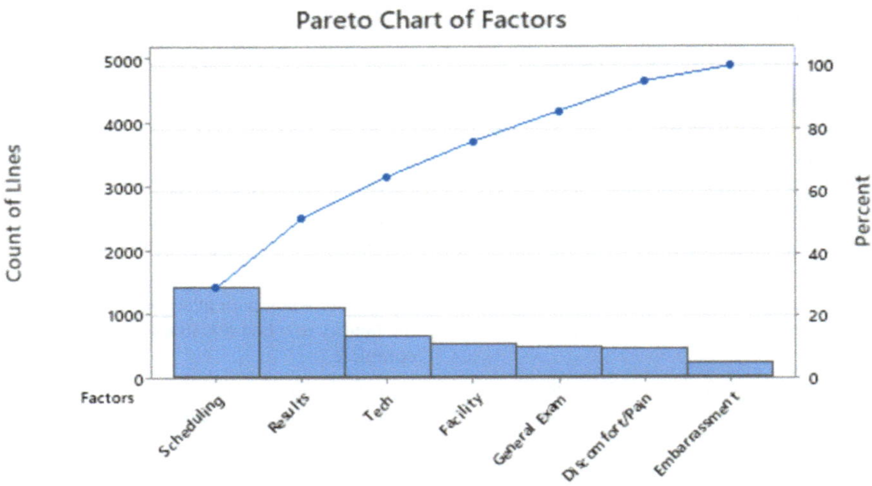

Count of Lines	1415	1093	657	541	473	462	238
Percent	29.0	22.4	13.5	11.1	9.7	9.5	4.9
Cum %	29.0	51.4	64.9	76.0	85.7	95.1	100.0

Fig. 2.6 Pareto chart of Women's Center factors

Check Sheet

A check sheet is a graphical tool that can be used to collect data on the process and types of defects so that root causes can be analyzed in the Analyze phase. The steps to create a check sheet are as follows:

Step 1 Choose a characteristic to track, i.e., defect types
Step 2 Set up the data collection check sheet
Step 3 Collect data using the check sheet

An example of a check sheet for potential errors when providing care to patients is shown in Fig. 2.7.

A Pareto Chart can then be created from the data collected on a check sheet.

Histogram

A histogram is a graphical tool that provides a picture of the centering, shape, and variance of a data distribution. Minitab or Excel is commonly used to create a histogram. It is always important to graph the data in a histogram as the first step to understanding the data.

Defect Type	Tally	Total
Incorrect patient name	𝍸	5
Incorrect patient ID	𝍷𝍷	2
Incorrect patient date of birth	𝍸 𝍷	6
Incorrect appointment time	𝍸 𝍷𝍷	7
Patient is late	𝍸 𝍸	10
Resource not available	𝍷𝍷	2
Bottlenecks in process	𝍷𝍷𝍷	3
Lack of system capacity	𝍸 𝍷𝍷𝍷	8

Fig. 2.7 Check sheet for errors when providing patient care

Statistics

Statistics can also be used to assess the VOP related to the metrics that are measured. Once the data is collected, it can be tested to see if the data distribution follows a Normal Distribution, using a test for normality. The null hypothesis is that the data is Normal. If the null hypothesis is not rejected, then the statistics that would describe the Normal distribution are the mean and the standard deviation. The mean is the average of the sample data. The mean describes the central location of a Normal distribution. The sample standard deviation is the square root of the sum of the differences between each data value and the mean divided by the sample size less one. Standard deviation (Sigma) is a measure of variation of the data. 99.997% of all data points within the normal distribution are within Six Sigma.

Validate the Measurement System

It is important to validate the measurement system to ensure that you are capturing the correct data and that the data reflects what is happening. It is also important to be able to assess a change in the process with our measuring system as well as the measurement system error. We must ensure that the measurement system is stable over time and collecting the data that allows us to make appropriate decisions.

Measurement Systems Analysis

A measurement systems analysis includes the following steps (Gitlow & Levine, 2005):

1. Prepare a flow chart of the ideal measurement system.
2. Prepare a flow chart of the current measurement system.
3. Identify the gaps between the ideal and current measurement systems.
4. Perform a Gage Repeatability & Reproducibility (R&R) study.

The measurement process variation is due to two main types of variation:

- Repeatability related to the gage.
- Reproducibility related to the operator.

The Gage R&R study assesses both the repeatability and the reproducibility. Minitab or other statistical software can be used to assess the measurement system error and improve the measurement system if necessary.

Define the Cost of Poor Quality (COPQ)

Cost of Poor Quality (COPQ)

The last step in the Measure phase can be to assess the COPQ related to your Six Sigma project. The COPQ identifies the cost related to poor quality or not doing things right the first time. The COPQ translates defects, errors, and wastes into the language of management which is cost or dollars. There are four categories of COPQ: (1) prevention, (2) appraisal, (3) internal failures, and (4) external failures.

Prevention costs are all the costs expended to prevent errors from being made or the costs involved in helping the employee do the job right every time. Appraisal costs are the results of evaluating already completed output and auditing the process to measure conformance to established criteria and procedures. Internal failure cost is defined as the cost incurred by the company as a result of errors detected before the output is accepted by the company's customer. External failure cost is incurred by the producer because the external customer is supplied with an unacceptable product or service.

Examples of prevention costs are as follows:

- Methods improvements.
- Training.
- Planning for improvement.
- Procedures.
- Quality improvement projects.
- Quality reporting.
- Data gathering and analysis.
- Preventive maintenance.
- SPC training costs.
- ISO 9000 training costs.

Examples of appraisal costs are as follows:

- Inspections.
- Process audits (SPC, ISO).
- Testing activity and equipment depreciation allowances.
- Product audits and reviews.
- Receiving inspections and testing.
- Reviews (meeting time).
- Data collection.
- Outside endorsements and certifications.

Examples of internal failure costs are as follows:

- Reaudit, retest, and rework.
- Defects and their impacts.
- Unscheduled lost time.
- Unscheduled overtime.
- Excess inventory.
- Obsolescence.
- Scrap.
- White collar mistakes.

Examples of external failures are as follows:

- Warranty.
- Technical support.
- Customer complaints.
- Customer bad will costs.
- Customer appeasement costs.
- Lost business (margin only) due to poor quality.
- Product liability.
- Return/Refunds.
- White collar mistakes.

The COPQ can help to identify potential categories of waste embedded in the process.

Analyze Phase

The purpose of the Analyze phase is to analyze the data collected related to the Voice of the Customer and the VOP to identify the root causes of the process problems, and to develop the capability of the process. The activities performed and tools applied during the Analyze phase are discussed below:

1. Develop cause-and-effect relationships.
2. Determine and validate root causes.
3. Develop process capability.

Table 2.16 Analyze phase activities and tools/deliverables

#	Analyze activities	Deliverables
7	Develop cause-and-effect relationships	Cause-and-effect diagrams Cause-and-effect matrix Why–Why diagram
8	Determine and validate root causes	Process analysis, histograms and graphical analysis (covered in measure), waste analysis, value analysis, 5S, FMEA, basic statistics (covered in measure)
9	Develop process capability	DPPM/DPMO, process capability

Table 2.16 shows the main activities mapped to the tools or deliverables most typically used during that step.

Develop Cause-And-Effect Relationships

There are several tools that can be used to generate the root causes of the problems identified in the Measure phase.

Cause-and-Effect Diagram

The cause-and-effect diagram can be used to brainstorm and document root causes of an effect or problem. It is helpful to group the causes into categories or use categories to brainstorm the causes. Typical categories are people, machine, materials, methods, measurements, and environments. Transactional categories are places, procedures, policies, people, and information systems. The steps for creating a cause-and-effect diagram are as follows:

1. Define problem.
2. Brainstorm all possible types of causes.
3. Brainstorm and organize causes by groups: people, machines, materials, methods, measurement, and environment. Can also add information systems.
4. Brainstorm/identify sub-causes for each main cause.

A Women's Center cause-and-effect diagram for patient receives results late is shown in Fig. 2.8.

Cause-and-Effect Matrix

The cause-and-effect matrix can be used to understand if the same root causes contribute to multiple effects. It is helpful to use the cause-and-effect matrix if you

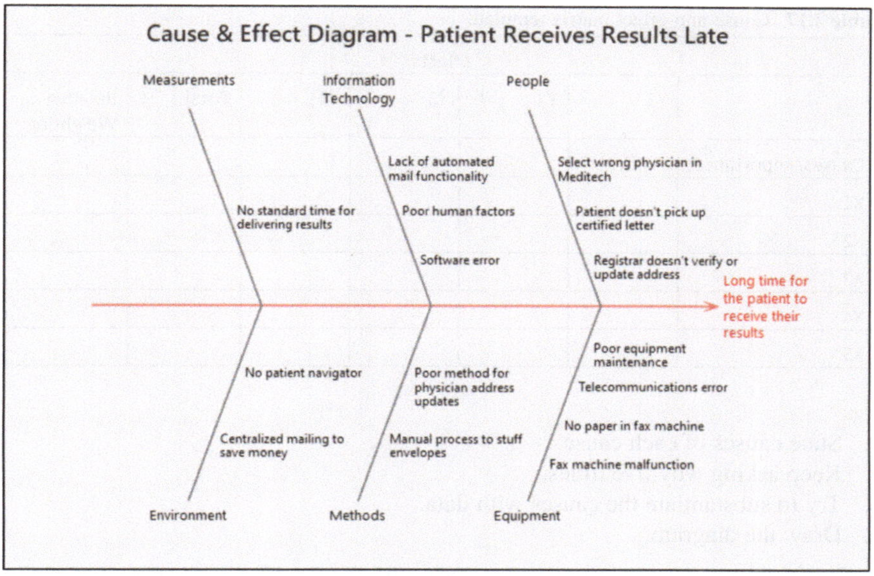

Fig. 2.8 Cause-and-effect diagram—patient receives results late

have multiple CTS characteristics or effects. The matrix establishes the relationship $Y = F(X)$, where Y equals the output variables and X represents the input/process variables or root causes. To create the cause-and-effect matrix, brainstorm the potential causes for the multiple CTS' or problems. The cause-and-effect matrix helps to relate the CTS' or output variables (Y's) to the process or input (X's) variables. The team can rate the strength of relationship between the CTS (effects) and the causes. A scale of 9 can be used for a high relationship, 3 a medium relationship, 1 a low relationship, and blank as no relationship. The customer should rate the importance of each CTS on a scale of 1–10, with 10 being the highest importance. This importance can be then multiplied by the relationship number to gain a total priority of the effects to understand where process improvement recommendations should be focused in the Improve phase. The relative weightings provide the order of importance, with 1 being the first effect to focus on related to the highest total score. A template for a cause-and-effect matrix is shown in Table 2.17.

Why–Why Diagram

The Why–Why diagram is also a powerful tool to generate root causes. It uses the concept of the five Why's, where you ask the question why several times until the root cause is revealed. Following are the steps to create a Why–Why diagram:

1. Start on left with problem statement.
2. State causes for the problem.

Table 2.17 Cause-and-effect matrix template

Effects						
	Y1	Y2	Y3	Y4	Total	Relative Weighting
Causes/Importance:						
X1						
X2						
X3						
X4						
X5						

3. State causes of each cause.
4. Keep asking why five times.
5. Try to substantiate the causes with data.
6. Draw the diagram.

Figures 2.9 and 2.10 show a sample Why–Why diagram for the Women's Center for Why does it take a long time for the patient to receive their results?

It is critical that once you brainstorm the potential root causes of the problems, that you collect additional data to substantiate the causes.

Determine and Validate Root Causes

Process Analysis

To determine and validate root causes, the Six Sigma team can perform a process analysis coupled with waste elimination. A process analysis consists of the following steps:

1. Document the process (using process maps from the Measure phase).
2. Identify nonvalue-added activities and waste.
3. Consider eliminating nonvalue-added activities and waste.
4. Identify and validate (collect more data if necessary) root causes of nonvalue-added activities and waste.
5. Begin generating improvement opportunities.

Value-added activities are those activities that the customer would pay-for, that add value for the customer. Nonvalue-added activities are those that the customer would not want to pay-for, or don't add value for the customer. Some are necessary, such as for legal, financial reporting, documentation reasons, others are unnecessary and should be reduced or eliminated. You can assess the percent of value-added activities as follows:

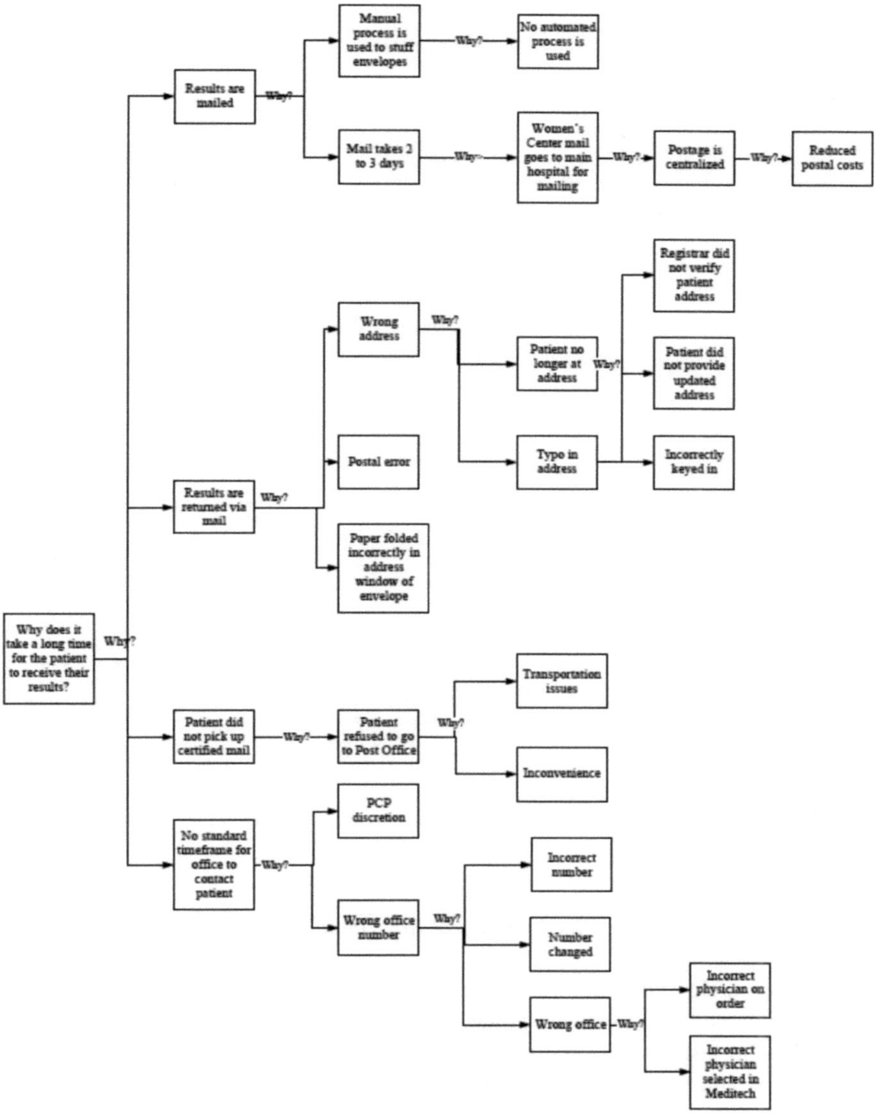

Fig. 2.9 Why–Why diagram for Women's Center Part I—why does it take a long time for the patient to receive their results?

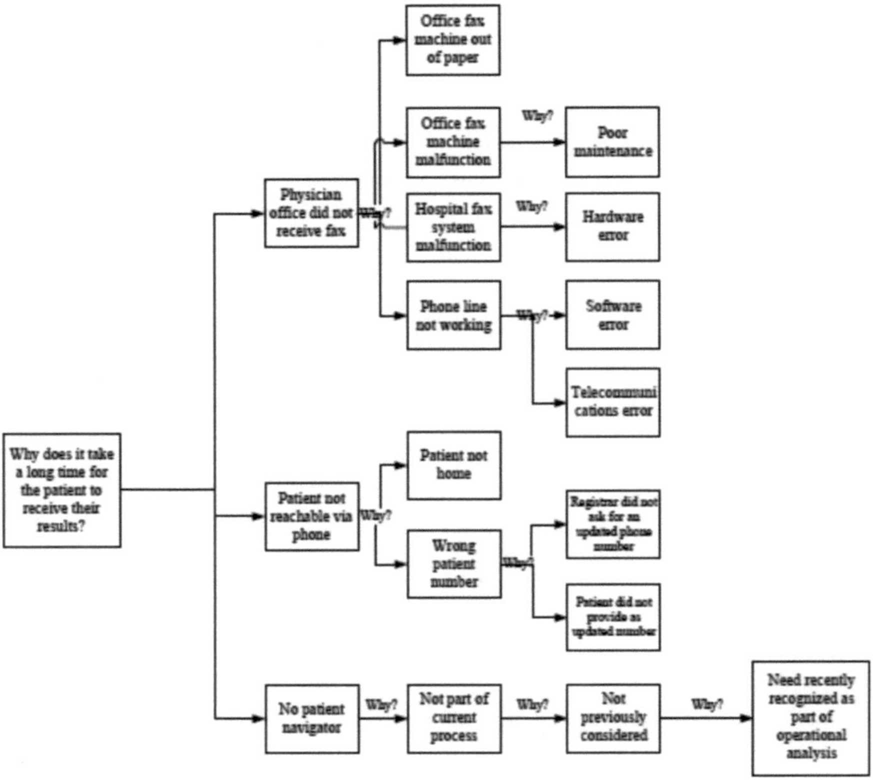

Fig. 2.10 Why–Why diagram for Women's Center Part 2—why does it take a long time for the patient to receive their results?

$$100 \times (\text{Number of value added activities}/\text{Number of total activities})\%$$

where value-added activities include operations that add value for the customer, and nonvalue-added activities include delays, storage of materials, movement of materials, and inspections. The number of total activities include the value-added activities and the nonvalue-added activities.

You can also calculate the percent of value-added time as follows:

$$100 \times (\text{Total time spent in value added activities}/\text{Total time for process})\%$$

Typically, the percent of value-added time is about 1 to 5%, with total nonvalue-added time equal to 95–99%.

During the process analysis, the team can focus on areas to identify inefficiencies in the following areas (Process Flow Analysis Training Manual, 1982):

- Can labor-intensive process be reduced, eliminated, or combined?
- Can delays be eliminated?
- Are all reviews and approvals necessary and value-added?
- Are decisions necessary?
- Why is rework required?
- Is all of the documentation, tracking, and reporting necessary?
- Duplicated process across the organization?
- What is slipping through the cracks causing customer dissatisfaction?
- Activities that require accessing multiple information systems.
- Travel—look at the layout requiring the travel.
- Is it necessary to store and retrieve all the information, do we need many copies?
- Are the inspections necessary?
- Is the sequence of activities or flow logical?
- Are standardization, training, and documentation needed?
- Are all of the inputs and outputs of a process necessary?
- How are the data and information stored and used?
- Are systems slow?
- Are systems usable?
- Are systems user-friendly?
- Can you combine tasks?
- Is the responsible person too high or low of a level?

Waste Analysis

Waste analysis is a Lean tool that identifies waste into eight different categories to help brainstorm and eliminate different types of wastes. The eight wastes are all considered nonvalue-added activities and should be reduced or eliminated when possible. Waste is defined as anything that adds cost to the product without adding value. The eight wastes are as follows:

- Transportation: Moving people, equipment, materials, and tools.
- Overproduction: Producing more product or material than is necessary to satisfy the customers' orders, or faster than is needed.
- Motion: Unnecessary motion, usually at a micro- or workplace level.
- Defects: Any errors in not making the product or delivering the service correctly the first time.
- Delay: Wait or delay for equipment or people.
- Inventory: Storing product or materials.
- Processing: Effort which adds no value to a product or service. Incorporating requirements not requested by the customer.
- People: Not using people's skills, mental, creative, and physical abilities.

5S Analysis

5S is a Lean tool that helps to organize a workplace. The 5S's are as follows:

- *Simplify*. Clearly distinguish between what is necessary and what is unnecessary, disposing of the unnecessary. A red tag is used to identify items that should be reviewed for disposal.
- *Straighten*. Organize the necessary items so they can be used and returned easily.
- *Scrub*. Fix the root cause of the dirt or disorganization.
- *Stabilize*. Maintain and improve the standards of the first three S's.
- *Sustain*. Achieving the discipline or habit of properly maintaining the correct 5S procedures.

Kaizen

Kaizen is a Lean tool that stands for Kai equals change and Zen equals for the good. It represents continuous incremental improvement of an activity to constantly create more value for the customer by eliminating waste. A Kaizen consists of short-term activities that focus on redesigning a particular process. A Kaizen event can be incorporated into the Analyze or Improve phase of the Six Sigma project to help to design and/or implement a focused improvement recommendation.

The Kaizen event follows the Plan-Do-Check-Act (PDCA) cycle, including the following steps:

Plan:

1. *Identify need*. Determine the purpose of the Kaizen.
2. *Form Kaizen team*. Typically, six to eight team members.
3. *Develop Kaizen objectives*. To document the scope of the project. The objectives should be SMART (Specific, Measurable, Attainable, Realistic, and Time-based).
4. *Collect current state baseline data*. From Measure phase or additional data as needed.
5. *Develop schedule and Kaizen event agenda*. Typically, 1 week or less.

Do:

6. Hold Kaizen event using DMAIC

Sample Kaizen Event Agenda:

- Review Kaizen Event Agenda.
- Review Kaizen objectives and approach.
- Develop Kaizen Event ground rules with team.
- Present baseline measure and background info.
- Hold event:
  - *Define*. Problem (derived from objectives), agree on scope for the event.
  - *Measure*. Review measure baseline collected.

  - *Analyze*. Identify root causes, wastes, and inefficiencies.
  - *Improve*. Create action item list and improvement recommendations.
  - *Control*. Create standard operating procedures to document and sustain improvements. Prepare summary report and present to sponsor.

- Identify and assign action items.
- Document findings and results.
- Discuss next steps and close meeting.

7. Implement: Implement recommendations, fine-tune, and train.

Check/act:

8. *Summarize*. Summarize results

   Kaizen summary report items:

- Team members.
- Project scope.
- Project goals.
- Before Kaizen description.
- Pictures (with captions).
- Key Kaizen breakthroughs.
- After Kaizen description.
- Results.
- Summary.
- Lessons learned.
- Kaizen report card with follow-up date.

9. Control: If targets are met, standardize the process. If targets are not met, or the process is not stabilized, restart Kaizen event PDCA cycle.

Failure Mode and Effect Analysis (FMEA)

A Failure Mode and Effect Analysis (FMEA) is a systemized group of activities intended to recognize and evaluate the potential failure of a product or process, identify actions that could eliminate or reduce the likelihood of the potential failure occurring and document the entire process.

The FMEA process includes the following steps:

1. Document process, define functions.
2. Identify potential failure modes.
3. List effects of each failure mode and causes.
4. Quantify effects: severity, occurrence, detection.
5. Define controls.
6. Calculate risk and loss.
7. Prioritize failure modes.

Table 2.18 Failure Mode and Effect Analysis (FMEA)

Process Step	Potential Failure Modes	Potential Effects of Failure	Severity	Occurrence	Current Process Control	Detection	RPN	Recommended Action

8. Take action.
9. Assess results.

A simple FMEA form is shown in Table 2.18.

The Risk Priority Number (RPN) is calculated by multiplying the Severity times, the Occurrence times, and the Detection ratings. The Severity is estimated for the failure and given a numerical rating on a scale of 1 (low severity) to 10 (high severity). The Occurrence is given a numerical rating on a scale of 1 (low probability of occurrence) to 10 (high probability of occurrence). The Detection scale is reversed, where a numerical rating is given on a scale of 1 (failure is easily detected) to 10 (failure is difficult to detect).

A Pareto diagram can be created based on the RPN values to identify the potential failures with the highest RPN values. Recommendations should be developed for the highest value RPN failures to ensure that they are incorporated into the improvement recommendations in the Improve phase.

Correlation Analysis

Correlation analysis measures the linear relationship between two quantitative variables which provides the relationship $Y = F(X)$ or the dependent variable (CTS) is a function of the independent variable(s). A correlation analysis can be run between any two variables, regardless of whether they both are independent or one is dependent and the other is independent. The correlation coefficient (r) is a measure that measures the strength of the relationship between the two variables. The r value falls between $+1.0$ and -1.0. The $+1.0$ signifies a positive strong correlation or positive linear relationship between the two variables. As one variable increases, so does the other variable. For example, as the number of customers increase, the sales increase. The -1.0 signifies a negative strong correlation or negative linear relationship between two variables. As one variable increases, the other variable decreases. A general rule of thumb is that an r value of $+0.80$ or greater, or -0.80 or less signifies a significant correlation between the two variables.

If we establish correlation between x and y that does NOT necessarily mean variation in x caused variation in y. There may be a third variable that is actually causing the variation that we have not accounted for. To conclude that there is a relationship between two variables does NOT mean that there is a cause-and-effect relationship.

Linear Regression Analysis

Linear regression analysis is a statistical tool that generates a prediction equation that allows us to relate independent factors to our response variable or dependent variable. We use the coefficient of determination, R^2, that allows us to identify the fit of the prediction equation. A simple linear regression relates a single x variable to a single y variable. A simple linear regression equation attempts to fit a line in with an equation of $Y = a + bx$. Here Y is the dependent or response variable, a is the Y-intercept value, and b is the slope of the line. A multiple linear regression relates multiple x's to a single y. The linear regression requires at least one independent variable and at least one dependent variable. The R-squared value will be between 0 and 1.0. Ideally, the R^2 value should be greater than 0.64 to represent a significant model. This means that our model accounts for 64% of the variation in the output.

Confidence Intervals

Confidence interval estimation or confidence intervals provide a range where there is some desired level of probability that the true parameter value is contained within it. It helps us to determine how well the subgroup average approximates the population mean. Confidence intervals help us understand the parameters between which the true population parameter, the mean, or standard deviation lies. As the sample size increases, the confidence interval width decreases. As the confidence level increases, say from 95 to 99%, the confidence interval will increase in width.

Hypothesis Testing

The purpose of hypothesis testing is to:

- Determine whether claims on process parameters are valid.
- To understand the variables of interest, the CTS'.

The hypothesis test begins with some theory, claim, or assertion about a particular characteristic (CTS) of one or more populations or levels of the X (independent variable).

The null hypothesis is designated as Ho (pronounced "H–O") and defined as there is no difference between a parameter and a specific value. The alternative hypothesis is designated as: there is a difference between a parameter and a specific value. The null hypothesis is assumed to be true, unless proven otherwise. If you fail to reject the null hypothesis, it is not proof that the null is true.

		Conclusion Drawn	
		Do not reject Ho	Reject Ho
Actual or True State	Ho True	Correct Conclusion	Type I error
	Ho False	Type II error	Correct conclusion

Fig. 2.11 Type I and Type II errors

In hypothesis testing, there are two types of error. Type I error or Alpha Risk is the risk of rejecting the null hypothesis when you should not. The probability of a Type I error is referred to as alpha. Type II error or Beta Risk is the risk of NOT rejecting the Null Hypothesis when you should. The probability of a Type II error is referred to as Beta.

When performing a hypothesis test, you select a level of significance. The level of significance is the probability of committing a Type I error or alpha and is typically a value of .05 or .01. Figure 2.11 shows the Type I and Type II errors.

The steps for performing a hypothesis test are as follows:

1. Formulate the null and alternative hypotheses.
2. Choose the level of significance, alpha, and the sample size, *n*.
3. Determine the test statistic.
4. Collect the data and compute the sample value of the test statistic.
5. Run the hypothesis test in Minitab or a statistical package.
6. Make the decision. If the *p*-value is less than our significance level, alpha, reject the null hypothesis, if not, then there is no data to support rejecting the null hypothesis. Remember, if *p* is low, Ho must go!

Some of the most common hypothesis tests are summarized in Table 2.19.

Table 2.19 Summary of hypothesis tests

Test statistics	Number of variables	Test	Parameters
Mean	1	1 sample Z	Variance
Mean	1	1 sample t	Variance unknown
Mean	2	2 sample t	Variance unknown, assume equal variances
Mean	2	2 sample t	Variance unknown, do not assume equal variances
Mean	2	Paired t-test	Paired by subject (before and after)
Proportion	1	1 proportion	
Proportion	2	2 proportion	
Variance	1	1 variance (chi square)	
Variance	2	Variance (F-test)	
Survey questions	1	Chi square goodness of fit test (one variable)	

ANOVA (Analysis of Variance)

Analysis of variance is another hypothesis test, when you are testing more than two variables of populations. Following are the steps for running ANOVA:

1. Formulate the null and alternative hypotheses.
2. Choose the level of significance, alpha, and the sample size, *n*.
3. Collect data.
4. Check for normality of data, using a test for normality.
5. Check for equal variances, using an F-test.
6. Run the ANOVA in Minitab or another statistical package.
7. Make the decision.

Common types of ANOVA:

- One-way ANOVA: Testing one variable at different levels, such as testing average grade point average for high school students for different ethnicities.
- Two-way ANOVA: Testing two variables at different levels, such as testing the average grade point average for high school students for different ethnicities and by grade level.

Customer Survey Analysis

Most surveys are attribute or qualitative data, where you are asking the respondent to answer questions using some type of Likert scale, asking importance, the level of agreement or perhaps, level of excellence.

Ways to analyze survey data:

1. Summarize percent or frequency of responses in each rating category using tables, histograms, or Pareto charts.
2. Perform attribute hypothesis testing using chi square analysis.

Unlike hypothesis testing with variable data, statistically with attribute data we are testing for dependence, not a difference, but you can think "makes a difference."

We formulate our hypotheses as follows:

- Ho: "{factor A} is independent of {factor B}."
- Ha: "{factor A} is dependent on {factor B}."

In addition to the p-value, we use contingency tables to help understand where the dependencies (differences) exist.

The customer survey analysis steps include the following:

1. State the practical problem.
2. Formulate the hypotheses.
3. Enter your data in MINITAB or another statistical package.
4. Run the chi square.

5. Translate the statistical conclusion into practical terms.

If *p*, the significance level is low, then reject Ho (if *p* is low, Ho must go). If you fail to reject the null hypothesis, Ho, that means that you fail to reject the hypothesis that the values are independent. If you reject Ho that means that they are dependent, or dependencies or differences exist.

Develop Process Capability

To develop the process capability, you can calculate the DPMO (Defects Per Million Opportunities) and related sigma level or you can calculate the capability indices. We'll first discuss DPMO.

DPMO

Six Sigma represents a stretch goal of six standard deviations from the process mean to the specification limits when the process is cantered but also allowing for a 1.5 sigma shift toward either specification limit. This represents a quality level of 3.4 defects per million. This is represented in Fig. 2.12.

The greater the number of sigmas, the smaller the variation (the tighter the distribution) around the average. DPMO provides a single measure to compare performance of very different operations, giving an apple to apples, comparison, not apples to oranges. Table 2.20 shows a sigma to DPMO conversion.

DPMO is calculated as follows:

$$\text{DPMO} = (\text{Defects} \times 1{,}000{,}000) / (\text{Units} \times \text{Opportunities})$$

where defects are the number of defects in the sample, units are the number of units in the sample, and opportunities are the number of opportunities for error. For example,

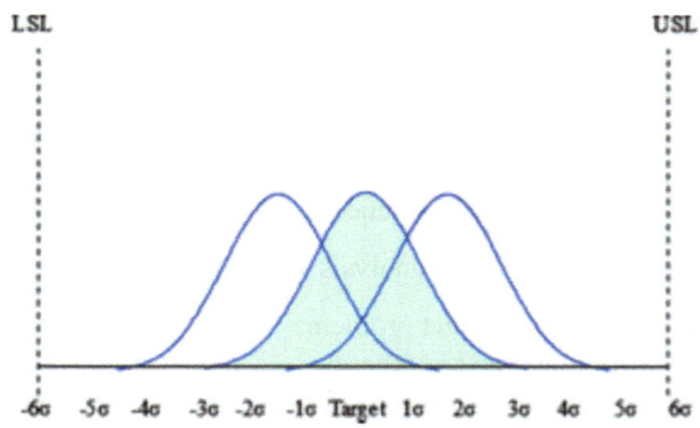

Fig. 2.12 A3.4 DPMO representing a 6 sigma quality level, allowing for a 1.5 sigma shift in the average

Table 2.20 Sigma to DPMO conversion (assuming 1.5 sigma shift)

Sigma level	DPMO
6	3.4
5	233
4	6,210
3	66,810
2	308,770
1	691,462

for a purchase order, for taking a sample of 100 purchase orders, finding 5 defects, with 30 fields on the purchase order (opportunities for errors), we calculate a DPMO of 1,667 or about 4.4 sigma.

Process Capability Study

Process capability is the ability of a process to produce products or provide services capable of meeting the specifications set by the customer or designer. You should only conduct a process capability study when the process is in a state of statistical control. Process capability is based on the performance of individual products or services against specifications. According to the central limit theorem, the spread or variation of the individual values will be greater than the spread of the averages of the values. Average values smooth out the highs and lows associated with individuals.

Following are the steps for performing a process capability study:

1. Define the metric or quality characteristic. Perform your process capability study for the metrics that measure your CTS characteristics defined in the Define and Measure phases.
2. Collect data on the process for the metric, take 25 to 50 samples.
3. Perform graphical analysis (histogram).
4. Perform a test for normality.
5. Determine if the process is in control and stable, using control charts. When the process is stable, continue to step 6.
6. Estimate the process mean and standard deviation.
7. Calculate the capability indices, usually Cp, Cpk.

$$\text{Cp} = (\text{Upper Specification Limit} - \text{Lower Specification Limit})/6\sigma$$
$$\text{Cpk} = \text{Minimum of } \{\text{CPU, CPL}\} \text{ where}$$
$$\text{CPU} = (\text{Upper Specification Limit} - \text{Process Mean})/3\sigma$$
$$\text{CPL} = (\text{Process Mean} - \text{Lower Specification Limit})/3\sigma$$

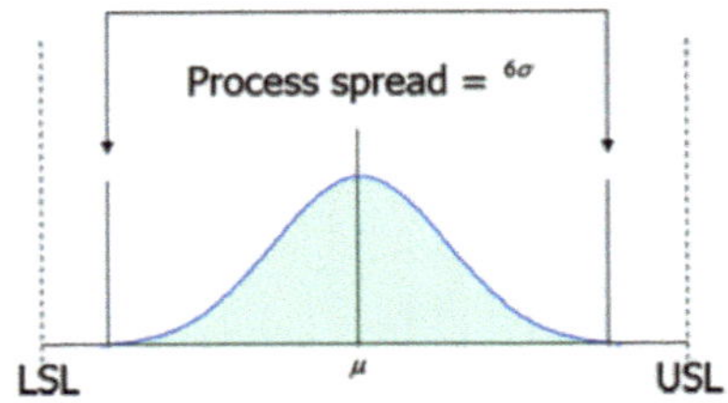

Fig. 2.13 Process is quite capable

A process can be in control but does NOT necessarily meet the specifications established by the customer or engineering. You can be in control and not capable. You can be out of control or unstable but still meet specifications. There is no relationship between control limits and specification limits. However, you must be in control before you use the estimates of standard deviation from your process to calculate process capability and your capability indices.

There are typically three scenarios regarding process capability:

1. Process spread is less than the specification spread. The process is quite capable. Figure 2.13 shows this scenario. Cp and Cpk are greater than 1.33.
2. Process spread is equal to specification spread, an acceptable situation, but there is no room for error. If the mean shifts or variation increases, there will be non-conforming product. Figure 2.14 shows this scenario:

$$\mathrm{Cp} = \mathrm{Cpk} = 1.$$

3. Process spread is greater than the specification spread. The process is NOT capable. Figure 2.15 shows this scenario. Cp and Cpk are less than 1.

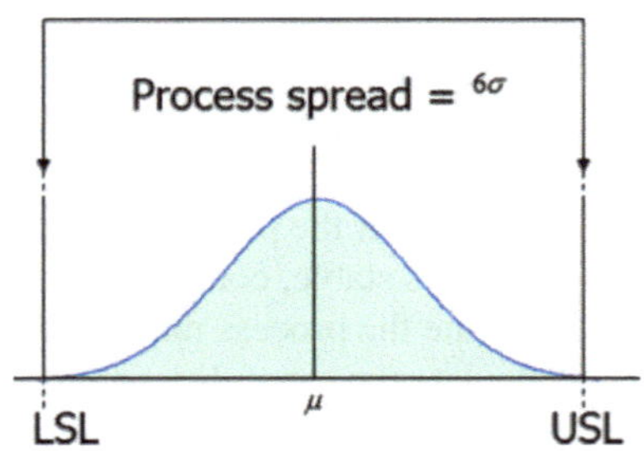

Fig. 2.14 Process is just capable

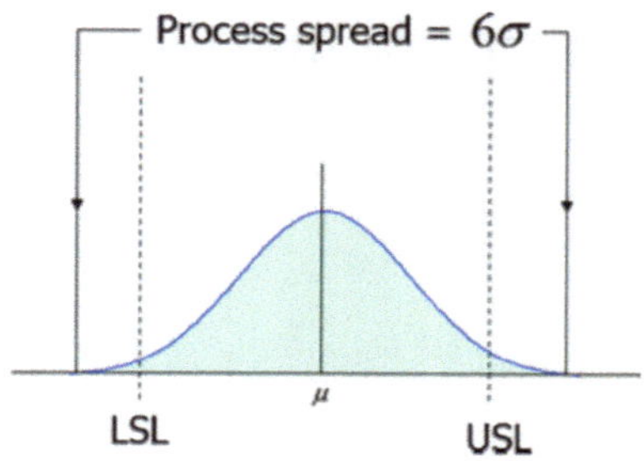

Fig. 2.15 Process is NOT capable

Improve Phase

The purpose of the Improve phase is to identify improvement recommendations, design the future state, implement pilot projects, train, and document the new processes. The activities performed and tools applied during the Improve phase are discussed below:

1. Identify improvement recommendations.
2. Perform cost/benefit analysis.
3. Design future state.
4. Establish performance targets and project scorecard.
5. Gain approval to implement, then implement.
6. Train and execute.

Table 2.21 shows the main Improve activities mapped to the tools or deliverables most typically used during that step.

Identify Improvement Recommendations

The Six Sigma team should use the data collected in the first three phases of the DMAIC to identify improvement recommendations. The recommendations should be designed to eliminate the root causes. The team should anticipate things that can go wrong or the type of resistance that you may have and design plans to work around these barriers.

Quality Function Deployment (Revised)

The QFD House of Quality can be used to map the CTS to the improvement recommendations, to ensure that the recommendations help to satisfy the CTS characteristics. The customer requirements on the left of the QFD matrix become the CTSs, and the technical requirements become the improvements.

Table 2.21 Improve phase activities and tools/deliverables

#	Improve activities	Deliverables
10	Design future state process and improvements	· Quality function deployment · Recommendations for improvement · Action plan · Cost/benefit analysis · Future state process map · Future state value stream map · Appreciative inquiry · Affinity diagram · Relationship diagram · Opportunity map · Strategies and strengths matrix
11	Establish performance targets and project scorecard	· Dashboards/Scorecards · Revised VOP matrix
12	Gain approval, train and pilot	· Training plan and materials · Procedures · Policies · RACI

The House of Quality can also be used to prioritize the recommendations, by weighting the CTS' and multiplying the weighting by the strength of the relationship in the relationship matrix of the QFD matrix.

Recommendations for Improvement

The recommendations for improvement can be a listing, with a description of the recommendation ideas. During the design of the recommendations, input should be obtained from the process owners to assess the reasonableness of the solutions. Information from the following previously used tools can be used to develop the improvement recommendations:

- QFD.
- Waste analysis.
- Process analysis.
- Cause-and-effect analysis (diagram and matrix).
- Why–Why diagram.
- 5S.
- Kaizen.
- Process map or value stream map.

Action Plan

The action plan for the improvement recommendations can be divided into short-term and long-term recommendations. Short term is 3 months or less, and long-term recommendations are those that can be implemented in greater than 3 months.

The action planning should include project plans, with resources, timelines, and risk analysis.

Change management is important when planning the improvement recommendations pilot projects. The following steps can be used to enhance the probability of success for the pilot projects:

1. Create a vision of the future state and communicate with the stakeholders.
2. Understand what is in it for each stakeholder and include them in the solution.
3. Identify who will be resistant to change, and who will be receptive. Gain some early adopters and engage them in the change process.
4. Implement the change, incorporating training and appropriate resources.
5. Monitor improvement, measure, and assess results.
6. Implement new processes, systems, and organizational structures if needed.

Perform Cost/Benefit Analysis

Cost/Benefit Analysis

You can perform a cost/benefit analysis or a COPQ assessment, to determine the return on investment or savings that are estimated by implementing the improvement recommendations.

In a cost/benefit analysis, we need to decide whether the benefits of the project outweigh the costs. We can use the benefit/cost ratio:

$$\text{B}/\text{C Ratio} = \frac{\text{Present value of benefits}}{\text{Present value of costs}}$$

If the B/C ratio is greater than 1, then you can accept the project, the benefits outweigh the costs.

Cost of Poor Quality

The COPQ categories can be used to assess the costs that can be eliminated by implementing the improvement recommendations. The team can either use this or the cost/benefit analysis or compare the improvement in the COPQ before and after the recommendations are implemented.

Design Future State

It is important to design the new future state by developing a future state process map. The team should challenge the boundaries and incorporate quality and Lean principles.

Future State Process Map

The future state process map is simply a process map of the new process incorporating the improvement recommendations.

Design of Experiments (DOE)

Design of experiments can be used to identify key variables and levels that optimize process performance and improvement quality. It helps to design a robust process that is insensitive to uncontrollable factors. It allows you to look at many factors simultaneously and to assess the interaction of variables. It enables the identification of the critical factors and the associated levels for the process design.

The steps for performing a design of experiment are as follows:

1. Set experimental objectives.
2. Select process variables.
3. Select an experimental design and identify hypotheses.
4. Execute the design.
5. Check that the data are consistent with the experimental assumptions.
6. Analyze and interpret the results.
7. Use/present the results and incorporate them into your future state design.

There are many types of experimental designs, the most common are as follows:

- One-factor experiments: Allows for the manipulation of one factor.
- Two-factor experiments: Allows for the manipulation of two factors.
- Full factorial experiments: Consists of all possible combinations of all selected levels of the factors to be investigated.
- Fractional factorial experiments: Studies only a fraction or subset of all the possible combinations of factors.

Establish Performance Targets and Project Scorecard

In this step, the team should identify the performance targets for the metrics identified in the Measure phase. They also should track pilot project status using project scorecards.

Dashboards/Scorecards

You should also create dashboards or scorecards to assess the performance of your process after trying out the improvement recommendations in the pilot projects.

The project scorecards or dashboards can be used to identify improvements in your process against your metrics that you have identified. The metrics should relate to your CTS' characteristics.

Scorecards should include the following ways to present your metrics (Pyzdek, 2003):

- Assessment of improvement of central tendency and variation overtime. SPC average and range charts can be used to meet this objective.
- Graphical distribution using a histogram for the most recent time period.
- Assessment of quality or number of defects, using SPC percent defective chart.
- Outliers showing a distribution of individual defectives using a dot plot.

Revised VOP Matrix

You can revise your VOP Matrix from the measure phase to relate your CTS', process factors, operational definition, metric, and more realistic targets for your metrics.

Gain Approval to Implement, then Implement

The Six Sigma project team should create a presentation and deliver it to the project sponsor, champion, and other management that must approve the improvement recommendations. The champion presentation should be a high-level executive summary.

The team can use the PDCA cycle to implement the pilot recommendation projects. Plan: you can plan your improvements; Do: implement your improvements, usually on a pilot scale; Check: verify that the improvement improved the process based on your metrics; and Act: if the improvements made a positive difference, implement them on a broader scale, if not, refine the improvements and try again. You may go through the PDCA cycle several times.

Train and Execute

The team should develop detailed procedures, as necessary to ensure consistency of the new process. They should develop and rollout training. The train the trainer concept is sometimes used to reduce the resources needed to train. A core group of people are trained on the new process and then they train others in the organization and become subject matter experts. The process owners should be included in the change process, and changes should be communicated to appropriate stakeholders. The team can use the future state process map as a training guide. They should assess the effectiveness of the training as part of the control plan in the next phase.

Control Phase

The purpose of the Control phase is to measure the results of the pilot projects and manage the change on a broader scale, report scorecard data and the control plan, identify replication opportunities, and develop future plans for improvement. The activities performed and tools applied during the Control phase are discussed below:

1. Measure results and manage change.
2. Report scorecard data and create process control plan.
3. Apply P-D-C-A process.
4. Identify replication opportunities.
5. Develop future plans.

Table 2.22 shows the main control activities mapped to the tools or deliverables most typically used during that step.

Measure Results and Manage Change

In the Control phase, the team should verify that the training and implementation was carried out correctly. They need to collect and analyze data to ensure process performance and improvements were made. The team needs to further manage the change for rollout of the pilot recommendations on a broader scale. The team needs

Table 2.22 Control phase activities and tools/deliverables

#	Control activities	Deliverables
13	Implement improvement recommendations and manage change	• Hypothesis testing
14	Incorporate process control plans and dashboards	• Dashboards • Basic statistics, graphical analysis, sampling, mistake proofing, FMEA, control plan, process capability, DPPM/DPMO • Statistical process control • Standard work • Revised VOP matrix
15	Implement continuous improvement cycle, plan do, check, act (PDCA)	• Prioritization matrices • Process map repository

to keep all of the stakeholders in the loop by developing and implementing a communication plan. We will collect data after we improve the process, for the same CTS' and metrics identified in the Measure phase. We will then assess whether the changes implemented made a "statistically" significant difference, using:

- Hypothesis testing

  Or

- Design of experiments.

Report Scorecard Data and Create Process Control Plan

In this step, the team should demonstrate the impact of the project's metrics and create or revise the process control plan. The Plan helps to deploy the Six Sigma approach across large areas and to coach groups through the major quality processes.

The purpose of the control plan is to maintain the gains. If a conscious plan and effort is not made to ensure that people continue to use the new process, the gains can slip, and when push comes to shove, and people get pressured and busy, they can very easily slip back to their old ways and old processes. The control plan can include the following:

- Deploying new policies and removing out-dated policies.
- Implement new standards.
- Modify procedures.
- Modify quality appraisal and audit criteria.
- Update prices and contract bid models.
- Change information systems.
- Revise budgets.
- Revise forecasts.
- Modify training.

Useful tools that can be used to derive the information to create a control plan include the following:

- Project planning for creating the control plan.
- Brainstorming.
- Failure mode and effects analysis.
- SPC.
- Process map.
- Training.
- Procedures.
- Mistake proofing.
- Statistics, graphical tools, sampling, FMEA, process capability, DPPM/DPMO.

A control plan format is provided in Table 2.23.

Table 2.23 Control plan template

Process Step	Control Mechanism	Measure/Metric	Criticality (High, Medium, Low)	Action to be taken if problems occur	Responsibility

For each major process step on the future process map, the control plan should identify how you will control the process step (control mechanism); how you will measure the process step; how critical it is to ensure control for that step; actions to be taken if problems occur; and who is responsible for monitoring control for each process step.

Mistake Proofing

Mistake proofing is a tool that helps to prevent errors in your process. Errors are inadvertent, unintentional, and accidental mistakes made by people because of the human sensitivity designed into our products and processes.

Mistake proofing, also called Poka Yoke, is the activity of awareness, detection, and prevention of errors which adversely affect our customers, our people, and result in Waste.

Some of the underlying mistake proofing concepts are as follows:

- You should have to think to do it wrong, instead of right.
- Easy to perform inspection at the source.
- Reduces the need for rework and prevents further work (and cost) on a process step that is already defective.
- Simplifies prevention and repair of defects by placing responsibility on the responsible worker.

Statistical Process Control (SPC) Charts

Statistical process control charts are an effective tool to monitor and control the process and ensure that the process is not out of control. SPC control charts are a graphical tool for monitoring the activity of an on-going process. The most commonly used control charts are also referred to as Shewhart control charts, because Walter A. Shewhart first proposed the general theory in the 1920s at AT&T Western Electric. Table 2.24 identifies the most commonly used control charts.

Table 2.24 Commonly used control charts

Most common variable charts	Most common attributes charts
X-bar and R-charts (average and range)	P-charts (proportion non-conforming)
X-bar and s-charts (average and standard deviation)	NP-charts (number non-conforming items)
X (individual) and moving range	C-charts (number non-conformities
	U-charts (number non-conforming per unit)

The following steps can be used to implement control charts:

1. Determine the type of chart, quality characteristic, sample size and frequency, and data collection mechanism.
2. Select the rules for out-of-control conditions.
3. Collect the data (10 to 25 subgroups).
4. Order data based on time order.
5. Calculate the trial control limits and create charts (Minitab).
6. Identify out-of-control conditions.
7. Remove points where you can assign causes.
8. Recompute the control limits.

Apply P-D-C-A Process

Applying the PDCA to help people continually improve the process. There is the need to focus on:

- What are we trying to accomplish?
- How will we know that change is an improvement?
- What change can we make that will result in improvement?
- If the process is performing to plan, then standardize the activities, if not, then study why not and develop a new plan for improvement.
- Focus on the next most important root cause and implement additional improvements.

Identify Replication Opportunities

In this step, it is important to identify opportunities where you can replicate the same process in the organization. This will leverage the improvement effort across the organization and potentially save additional money for the company. Identifying replication opportunities can help to support organizational learning.

Develop Future Plans

The purpose of developing future plans is to recognize the time and effort that went into the LSS project by reflecting on the lessons learned and incorporating these into future projects. Some important questions are as follows:

- Have you identified lessons learned?
- Have you identified the next opportunity for improvement?
- Have you shared the learnings with others?
- Have you documented the new procedures?
- Has everyone been trained that needs to be?
- Take time to celebrate.

Dashboards and scorecards can be used to assess where you need to focus improvement efforts in the future. Also, the cause-and-effect analysis can be used to identify the next root cause to focus improvements on.

Summary

This chapter provided a comprehensive overview of the LSS DMAIC problem-solving approach along with key tools of each phase.

References

Alukal, G. (2003). *Create a lean, mean machine.* Quality progress. ASQ.

Brassard, M., & Ritter, D. (2001). *Sailing through Six Sigma: How the power of people can perfect processes and drive down costs.* Brassard & Ritter.

Dubai Quality Group. (2003). *The birth of Lean Sigma.* The Manage Mentor.

Engelman, K., Cizik, A., & Ellerbeck, E. (2005). Women's satisfaction with their mammography experience: Results of a qualitative study. *Women and Health, 42*(4), 17–35.

Evans, J., & Lindsay, M. (2002). *The management and control of quality* (5th ed.). South-Western Thomson Learning.

Furterer, S. (2009). *Lean six sigma in service: Applications and case studies.* CRC Press.

George, M., Rowlands, D., Price, M., & Maxey, J. (2005). *Lean six sigma pocket toolbook.* McGraw-Hill.

Gitlow, H. S., & Levine, D. M. (2005). *Six sigma for green belts and champions: Foundations, DMAIC, tools, cases, and certification.* Pearson/Prentice Hall.

Kandebo, S. (1999). *Lean, six sigma yield dividends for C-130J.* Aviation Week & Space Technology.

Malone, L. (2005). *Class notes guest lecture, ESI 5227.* University of Central Florida

McIlroy, J., & Silverstein, D. (2002). *Six Sigma deployment in one aerospace company.* Six Sigma Forum website. www.sixsigmaforum.com

Process Flow Analysis Training Manual, Control Data, 1982.

Pyzdek, T. (2003). *The six sigma handbook: A complete guide for green belts, black belts, and managers at all levels.* McGraw-Hill.

Sheridan, J. (2000). *Aircraft-controls firm combines strategies to improve speed, flexibility and quality*. Gale Group, Penton Media Inc.
Womack, J. P., & Jones, D. T. (1996). *Lean thinking: Banish waste and create wealth in your corporation*. Simon & Shuster.

Bibliography

Certified Six Sigma Primer, Quality Council of Indiana, West Terre Haute, IN, 2001.

Chapter 3
Hospital and Emergency Department Throughput Improvement

Abstract Visits to emergency departments (ED) are increasing, and overcrowding has become a serious problem for hospitals, patients, physicians, and clinical staff. In this case study, a cross-functional team representing the ED, inpatient floors, and units including clinical, ancillary support staff, and nonclinical members applied Lean Six Sigma methodologies and tools to drastically reduce the patient length of stay (LOS) and the percent of patients Left Without Being Seen. There was a 31% reduction in LOS which is closely tied to patient satisfaction. The percent of Left Without Being Seen was reduced from 6.5 to 0.3%. The door-to-doctor time was reduced by 78% to an average of 23 min. Patient Satisfaction scores increased by 24% to 89.9% which placed this EDs department in the top 1% of EDs nationally.

Keywords Emergency Department · Chief Executive Officer · Emergency Department Patient · Emergency Department Physician · Against Medical Advice

Introduction

Emergency departments (ED) across the United States are experiencing increased volume and overcrowding. There are many reasons for this crowding in the ED including higher patient acuity, prolonged ED evaluations, inadequate bed capacity, a severe nursing shortage, problems with access to on-call specialists, and the use of the ED by those with no other alternative to medical care, such as the uninsured. Even those with health insurance need to access the ED as primary care networks have been unable to meet demands of unscheduled healthcare (American College of Emergency Physicians 2002).

The hospital is an approximately 500-bed acute care facility, accredited by the Joint Commission. It is affiliated with over 700 physicians representing 27 medical specialties. Its Centers of Excellence include: The Heart Center, The Cancer Center, Minimally Invasive Surgical Institute, and the Orthopedic Institute.

Our hospital-wide cross-functional team worked together to apply Lean Six Sigma tools and methods to significantly improve the patient throughput and time spent in

S. Furterer, *Lean Six Sigma for the Healthcare Enterprise*,
https://doi.org/10.1007/978-1-4471-7617-6_3

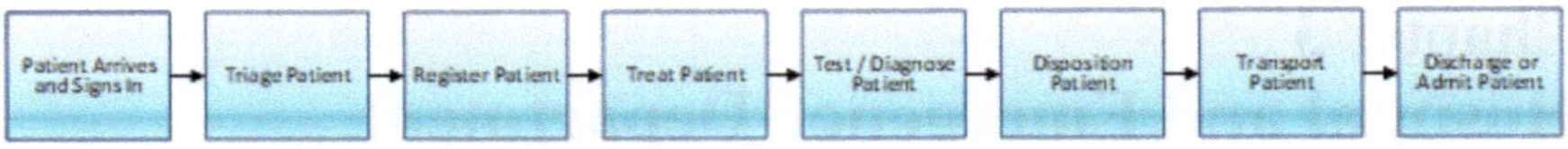

Fig. 3.1 Emergency services value chain

the ED when they seek emergency medical care. To ensure a system and hospital-wide view of the processes, the team created the Emergency Services Value Chain. This Value Chain consisted of providing emergency services to patients that came to the ED and were either discharged after assessment, testing, and medical care and disposition, or were admitted to the hospital due to their critical condition. The Emergency Services Value Chain is shown in Fig. 3.1.

The Emergency Services Value Chain consists of the following activities:

- Patient Arrives and Signs In: The patient arrives to the ED and signs into the patient kiosk.
- Triage Patient: Assess patient condition and acuity.
- Register Patient: Collect patient personal, demographic, and insurance information.
- Treat Patient: Nurse and physician stabilize the patient, provide treatment, and order diagnostics tests.
- Test/Diagnose Patient: Perform lab and radiology diagnostic tests.
- Disposition Patient: Determine patient disposition (discharge or admit to hospital).
- Transport Patient: Move patient to room if admitted.
- Discharge or Admit Patient: Discharge patient from ED or admit patient to hospital.

We used the DMAIC (define, measure, analyze, improve, control) problem-solving methodology to improve the throughput, measured by patient (LOS) and the percent of patients who leave without being seen (LWBS) treated. A truly cross-functional team was engaged on the project. The team represented clinical and nonclinical team members from the ED, administration, the hospital floors, Information Technology, Case Management, Environmental Services, Radiology, and Lab Services.

The average patient ED LOS was reduced by 86 min and 21% within the first month of the piloted improvements, while the standard deviation was reduced by 43 min and 19% in the same time period. This showed statistically significant improvement with a *p*-value of zero.

Define Phase

The Hospital and ED Throughput Improvement Project was kicked off in May 2009. This project was identified as a strategic project for the hospital. The chief executive officer (CEO) and the chief operating officer (COO) identified the project team participants by selecting key Vice Presidents and Directors in the organization that represented the critical cross-functional areas in the Emergency Services Value Chain.

The author, a certified Six Sigma Master Black Belt, was assigned as the Lean Six Sigma project leader and facilitator for the project. As the project lead, she worked with the CEO and COO to understand the desired scope of the project. The CEO and COO wanted a very quick turnaround time in reducing the LOS. The project time frame to piloting the improvements was reduced from a proposed five months to just three months.

The Lean Six Sigma DMAIC problem-solving approach was combined with the concepts of Theory of Constraints and Systems Engineering to identify the critical constraints impeding ED patient flow. The goal was to implement improvements that could quickly improve patient flow and then form additional Lean Six Sigma teams that targeted the critical constraints.

The following steps were executed in the Define Phase:

1. Develop project charter.
2. Perform stakeholder analysis.
3. Perform initial voice of customer (VOC) and identify critical to satisfaction (CTS).
4. Select team and launch the project.
5. Create project plan.

Develop Project Charter

The team reviewed the initial draft of the project charter in the first team meeting. The project charter is shown in Table 3.1, Project Charter.

Project Charter

This project is focused on improving patient throughput in the ED. About 40% of the patients seen through the ED are admitted to the hospital, versus 60% that are seen are discharged.

The Emergency Department had been experiencing delays in moving the patients through the ED in a timely manner. They recently built a larger ED to be able to better handle anticipated patient volume and better manage patient flow. However, the patient LOS was still not meeting expectations of ED physicians, nurses, staff, and administration. For January through April 2009 patients, it took an average of 5.8 h for a patient to be seen, treated, tested, and discharged, and an average of 8.7 h for a patient to be seen, treated, tested, and admitted to the hospital. There were excessive delays, and an average 6.5% of the patients left without being treated or seen by an emergency department physician (EDP), due mainly to long delays in the ED.

Table 3.1 Project charter

Project Charter
Project Overview, Problem Statement, & Goals
Provides a description of the problem to be solved. Be as descriptive as possible. Include improvement goals that are SMART (Specific, Measurable, Attainable, Realistic, and Time-Based).
Project Overview: This project is focused on improving patient throughput in the ED. About 40% of the patients seen are admitted to the hospital, vs. 60% seen are discharged. Problem Statement: The Emergency Department is experiencing delays in moving the patient through the ED in a timely manner. For January through April, patients took an average of 5.8 hours for a patient to be seen, treated, tested and discharged and an average of 8.7 hours for a patient to be seen, treated, tested and admitted. There are excessive delays and an average of 6.5% of the patients left without being seen. The total bypass hours for the past year were 341 hours. Goal(s): Improve ED throughput time to 3 average hours for discharged patients and 5 average hours for admitted patients. There could be an improvement in patient satisfaction and quality of care due to the synergistic relationship between throughput, quality and satisfaction. Scope Statement: The scope includes the ED processes starting from patient entrance, to triage, treat, transport, test/diagnose, disposition and discharge/admit. Projected Financial Benefits: Improved revenue; increased volume (reduction of Left Without Being Seen (LWBS), increased volume through increased performance; reduced costs through improved efficiency (time per patient).

The primary stakeholders are the ED Patients, Medical Associates (Doctors, Nurses, Technicians), and Ancillary Services (lab, radiology, pharmacy, transportation) that provide care to the patients. Secondary stakeholders are emergency medical services (EMS), Registration, Admitting Physicians and Consultants, Regulatory Agencies, and Hospital Administration.

The goals of the project were to improve ED throughput time to 3 average hours for discharged patients and 5 average hours for admitted patients. There could be an improvement in patient satisfaction and quality of care due to the synergistic relationship between throughput, quality, and satisfaction.

The scope includes the ED processes starting from patient arrival, triage, test and diagnosis, treatment, transporting the patient, to finally disposition and discharge or admit to the hospital. This project also included identifying the major constraints to throughput, the root causes, and improvement recommendations.

Projected Financial Benefits include improved revenue, increased volume (reduction of Left Without Being Seen (LWBS)), increased volume through increased performance, and reduced costs through improved efficiency (time per patient).

Perform Stakeholder Analysis

The primary stakeholders are the ED Patients, Medical Associates (Doctors, Nurses, Technicians), and Ancillary Services that provide care to the patients, Table 3.2. The primary stakeholders have a critical stake in the success of the project, and the processes to be improved. The ED patients are those that come to the ED either as walk-ins or are brought in by ambulance. The Medical Associates include the ED clinical staff, including the physicians, physician assistants, nurses, and medical technicians. The Ancillary Services staff includes Radiology, Transportation, Lab, and Pharmacy who provide services for the ED patients.

The secondary stakeholders are involved in the process, but at a less critical level than the primary stakeholders. The secondary stakeholders, shown in Table 3.2 are: EMS who bring the patients to the hospital via ambulance, Registration who registers the patients, Admitting Physicians and Consultants who admit patients to the hospital from the ED, Regulatory Agencies who provide standards for compliance, and the hospital's Administration who administer the hospital.

The suppliers-inputs-processes-outputs-customers (SIPOC) is used to ensure the scope of the project is well understood and is shown in Table 3.3. It is also a check and balance for the customer and stakeholder analysis. Any customer and supplier should also be identified as a project stakeholder.

The high-level process steps are triage, patient registration, treat, diagnose/test, disposition, transport, and discharge or admit. The patients come into the ED either by EMS or by private vehicle. The patients request ED care, and some are referred by their primary care physicians to go to the ED. In triage, the initial acuity of the patient is assessed. The patients are also registered in the medical information system by registration personnel. The nurse will see the patient and ensure that they are stabilized. The physician or physician assistant will also see the patient and determine the disposition. Lab and imaging diagnostic tests are ordered and performed on the patients. The patient may be further treated based on the results of the tests. The physician determines whether the patient will be discharged from the ED or admitted to the hospital. The patient is then either discharged or admitted.

Perform Initial Voice of Customer and Identify Critical to Satisfaction

The Critical to Satisfaction Criteria for this project are:

- Patient throughput time measured by LOS.
- Patient Satisfaction assessed by the percent of patients leaving without treatment.
- Quality of Care measured by quality of care measures.
- Patient waiting times, such as waiting to be seen by the EDP, waiting to be triaged, waiting for tests or test results, waiting for transportation, and waiting to be admitted or discharged.

Table 3.2 Stakeholder analysis

Stakeholder analysis					
Stakeholder	Type	Primary role	Potential impacts/ Concerns	Initial receptivity	Future receptivity
ED patients	Primary	Patients that go through the Emergency Department. Includes patients who are seen and discharged, admitted, or who leave without being seen	• Quality of Care • Low waiting time • Patient Satisfaction	Neutral	Neutral
ED physicians	Primary	Physicians who provide care for ED patients	• Efficient processes • Patient Satisfaction • Patient Throughput • Patient Capacity	Strongly support	Strongly support
Medical staff	Primary	Nurses, Technicians, Transport, Lab, Radiology, Pharmacy, other inpatient areas, who provide care for ED patients	• Efficient processes • Patient satisfaction	Moderate support	Strongly support
Administration	Primary	Administration of the Hospital	• Efficient use of resources • Patient Satisfaction • Patient Throughput	Strongly support	Strongly support
EMS	Secondary	Emergency Medical Services who transport patients to the ED from outside the hospital	• Quality of Care • Low waiting time • Patient Satisfaction	Neutral	Neutral
Registration	Primary	Register the patient	• Correct registration • Accurate billing	Moderate support	Strongly support

(continued)

Table 3.2 (continued)

Stakeholder analysis					
Stakeholder	Type	Primary role	Potential impacts/ Concerns	Initial receptivity	Future receptivity
Admitting physicians and consultant	Secondary	Physicians and Specialty Consultants who admit patients to the hospital, or provide care for the patients that go through the ED	• Patient satisfaction • Quality of Care	Neutral	Strongly support
Regulatory agencies	Secondary	Regulatory Agencies who define regulatory criteria	• Quality of Care • Revenue Integrity	Neutral	Neutral
Ancillary support	Primary	Support Staff, Environmental Services, Central Supply, Dietary, insurance, payors	• Efficient processes • Patient satisfaction • Patient throughput	Moderate support	Strongly support

Select Team and Launch the Project

The cross-functional team was identified and formed. The following areas were included as team members:

- ER nurses and physicians.
- Admitting and registration.
- Compliance.
- Imaging, Lab Diagnostics.
- Floor, unit nursing.
- Chief Nursing Officer.
- Information Technology.
- Decision Support/Finance.
- Pharmacy.
- Environmental Services.
- Administration.

Rules of Engagement:
The following rules of engagement were embraced by the project team:

- Be present and be committed at the meetings.
- Look at issues, not people; not to be defensive.
- Be open to new ideas.

Table 3.3 SIPOC

SIPOC				
Suppliers	Inputs	Process	Outputs	Customers
Patient EMS Physicians Other facilities	Request for ED Care Referrals Patient information	Triage patient	Acuity level Triage decision	Medical staff Patients
Patient Medical staff	Patient information Acuity level Triage decision	Register patient	Registered patient	Registration Medical staff ED Physicians
Patient Medical staff ED Physicians	Patient information Acuity level Triage decision	Treat patient	Patient information Physician orders	Patient Medical staff Ancillary staff ED Physicians
Patient Medical staff ED Physicians Ancillary staff	Patient information Physician orders Patient medical record	Test/Diagnose patient	Test results Specimens Diagnosis Patient medical record	Patient Medical staff Ancillary staff
Patient Medical staff Ancillary staff	Test results Diagnosis	Disposition patient	Diagnosis Patient medical record	Patient Medical staff ED Physicians Admitting and Consulting Physicians Ancillary staff
Patient Medical staff ED Physicians Ancillary staff	Diagnosis Patient medical record	Transport patient	Transported patient Patient medical record	Patient Ancillary staff Medical staff
Patient Ancillary staff Medical staff	Patient Patient medical record	Discharge or admit patient	Discharged or admitted patient Bill	Patient Insurance companies Medicare/ Medicaid

- Think outside of the box.
- Help each other work together as a team.
- Be respectful for people to finish their thoughts.
- Focused structure: agendas, outcomes, and parking lot issues.
- Limit our time; keep our meetings to the time; and finish meetings.
- Fix the process, do not fix blame.

The rules of engagement are critical to ensure the smooth running of the team meetings, so that the team can stay on topic, and change management can begin.

	Estimated Finish Date	Activities
D	5/29	Define project charter and scope, stakeholders and initial Critical to Satisfaction (CTS); document value chain, identify constraint processes
M	6/5	Measure and document current As-Is process; baseline throughput time; collect data on current process; identify constraints & bottlenecks
A	7/15	Analyze problems and their root causes; identify waste and non-value added activities.
I	7/20	Identify improvement ideas; pilot improvements; assess improvement to throughput time
C	11/9	Identify and implement control mechanisms; create procedures and training for new improvements

Fig. 3.2 Initial project milestones

Create Project Plan

The aggressive project milestones are shown in Fig. 3.2. The project was started at the end of May, and the initial pilot improvement ideas were designed by mid-July. The pilot period started on July 22nd. The pilot activities continued for several months, and the control plan was implemented in November. The project scope was much broader than was feasible to fully implement in only two months as initially proposed.

Measure Phase

The Measure Phase included the following activities:

- Define the current process.
- Define detailed Voice of Customer (VOC).
- Define the voice of process (VOP) and current performance.
- Validate measurement system.

Define the Current Process

The first step in the Measure phase was to map the current process by creating detailed process maps of the current processes in the Emergency Services Value Chain. The patient arrival, triage assessment, and assigning an ED bed is shown in Fig. 3.3. There were several additional process maps that were also generated, including, lab test processing, radiology diagnostic testing, registration and admitting, patient

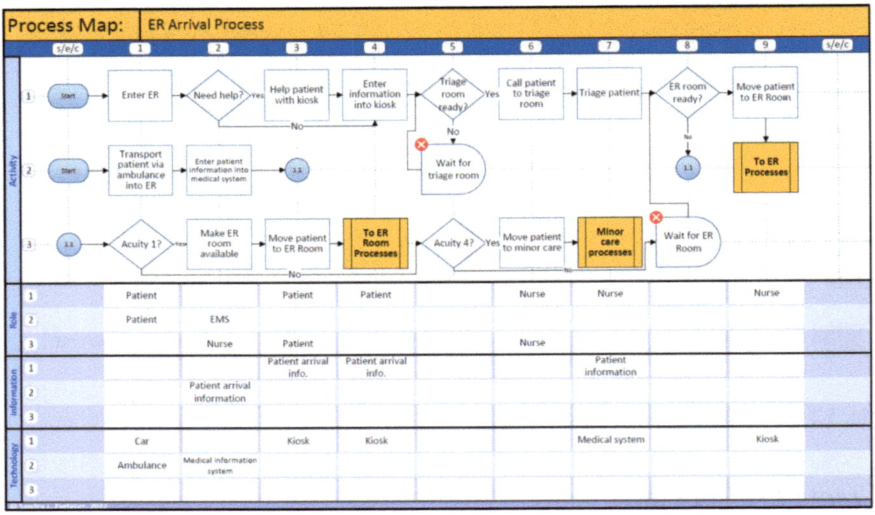

Fig. 3.3 ER Arrival current state process map

transportation, and hospital bed assignment for patients admitted to the hospital from the ED.

Define Detailed Voice of Customer

The internal VOC was gathered during the weekly team meetings as the team developed the current state process maps and later identified the performance targets and goals. There was no direct collection of ED patient voice of customer beyond the existing patient survey process, and collection of patient complaints on as needed basis.

Define the Voice of Process and Current Performance

The VOP and current performance were an extensive data collection experience. The data collection plan was developed by the team and shown in Table 3.4. The goal is to measure the Critical to Satisfaction criteria identified in the Define phase, including patient throughput time, patient wait time, and patient satisfaction.

The operational definitions were developed for each metric.

The purpose of the first set of metrics was to measure the overall ED patient throughput time, and then the detailed process times.

Table 3.4 Data collection plan

Data collection plan						
Critical to satisfaction (CTS)	Metric (short title)	Operational definition (metric description)	Data collection source	Analysis mechanism	Sampling plan (size, frequency)	Process to collect and report
Patient throughput time	Total LOS	Time from patient arrives to discharged from ED (physically leaves)	ED Bed Board System	Statistics	Jan–April	Verify and validate data with team and subject matter experts
Patient throughput time	Triage time	Time from arrival until patient is triaged				
Patient throughput time	EDP Time	Time from when the patient is seen by the ED Physician to when they complete their disposition				
Patient wait time during visit	Wait time for ED	Time after Triage until the ED physician sees the patient				
Patient wait time during visit	Orders to leave ED	Time from when orders are received to when the patient physically leaves the ED				
Patient satisfaction	% Left Without Being Seen	The number of patients left without being seen / Total number of patients signed into ED				
Quality of care	Lab time	Time to perform labs: time from ordered by physician to results provided to physician				

(continued)

Table 3.4 (continued)

Data collection plan						
Critical to satisfaction (CTS)	Metric (short title)	Operational definition (metric description)	Data collection source	Analysis mechanism	Sampling plan (size, frequency)	Process to collect and report
Quality of care	Diagnostic time	Time to perform diagnostic tests: time from ordered by physician to results provided to physician				
Quality of care	Register time	Time from arriving in ED to the patient is registered				
Quality of care	Admit time	Time from diagnosis to when the patient is admitted to inpatient floor				

Process Throughput and Time Metrics

The Patient LOS is the time from when the patient arrives at the ED (signs in at the Kiosk) to when they leave the ED, either being discharged and leaving in a private vehicle or being moved by patient transport to the hospital to be admitted. The time when the patient arrives is captured in the ED Bed Board when the patient signs in at the Kiosk in the ED waiting room. The nurse selects a field on the ED Bed Board when the patient physically leaves the ED, which is logged in the electronic ED Bed Board system. This ED Bed Board system tracks the major process steps and records the time of each step. It also provides alerts if the patients are waiting in any one step beyond identified target times. The following times are measured:

ED Patient Throughput Time (Length of Stay): Time from arrival to leaving the ED.

Wait for ED Physician: The time from patient arrival to when the physician first sees the patient.

Triage Time: The time from patient arrival to when the nurse completes the triage assessment and defines the patient acuity level.

Time to ED Bed: The time from when the patient arrives to when they are placed in an ED Bed.

Time from Bed to Physician: The time from when the patient is placed in the ED Bed to when the physician first sees them.

EDP Visit Time: The time from when the EDP initially sees the patient to when they determine the disposition (whether to discharge or admit the patient). The EDP visit time includes the time to receive the diagnostic (typically lab and radiology) tests and results.

Lab Turnaround Time: Time from when the lab order is placed to when the results are available to the EDP.

Radiology Turnaround Time: Time from when the radiology tests are ordered to when the results are available to the EDP.

Disposition to Leave ED: The time from when the EDP provides the patient disposition (discharge or admit) to when they leave the ED.

Disposition to Receive Admitting Orders: The time from when the physician provides the disposition to admit to the time when the nurse receives the admitting orders from the admitting physician.

Admit Orders to Assign Bed: The time from when the nurse receives the admit orders to when they assign the hospital bed.

Assign Bed to Leave ED: The time from when the hospital bed is assigned to when the patient leaves the ED.

Disposition to Discharge: The time from when the patient receives the admitting orders to when they leave the ED.

Figure 3.4 shows a process timeline of the process steps and relative times of each step in the current baseline performance.

To assess the baseline process performance, we collected data from the ED Bed Board electronic system for data from January through April. We validated the data and removed any data inconsistencies, such as process times that were not selected at the appropriate process step that would cause negative time values.

The second set of metrics collected was the patient volume metrics and the percent of patients leaving the ED without being treated.

We tracked the number of ED visits on a daily basis, as well as the percent of patients that were admitted to the Hospital from the ED. The other statistic that we

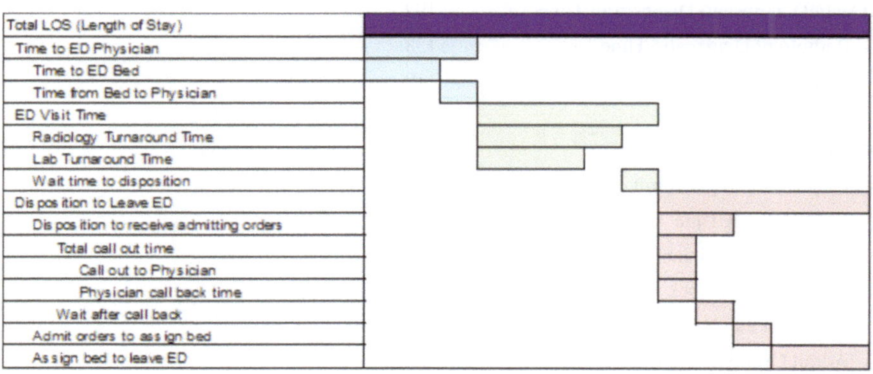

Fig. 3.4 Baseline ED process timeline

tracked was the Percent of Patients leaving the ED without being seen or treated by an EDP. The patient volume metrics are discussed in the next section.

Patient Volume Metrics

ED Visits The number of patients that came to the ED on a daily basis.

ED Admissions The number of patients that were admitted to the hospital from the ED.

LWBS The Percent LWBS patients is the number of patients LWBS divided by the Total ED visits for that time period.

The baseline metrics are shown in Table 3.5. The baseline metrics showed that the team had a tough road in front of them to get to the desired metrics of 3 and 5 h for discharged and admitted patients, respectively. The current baseline was 5.8 h for discharged patients and 8.7 h for admitted patients. For all patients combined, the baseline LOS was 6.9 h.

Table 3.5 Baseline process time and LWBS metrics

Baseline		
Process element	ED admitted	ED discharged
Total LOS	8.7 h	5.8 h
% LWBS	6.50%	
All times below are in minutes		
Total Time to EDP	93	109
– Triage Time	34	34
– Wait time triage to EDP	65	79
– Time to ED Bed	61	81
– Bed to Physician	34	55
Treat/éDiagnose to Disposition Time	192	175
– Radiology Diagnostic Time	138	270
– Lab Time	119	
– EDP Time to Visit	192	175
Disposition to leave ED	223	64
– Disposition to get admit orders	103	
– Total Call out Time	41	
– Call out to Physician	13	
– Physician Call Back	29	
– Wait after call back	27	
– Admit Orders to Assign Bed	42	
– Assign Bed to Leave ED	112	

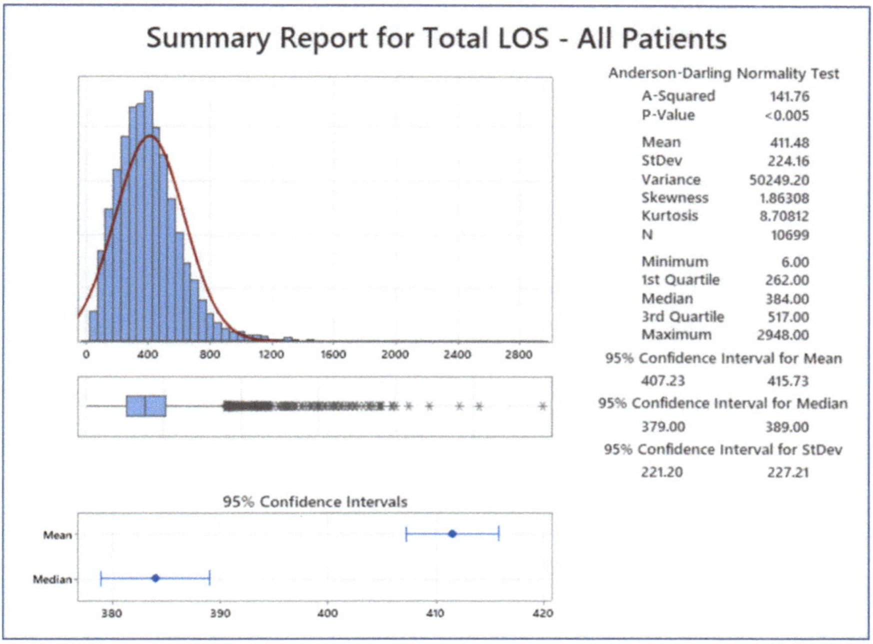

Fig. 3.5 Histogram of baseline length of stay—all patients

We created histograms of the LOS for all patients using Minitab®. The histogram is shown in Fig. 3.5, where the mean for all patients is LOS is 411.48 min. The histogram for LOS for admitted patients is shown in Fig. 3.6, and the histogram for LOS for discharged patients is shown in Fig. 3.7. The mean LOS for the admitted patients is 521.21 min, and the mean LOS for the discharged patients is 345.80 min.

The histograms illustrate non-normal distributions with fairly long tails indicating a few patients with quite long length of stays. These are typically patients that are held in the ED when critical care beds are not available in the hospital due to increased volumes. The standard deviation for the admitted patients was 3.1 h, and for discharged patients was 3.0 h. The standard deviation across all patients was 3.1 h.

Validate Measurement System

We reviewed the electronic data in the ED Bed Board to validate the data collected electronically in the ED. We observed the processes to understand how the data are collected, and whether they are consistent and represent the process times.

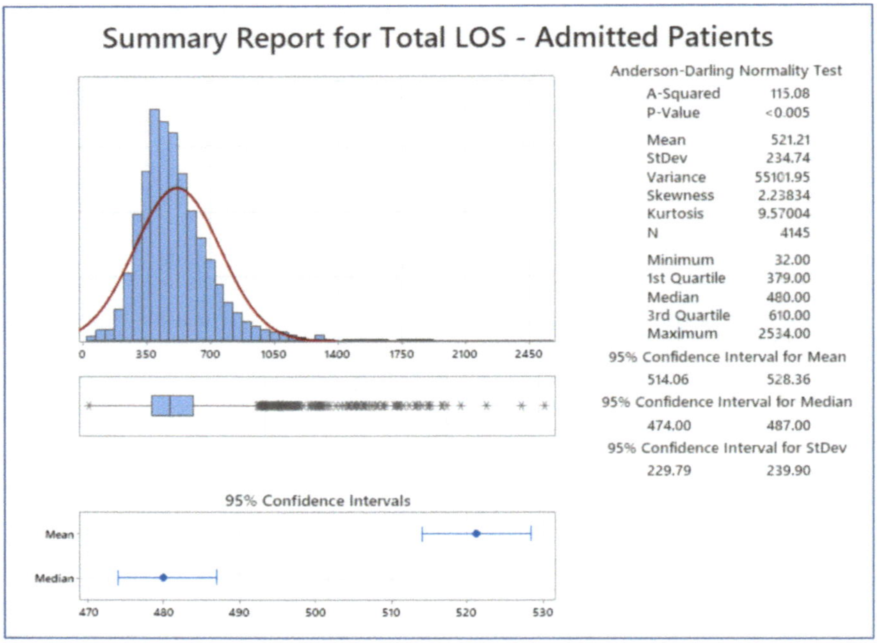

Fig. 3.6 Histogram of total LOS (minutes) for admitted patients

Analyze Phase

The following activities were performed for the Analyze Phase to identify and validate the root causes and to assess whether the process is currently capable of meeting the desired specifications for the LOS:

1. Develop cause-and-effect relationships.
2. Determine and validate root causes.
3. Develop process capability.

Develop Cause-And-Effect Relationships and to Determine and Validate Root Causes

We collected and analyzed data from the electronic ED Bed Board system regarding the metrics identified in the measure phase. We stratified the data in many ways to determine the relationships between different process factors.

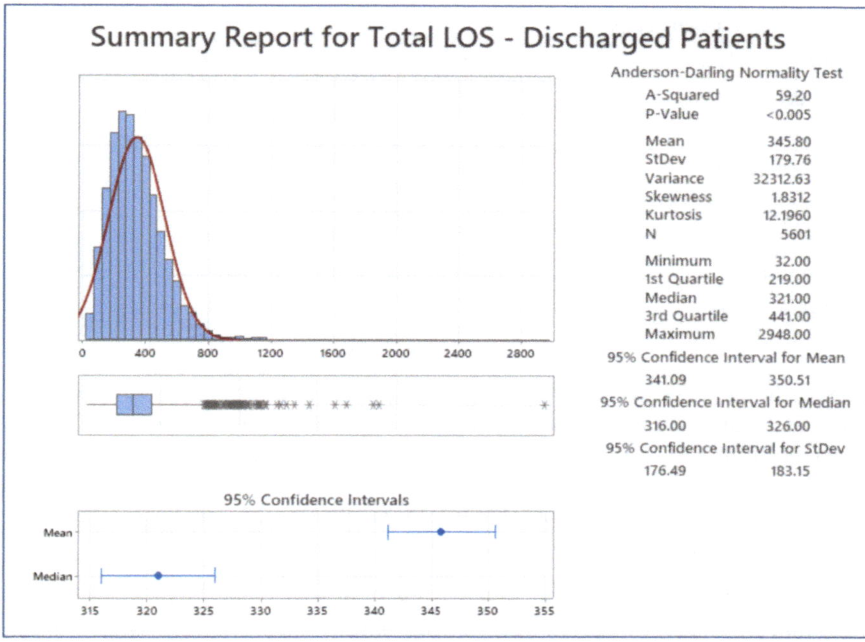

Fig. 3.7 Histogram of total LOS (minutes) for discharged patients

Patient Demographics

We first provide some patient demographics:

The average patient age is 55 years of age. There are 56% females and 44% males.

For the patients that were admitted to the hospital 54% of the patients are females, and 46% are males. Of the discharged patients, 58% of the patients are female and 42% are males. Therefore, the sicker patients tend to be males. Males are also more likely to leave against medical advice (AMA), with 56% of the AMA patients being males, and 44% are females. Of the expired patients, 55% were males and 45% were females. Of the patients who leave without being seen by an EDP, 58% are female, and 42% are male. Therefore, the women who tend to be less sick will also leave quicker if there is a longer wait to be seen by the EDP. Twenty-two percent of the patients are self-pay (do not have insurance) that come through the ED. The demographics statistics are summarized in Table 3.6.

Statistical Analysis

We then performed our statistical analysis:

The first analysis was a correlation analysis to determine if there were any significant correlations between the following variables, with the R (correlation coefficient),

Table 3.6 Patient Demographics

Demographic characteristic	Percent males (%)	Percent females (%)
Percentage of patients overall	44	56
Admitted patients	46	54
Discharged patients	42	58
Against Medical Advice (AMA)	56	44
Left Without Being Seen (LWBS)	42	58
Expired patients	55	45
Average Age: 55	Self-Pay: 22	

shown in Table 3.7. Generally, to show a strong linear relationship between two variables, the R value should be close to $+\ 0.80$ or above, or -0.80 or below. The R value can range from 1 to $-\ 1$. An R of 1 shows a positive linear relationship, as the one variable increases, the other variable also increases. An R of $-\ 1$ shows a negative linear relationship, as the one variable increases, the other variable decreases.

The variables that had a very high positive correlation were:

1. Time to EDP and time to ED Bed. This demonstrates that the longer it takes to get a patient placed in an ED Bed, the longer the time is until the patient first sees an EDP.
2. Time to Assign Hospital Room and Time from ED Disposition to Admit Orders Received. The time after the EDP provides the disposition to admit the patient, and the time that it takes to receive the admitting physician's admitting orders is correlated to the time it takes to assign the hospital room. These times are both in the back-end of the process in the steps to get the patient admitted to the hospital.
3. LOS and Time from ED Disposition to Admit Orders Received. The time it takes to call the admitting physician and get the admitting orders back from them is highly correlated to the LOS, meaning that this time tends to be fairly lengthy and extends the entire patient's LOS.
4. ED Patient Census and Hour of Day. The ED Patient Census is fairly well correlated to the time of day.
5. Time from ED Disposition to Admit Orders to Time Disposition to Call. It makes sense that the time to call out for admitting physician orders to when the orders are received is correlated.
6. LOS and Time to Assign Hospital Room. The total LOS is somewhat correlated to the time to assign the hospital room.
7. Imaging tests and lab tests. The number of imaging tests and lab tests is correlated. Many imaging tests also require lab tests, such as a pregnancy test being required for women of childbearing age prior to performing a CT Scan of the abdomen. Additionally, 72% of the patients get at least one lab test, while there is a median of 1 radiology test per patient.
8. Clinical Staffing in the ED (Nurses and Medical Technicians) and ED Census by hour of the day were fairly well correlated.

Table 3.7 Correlation analysis with R (correlation coefficient)

Variable 1	Variable 2	Correlation Coefficient (R)
Relatively strong correlations		
Time to ED Physician	Time to ED Bed	0.931
Time to Assign Hospital Room	Time from ED Disposition to Admit Orders Received	0.722
Length of Stay	Time from ED Disposition to Admit Orders Received	0.614
ED Census	Hour of Day	0.596
ED Census	Clinical Staffing	0.594
Time from ED Disposition to Admit Orders	Time Disposition to Call	0.564
Length of Stay	Time to Assign Hospital Room	0.523
Imaging tests	Lab tests	0.501
Relatively low positive correlations		
Length of Stay	ED Physician Visit Time	0.45
Length of Stay	Time from Hospital Room Assigned to Leave ED	0.449
Length of Stay	Imaging tests	0.407
Length of Stay	Lab tests	0.407
Length of Stay	Total Time to EDP	0.41
Staffing	Patient arrival patterns	0.33
Length of Stay	Time to ED Bed	0.264
Length of Stay	Arrival Time of Day	0.05
Low negative correlation		
Length of Stay	ED Physician	−0.024
Length of Stay	Day of Week	−0.094
Length of Stay	ED Nurse	−0.074
Length of Stay	Acuity Level	−0.212
Acuity Level	Number of Imaging Tests	−0.287

The remaining variables that we tested did not have a very high positive or negative correlation.

Graphical Analysis

We graphed the patients' census, arrival, and departure patterns by hour of the day, shown in Figs. 3.8 and 3.9. The patient arrival patterns are fairly predictable and peak at around 10 am. The patient census is also fairly predictable, and mimics the arrival patterns phased approximately by the average LOS of 6 h. The ED patient census tends to peak about 5 pm. The patients get "clogged" in the system due to the peak arrivals in the middle of the day, and it takes 6 to 7 h to finally move them through the system.

The departures per hour are roughly an inverse of the patient arrival patterns by hour. The patients' departures peak about 10 pm, shown in Fig. 9. So, the arrival peak is 10 am, and the patients don't clear out until roughly 12 h later.

The mean patient LWBS pattern is similar to overall patient departure patterns, shown in Fig. 3.10. The patient LWBS pattern also appears to peak with the patient departure rates. However, this is somewhat misleading, because the ED staff only records the patient as an LWBS when they are called from the waiting room, when they would have been triaged, but the patient could have left hours earlier and just did not notify anyone of their departure.

Another interesting finding is that the average triage time for the patients that were LWBS is twice the average of the admitted and discharged patients (53 min compared to 34 min), as shown in Fig. 3.11. So, it appears that the patients that wait almost an hour for triage get tired of waiting early in the process, and if we can improve the patients' experience right at the beginning and get them triaged faster, we may be able to drastically reduce the LWBS percentage.

The average LOS differs slightly by day of the week, as does the average ED Census, so the more patients in the system, the longer the LOS. The Average LOS by day and the average ED Census by day are shown in Fig. 3.12.

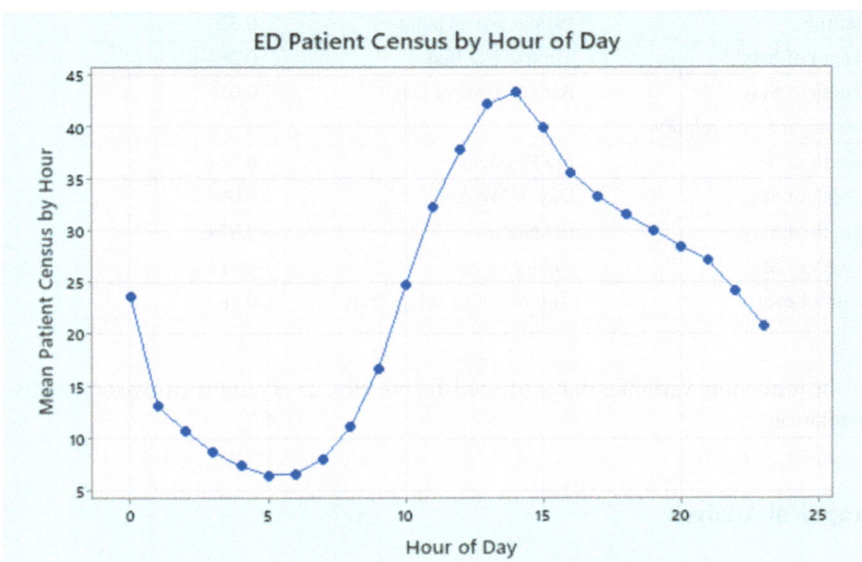

Fig. 3.8 ED patient census by hour of day

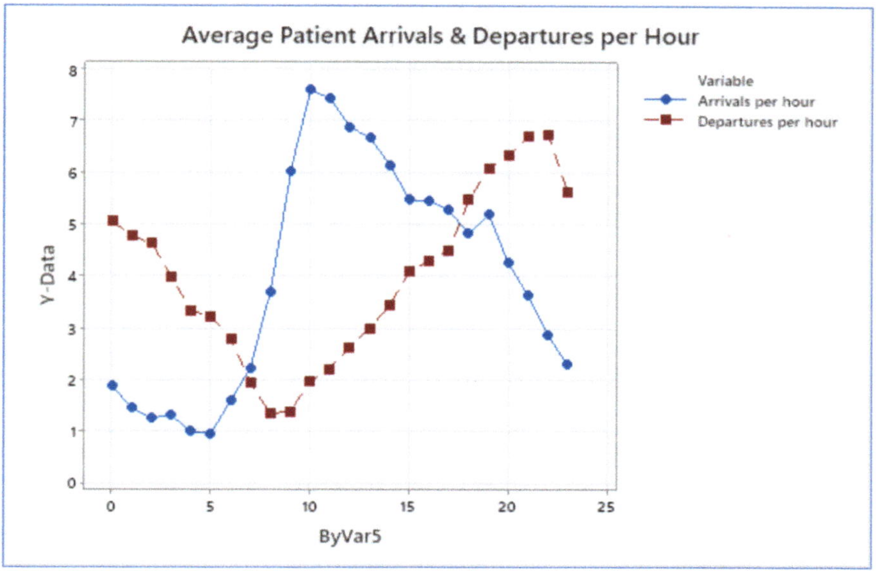

Fig. 3.9 Patient average arrivals and departures by hour of day

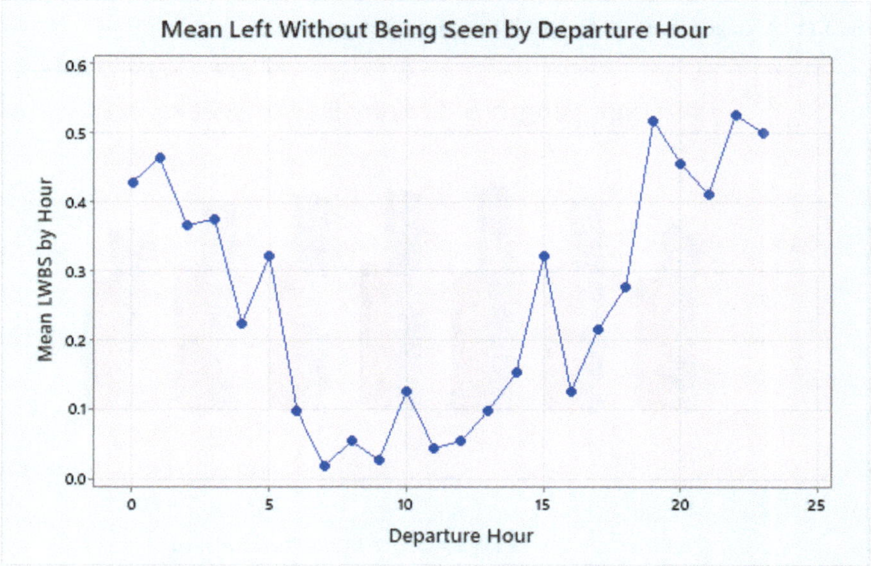

Fig. 3.10 Mean patients left without being seen by departure hour

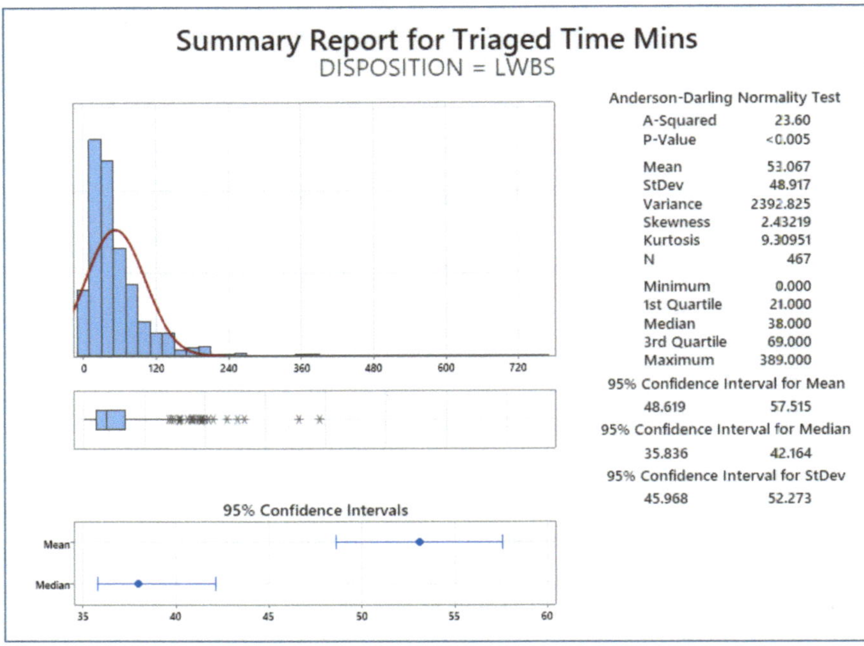

Fig. 3.11 Average triage time for left without being seen patients

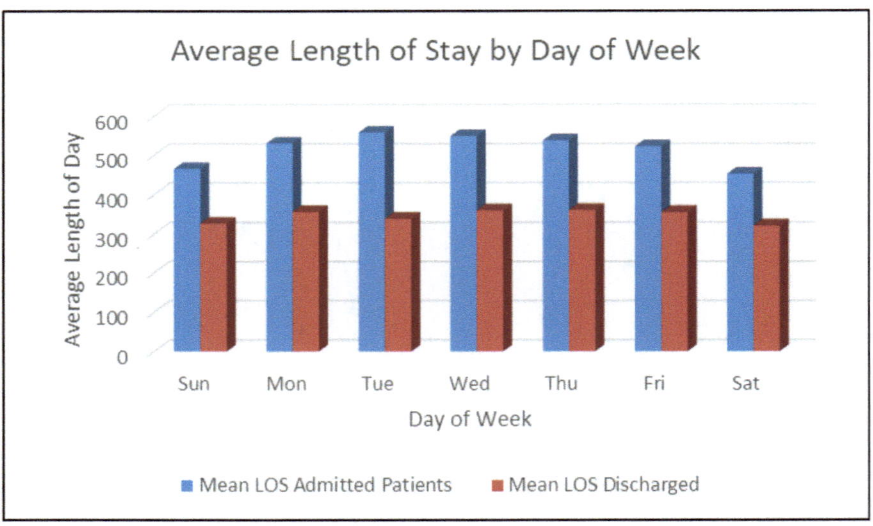

Fig. 3.12 Average length of stay by day of week

Takt Time, Identifying Constraints/Bottlenecks, and System Capacity

We calculated the takt time for the ED Value Chain to assess the required LOS based on the current volumes. With the complexity of the value chain that we are trying to improve, it is important to identify the key bottlenecks so that it focuses our improvement efforts. The takt time describes the time that is required to complete each unit to be able to meet the customer demand.

Takt Time = Total Available Time/Number of ED Visits.

Total Available Time = 32 hospital rooms * 1,440 min per day = 46,080.

Number of ED visits per day = 90.

Takt time = 46,080/90 = 512 min per day = 8.5 h per day.

We then calculated the patient volume capacity.

Patient volume capacity = Total Time Available/System cycle time.

Patient volume capacity = 46,080/409 = 113 patients.

Since the average number of patients per day is 90, the ED is working at only 80% of capacity (90/113). Once the census gets above about 85 per day, the LWBS rate skyrockets.

The process with the lowest needed takt time will constrain the process the most. The takt time for each major process is shown in Table 3.8. The triage and registration processes are the most constraining bottlenecks in the ED process. Entering orders and then performing the labs are the next most constraining bottlenecks.

Table 3.8 Takt time for major processes

Process	Staff	Available Time	Demand	Takt Time
Triage	1	1440	91	15.8
Register	1	1440	91	15.8
Enter orders into system	1	1440	74	19.5
Perform labs	1	1440	66	21.8
Perform x-rays	2	2880	53	54.3
Perform CT scans	1	1440	36	40.0
Perform Ultrasound	1	1440	12	120.0
Perform Disposition	2	2880	91	31.6
Call for admit orders	1	1440	34	42.4
Take report	8	11,520	34	338.8
Move patient	1	1440	34	42.4
Discharge	8	11,520	57	202.1

Develop Process Capability

The process is not stable, as shown in the X-bar and S control charts in Fig. 3.13. Therefore, a long-term process capability should not be assessed until control and stability is attained. A process capability analysis was performed, in Fig. 3.14, showing that the short-term process capability index Cp is 0.14, with a Cpk of –0.14. The existing process, as designed, is not capable of meeting the target LOS for admitted patients of 5 to 7 h.

Cause-And-Effect Analysis

The team performed a cause-and-effect analysis using the five Whys and Why–Why Diagrams. The team started with asking the following questions and asked Why at least five times until they believed they reached the root causes of each question.

- Why is the patient flow not continuous?
- Why are admitting physician call backs taking on average 43 min?
- Why are patients waiting in minor care?
- Why is triage time an average of 34 min for admitted and discharged patients, but an average of 53 min for patients that are LWBS (treated)?
- Why are there delays in registering patients?
- Why are there delays in the nurse seeing patients?

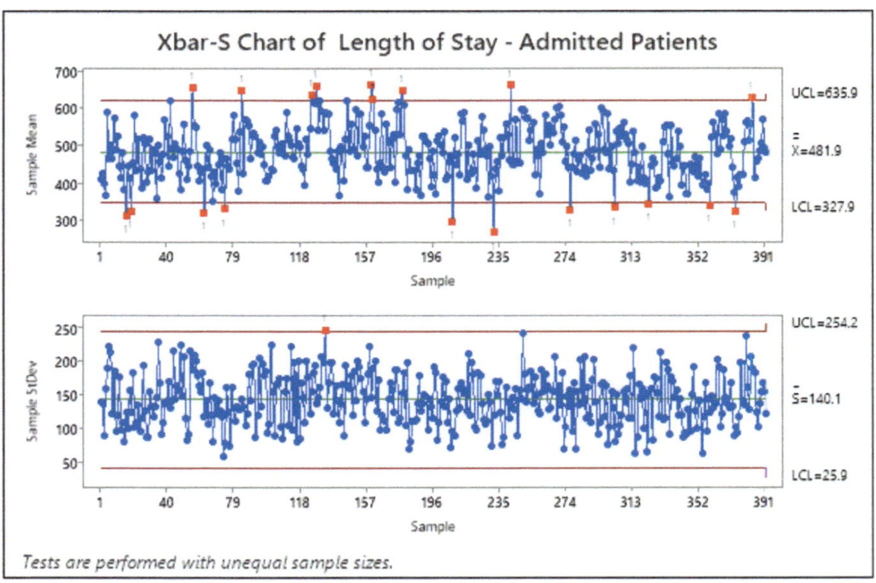

Fig. 3.13 Length of Stay X-bar and S control charts

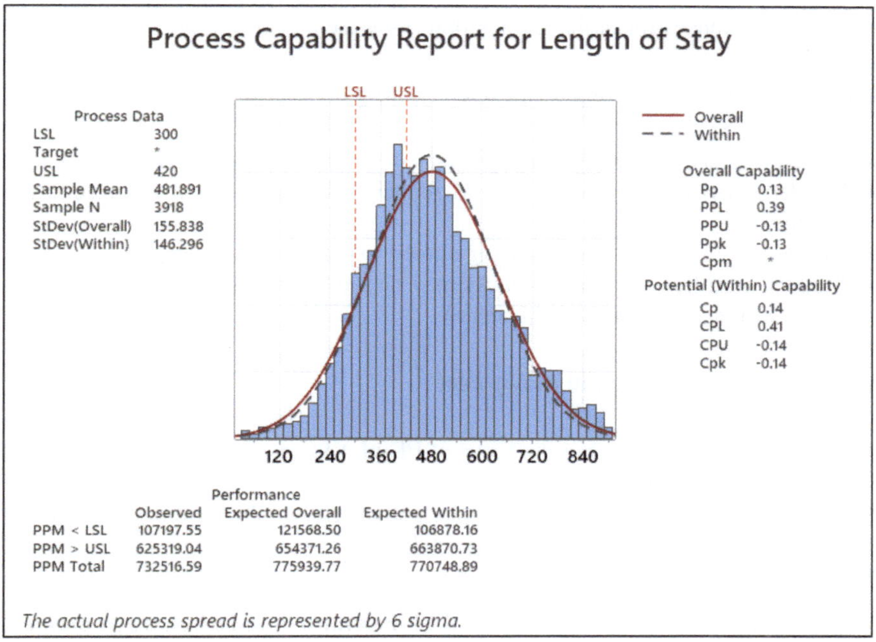

Fig. 3.14 Process capability analysis

- Why are there delays in expediting time to physicians?
- Why are there delays in radiology diagnostic testing?
- Why are there delays in lab tests?
- Why are there delays in discharging patients?
- Why are there delays in assigning beds for admitted patients?
- Why are there delays in moving patients to rooms?

The following root causes were identified for the key questions asked.

Why is the patient flow not continuous?

- More discharges occur in the PM, causing batches of patients to be discharged.
- Unit clerks on the floors batch discharges and schedule discharge order entry around lunch and end of shifts, which clogs the hospital rooms and prevents patients from being able to be admitted from the ED.
- Not prioritizing ED admissions or separating them from other admissions.
- Charts are not organized in one place.
- Staff is not always available.
- No training available to enter rooms as clean by cleaning staff.
- No expectations for cleaning personnel to enter rooms as cleaned directly into system.
- Order entry is not user friendly.

- No consequences for hiding beds, not turning them over quickly.

Why are admitting physician call backs taking on average 43 min?

- Not calling out fast enough to physicians or not calling back frequently enough.
- No dedicated call person to call to request admitting orders from admitting physicians.
- Not calling on-call physicians using agreed to protocols.
- Medical staff by-laws for call backs may be too liberal (three call backs 15 min apart before calling on-call physician).

Why are patients waiting in minor care?

- Minor care patients are not the focus of ED physicians, due to lower acuity, they are seen first by physician assistants.
- Physicians batching several minor care patients to see them all at once after the physician assistants have seen the patients.
- Lack of flow and staffing.

Why is triage time an average of 34 min for admitted and discharged patients, but an average of 53 min for patients that are LWBS (treated)?

- Only one triage nurse, even when volume increases.
- No surge policies or procedures to address intense volume increases in the ED.
- Too busy and forgot to triage.
- Triaging is not a priority.
- No one "owns" patients in waiting room.
- No clinical visibility of patients in waiting room.
- Staffing is not based on potential ED admits from actual ED volume and admission trends.

Why are there delays in registering patients?

- Registration screens are not organized efficiently.
- Quick registration requires screens several layers deep.
- The way registration staffing is scheduled, and registration is done.
- Wrist bands print only in one place.

Why are there delays in the nurse seeing patients?

- Nurses must do what the techs would do because the techs are transporting patients.
- No script for communicating expectations to patients.
- No one in charge or focused on patient flow.

- Nursing coverage is down due to call outs and staffing.
- Nurses call in sick, can't schedule on short notice, and not enough ED nurses as backup.
- Staffing not based on actual ED volume.
- Transport not trained to transport telemetry patients, so nurse must transport them.
- Transportation prioritization not always aligned to ED needs.
- If patient is already moved when transport arrives, they won't take the next patient.
- No paging or easy way to contact the nurse.
- Admitting several patients in a batch at the same time, due to no floor nurse available to take report.

Why are there delays in expediting time to physicians?

- Staffing not based on actual ED volume and patient arrival patterns.
- No process to deal with ED surge in volume.
- Number of tests ordered by physicians varies.
- Some ED physicians do not see patients near the end of the shift.
- Not having prompt discharge instructions.
- Physicians are batching to see several minor care patients at the same time.
- Twelve-hour shifts create fatigue and slowness toward end of shift.
- Staff works multiple shifts during week.
- Radiologists go to lunch together.
- Delays at the end of shift in reading radiology results.

Why are there delays in radiology diagnostic testing?

- Shorter contrast is not used; it takes 1½ hours to prep patient with contrast.
- Not always prioritizing, planning, or sequencing tests.
- Competing modalities for tests.
- After hours tests may not be prioritized.
- After hours radiology services don't read directly on PACS (in-house radiology system).
- Sometimes radiologists leave early.
- Radiologists go to lunch together.

Why are there delays in lab tests?

- Pregnancy tests are not always ordered right away.
- Not always planning, prioritizing, or sequencing tests.
- Label printers go down.
- Specificity of some tests.
- High volume of diagnostic tests.

Why are there delays in discharging patients?

- Wait on physician to do discharge instructions and prescriptions.
- Wait on nurse to discharge patient physically and from the system.
- Private physicians want to see their patients before they are discharged.
- Physician Assistants batch to physicians from minor care.
- Homeless patients need shelter and medications.
- Patient may not have a ride.
- Patients eat before they leave.

Why are there delays in assigning beds for admitted patients?

- Inefficient admit and discharge processes.
- Unit/floor staffing is not integrated with ED staffing.
- Late physician rounding.
- Too many steps and ways to communicate.
- Admitting must have order to enter bed.
- Not updating the system with potential and actual discharges.
- No way to identify pending discharges in the bed board system.
- When Bed Board System person is not there, no one assigned to cover.

Why are there delays in moving patients to rooms?

- Cleaning personnel are not trained to enter cleaned rooms directly into the system.
- No status in system for pending versus actual discharges.
- No clear definition for pending versus potential discharges.
- No pending information from Assistant Nurse Manager who knows pending discharge information.
- ED not calling physicians promptly.
- Liberal call back policies.
- Practices of not giving the physician the messages.
- Not leveraging the Bed Board System.
- No easy communication method to give report to floor/unit.
- More discharges in PM versus AM, but staffing assigned to AM.
- Transports not prioritized based on flow between the ED and floors/units.

Figure 3.15 shows one example of a Why–Why Diagram for asking the question, "Why are there delays in moving patients to rooms?".

The next step in the cause-and-effect analysis was to align the causes by the process steps and then prioritize them based on the CTS Criteria of LOS, wait times, and patients LWBS.

An example of one of the Cause-and-Effect Matrices is shown in Table 3.9.

To derive the high-priority root causes, we first prioritized the critical to satisfaction (CTS) criteria, with the LOS being given the highest priority of a nine, the

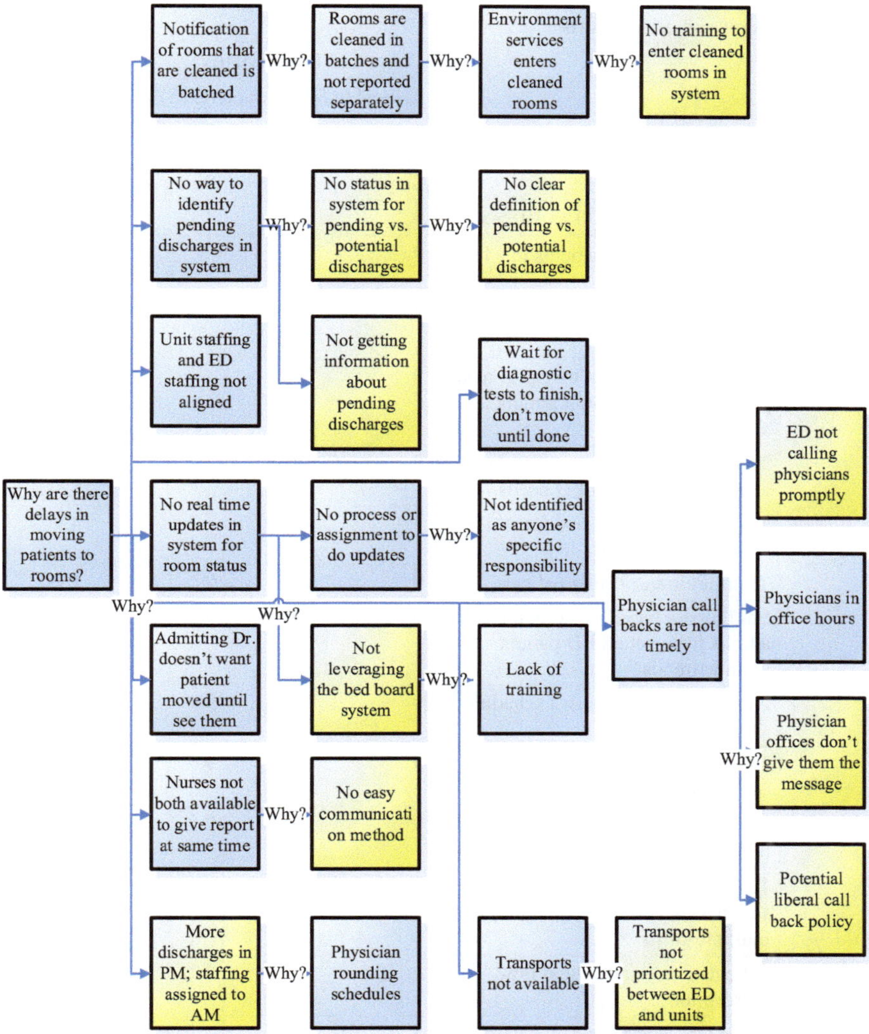

Fig. 3.15 Why–Why diagram for "Why are there delays in moving patients to rooms?"

wait times being given a five, and LWBS as a three. It is important to not have the patients leave, but the team believed that if the LOS and wait times were decreased that the patients would not leave as frequently. Each cause was assigned a strength of relationship to each CTS, with a nine being a very strong relationship, a five being a medium strength, three being a low strength, and blank being no relationship. An example of the strength of relationship is that one root cause for the Triage process step is that only one triage nurse is assigned even with patient surges in volume. This has a very high impact, and therefore a very high strength of relationship (9 rating)

Table 3.9 Cause-and-effect matrix

		Cause-and-Effect Matrix			
		Critical to Satisfaction (CTS) Criteria			
Process Steps	Root Causes	Total Length of Stay	Wait Times	% Left Without Being Seen	Total Impact Score
	Importance of CTS (1 to 10) > >	9	5	3	
Treat Patient	Admitting several patients in a batch at the same time	9	9		126
	No paging or easy way to contact nurse for report	5	5		70
	Transporter does not transport next person unless on list, if patient already transported	9	9		126
	Transporter prioritization is not always aligned with ED needs	9	9		126
	Transporters not trained for telemetry transport	3	3		42
	Techs are transporting patients	9	9	9	153
	Staffing not based on potential ER admits from actual ED patient volume trends	9	9	9	153
	Nurses call in sick, can't schedule at short notice	9	9	9	153
	No one is in charge of patient flow (to get patient to next step)	9	9	9	153
	No surge process to re-align staff for high patient volumes	9	9	9	153
	12-h shifts create fatigue and slowness at end of shift; nurses work multiple shifts	9	9	9	153
	Patients don't understand flow (perception)			9	27
	Some Drs. don't see new patients near end of shift	5	5	3	79
	Physicians batching (waiting) to see several minor care patients	9	9	3	135
	Minor care patient are not the focus of the ED physicians	9	9	3	135
	8 beds for minor care, but not always staffed	9	9	9	153
	Lack of flow, staffing	9	9	9	153

with all three CTS'. This results in an overall weighting of 153, derived by multiplying each CTS importance rating by the strength of relationship weighting: $((9 * 9) + (9 * 5) + (9 * 3)) = (81 + 45 + 27) = 153$. Assigning the root cause weighting gives us a focus on the high-priority improvement areas as we move into the Improve phase of the project. The highest priority root causes are shown in Table 3.10.

Improve Phase

The following steps were executed in the Improve Phase:

- Identify improvement opportunities and plans.
- Perform cost/benefit analysis.
- Design future state.
- Establish performance targets and project scorecard.
- Pilot improvements.
- Train and execute.

Identify Improvement Opportunities and Plans

The team developed an improvement plan based on eliminating the identified root causes. Improvements were categorized into the following themes:

- Hospital Bed Assignment.
- Triage and Waiting Room.
- Admitting Physician Call Out and Call Back.
- Patient Transport.
- Patient Flow.
- Minor Care Staffing.
- Diagnostic Tests.
- ED and Floor Reporting Process.
- Discharge Process.
- Dashboards, Metrics, and Performance Monitoring.

Hospital Bed Assignment

The first improvement category was focused on the bed board process and system. The electronic bed board was not being utilized to its full capacity. They needed to better utilize the bed board system to provide hospital-wide bed availability and visibility of beds. This would improve the time to assign beds for admitted ED patients. The Assistant Nurse Managers would need to provide the information of the bed availability on a real-time basis. Prior to the pilot, the Admitting Department assigned beds. The team proposed that the Nurse Supervisors should assign the

Table 3.10 Cause-and-effect matrix of high-priority causes

		Cause-and-Effect Matrix			
		Critical to Satisfaction (CTS) Criteria			
Process Steps	Root Causes	Total Length of Stay	Wait Times	% Left Without Being Seen	Total Impact Score
	Importance of CTS (1 to 10) > >	9	5	3	
Triage	Only 1 triage nurse even when have patient volume surges	9	9	9	153
	No surge policies or procedures	9	9	9	153
	Staffing not based on potential ER admits from actual ED patient volume trends	9	9	9	153
	No clinical visibility to waiting room	9	9	9	153
	No one "owns" patients in waiting room	9	9	9	153
Treat Patient	Techs are transporting patients	9	9	9	153
	Staffing not based on potential ER admits from actual ED patient volume trends	9	9	9	153
	Nurses call in sick, can't schedule at short notice	9	9	9	153
	No one is in charge of patient flow (to get patient to next step)	9	9	9	153
	No surge process to re-align staff for high patient volumes	9	9	9	153
	12-h shifts create fatigue and slowness at end of shift; nurses work multiple shifts	9	9	9	153
	8 beds for minor care, but not always staffed	9	9	9	153
	Lack of flow, staffing	9	9	9	153
	Not always planning, prioritizing or sequencing tests, and adjusting for competing tests	9	9	9	153
	High volume of diagnostic tests	9	9	9	153
Assign bed	Unit (floor) staffing is not integrated with ED, surgery (hospital-wide) needs staffing	9	9	9	153
	Late physician rounding	9	9	9	153
	Need a way to define and identify pending discharges in the bed board system	9	9	9	153

(continued)

Table 3.10 (continued)

		Cause-and-Effect Matrix			
		Critical to Satisfaction (CTS) Criteria			
Process Steps	Root Causes	Total Length of Stay	Wait Times	% Left Without Being Seen	Total Impact Score
	Importance of CTS (1 to 10) > >	9	5	3	
Admit patient	No training to have cleaning update clean room in system when done cleaning each room	9	9	9	153
	Not getting info from Assistant Nurse Managers who know pending discharges	9	9	9	153
	No real-time updates in bed board	9	9	9	153
	ED not calling physicians promptly	9	9	9	153

hospital beds to reduce the amount of time and non-value-added activities involved in the Admitting Clerk calling to the floors to request open beds.

They also proposed implementing pre-emptive bed assignment for those patients most likely to be admitted. This would enable the earlier assignment of the beds by requesting beds when the admitting orders were initially received by the nurses.

Triage and Waiting Room

When a surge in patient volume occurs, they propose to add a triage nurse to more quickly triage the patients. When there are ED beds available, they will place the incoming patients into an ED bed immediately, and triage the patient in the ED bed, instead of triaging them in the triage room and then sending them back to wait in the waiting room. They also will do the medication reconciliation in the ED room, not in the triage room.

Admitting Physician Call Out and Call Back

The team proposed to add a dedicated communication secretary during the peak ED census times of 2 to 10 pm. This dedicated person would focus on calling the admitting physicians for admitting orders. The medical policy requires the admitting physician to be called every 15 min for three times, and then the on-call physician can be called. Previously, the admitting physicians were not called consistently or quickly. The average time to call out to the admitting physicians was 13 min with a standard deviation of 20 min. The average time for the physician to call back was 29 min with a standard deviation of 34 min. There was also an average of 1.8 call attempts before the admitting physicians called back.

Patient Transport

The team added an ED-dedicated patient escort during the peak admitting times of 3 to 11 pm. We prepared a cost/benefit analysis estimating a rate of return based on proposed increased ED volumes from gaining additional patients and reducing the number of patients leaving the ED without treatment. The benefit to cost ratio was 13.4 to 1 (benefits to cost). The desired benefit to cost ratio is greater than 1. We showed that there was a positive return on investment when adding additional staff for the dedicated communication secretary, the patient escort, and the Nurse Supervisors for assigning beds.

Patient Flow

The team proposed to add additional electronic alerts on the ED Bed Board to improve visibility of the status of each patient and which step in the process they are at in the Emergency Services Value Chain. The team also developed an ED surge plan with volume and ED room availability triggers.

Minor Care Staffing

The team realigned minor care staffing to add a physician for minor care when ED volume increases.

Diagnostic Tests

The team implemented both lab and imaging diagnostic testing improvements. Since many of the imaging exams that expose a patient to radiation (X-rays, CT scans) require a pregnancy test for women of childbearing age, we decided to fast-track pregnancy and other tests required for radiology exams. The team also created another Lean Six Sigma team to investigate and reduce the number of imaging exams and to provide appropriate CT scans. The amount of radiation exposure that CT exams give can be equivalent to 1,000 X-rays, so it is critical to ensure that all tests are medically appropriate.

To decrease the time to draw the specimens for tests, we added an alert to the electronic ED Bed Board to track the time it takes for the phlebotomist to draw the specimens.

ED and Floor Reporting Process

We redesigned the reporting process to streamline the communication and handoff for admitted patients from the ED to the floors or units. We also automated the transfer form to reduce the time it took for the ED nurses to re-write information that was already electronically captured in the ED Electronic Medical Record.

Discharge Process

We formed another Lean Six Sigma team to investigate and improve the discharge process for the floors, so that more beds are available when they are needed to admit patients from the ED.

Dashboards, Metrics and Performance Monitoring

We developed a dashboard, metrics, and performance monitoring processes to measure performance aligned to flow and throughput. We incorporated necessary performance measures to ensure metrics and performance is met. A sample dashboard is provided in Fig. 3.16. We also developed a Flow Dashboard that identified where each patient is in the ED Value Chain, and which process is holding the patient from moving through the value chain and being released. An example of the Flow Dashboard is shown in Fig. 3.17.

Perform Cost/Benefit Analysis

We performed a cost/benefit analysis to determine if it was c ost beneficial to add three needed people, a communication secretary, an ED-dedicated patient escort, and an additional ED nurse. We justified the additional resources based on the additional revenue that would be realized by reducing the patients leaving without treatment.

Metrics	Baseline		Target		Improved		% Improvement	
	Admitted	Discharged	Admitted	Discharged	Admitted	Discharged	Admitted	Discharged
Total LOS:	8.7 hrs	5.8 hrs	5 hours (43%)	3 (48%)	5.6 hrs	3.9 hrs	36%	33%
% LWBS:	6.50%		3.50%		0.51%		92%	
Total Time to EDP:	93	109	35	35	26		75%	
Triage Time:	34	34	15	15	11		68%	
Wait time triage to EDP:	65	79	20	20	20		73%	
Time to ED Bed:	61	81	35	42	13		82%	
Bed to Physician:	34	55	20	20	13		72%	
Treat/Diagnose to Disposition Time:	192	175	109	91	161		11%	
Radiology Diagnostic Time:	138	270	154	154				
Lab Time:	119							
EDP Time Visit:	192	175	109	91	160		13%	
Disposition to Leave ED	223	64	127	33	155	47	30%	27%
Disposition to get admit orders:	103		59		77		25%	
Total Call Out Time:	41		23					
Call Out to Physician	13		7					
Physician Call Back	29		17					
Wait After Call Back:	37		21					
Admit Orders to Assign Bed:	42		24		25		40%	
Assign Bed to Leave ED:	112		64		64		43%	

Fig. 3.16 Sample Dashboard

Bed	Status	Triage	Regist-ration	Waiting EDP	Orders Pendin g	Phleb Pendin g	EKG	Meds Pendin g	Repira-tory	Lab Results	Xray	CT	US	To the DC/ Admit	Callout Pend	Bed Reques t	Bed Assigned	Report
ED-01																	42 mins	
ED-02	Labs not resulted									> 70 mins								
ED-03	Empty																	
ED-04	Waiting to be discharged																50 mins	
HW-1N	Empty																	
ED-05	Empty																	
ED-06	Empty																	
ED-07	Empty																	
ED-08	Empty																	
HW-5H	Empty																	
ED-09														> 2 H				
ED-10														> 3 H				
ED-11	Waiting EDP disposition													> 4 H				
ED-32	Empty																	
HW-32W	Empty																	
ED-12	Empty																	
ED-13																		
ED-14										57 mins								
ED-31	Waiting EDP disposition													> 6 H				
NW-13H	Empty																	
ED-15																		
ED-16												45 mins					60 mins	

Fig. 3.17 Flow status board

Metrics	Baseline		Target		Improved		% Improvement	
	Admitted	Discharged	Admitted	Discharged	Admitted	Discharged	Admitted	Discharged
Total LOS:	8.7 hrs	5.8 hrs	5 hours (43%)	3 (48%)	6.1 hrs	4.1 hrs	30%	29%
% LWBS:	6.50%		3.50%		0.60%		91%	
Total Time to EDP:	93	109	35	35	32		56%	
Triage Time:	34	34	15	15	13		62%	
Wait time triage to EDP:	65	79	20	20	25		62%	
Time to ED Bed:	61	81	35	42	16		74%	
Bed to Physician:	34	55	20	20	16		53%	
Treat/Diagnose to Disposition Time:	192	175	109	91	166		14%	
Radiology Diagnostic Time:	138	270	154	154	170		-23%	
Lab Time:	119				57		109%	
EDP Time Visit:	192	175	109	91	175		9%	
Disposition to Leave ED	223	64	127	33	171	47	23%	27%
Disposition to get admit orders:	103		59		70		32%	
Total Call Out Time:	41		23		24			
Call Out to Physician	13		7		5			
Physician Call Back	29		17		19			
Wait After Call Back:	37		21		46			
Admit Orders to Assign Bed:	42		24		44		-5%	
Assign Bed to Leave ED:	112		64		67		40%	

Fig. 3.18 Metrics dashboard three-month improvement

Design Future State

We designed the new process and updated the original process maps.

Establish Performance Targets and Project Scorecard

The metrics dashboard with the targets is shown in Fig. 18.

Pilot Improvements

The team began the pilot on July 20th. We tracked improvement on a weekly basis, and further tweaked the improvement ideas, and continued to generate additional ideas. We also started a detailed ED process Lean Six Sigma improvement project to continue additional process improvements. We created revised process maps based on implementing the improvement ideas.

In the first week of the pilot, the percentage of patients that LWBS was reduced from 6.5 to 3.5%, meeting our initial target. We also met our targets for the time to get the patient to the ED bed, reducing the time for admitted patients by 30 min (50%) and reducing the time for discharged patients by 42 min (51%).

This was a great way to start the pilot activities.

Within the first month of the pilot, we met the following targets:

- Percent LWBS: 7 to 0.64% (91% improvement!)
- Time to ED Bed: (78% improvement!)
  - Admitted: 61 to 13 min (target 35)
  - Discharge: 81 to 18 min (target 42)
- Triage time: (~56% improvement)
  - Admitted: 14 min (target 15)
  - Discharge: 15 min (target 15)

Other improvements in the first month were:

- Total LOS—Improved by 24 to 28%
  - Admitted: improved by 2.4 h (8.7 to 6.3 h)
  - Discharged: improved by 1.4 h (5.8 to 4.4 h)
- Cut Triage Time by more than Half:
  34 to 15 min
- Improved Time from Bed Assigned to Leaving ED by 40%
  (112 to 67 min)
- Total Time to EDP improved by 35%
  - Admitted: 93 to 60 min
  - Discharged: 109 to 68 min
- Improved Admit Orders to Assign Bedtime by 24%
  (42 to 32 min)

Culture change was also quite significant. We had an intense focus on wanting to reduce and prevent patients from leaving without treatment. One of the physician assistants even called a patient on their cell phone when they found out that they had just left the ED because they could not wait any longer. The patient was still in the parking lot, returned to the ED, and was treated. One of the ED physicians that had drastically reduced their schedule due to the chaos in the ED returned to the ED full time after seeing the drastic improvements, and reduction in the LOS, and reduction in the chaos in the ED.

The improvements continued to be quite significant 3 months after the pilot improvements began. A histogram of the improvements in LOS is shown in Fig. 3.19.

The following key improvements were observed:

- Total LOS—Improved from 26 to 30%
  - Admitted: improved by 2.3 h (8.7 to 6. 4 h)
  - Discharged: improved by 1.8 h (5.8 to 4.3 h)
- Percent LWBS: 7 to 0.6% (91% improvement!)
- Front end time:
  - Triage Time = 13 min (from 34)
  - Time to EDP = 32 min (from 93)
- Diagnostic time: 166 min (from 184)
- Back-end time:
  - Disposition to Leave ED 171, 47 (from 223 to 64).
  - Disposition to get admit orders 70 (from 103).
  - Time to assign bed 44 (from 42).
  - Assign bed to leave ED 67 (from 112).

The improved ED timeline is shown in Fig. 3.20. The improved census pattern by hour of the day also improved by getting patients through the ED more quickly, as shown in Fig. 3.21. The revised metrics after 3 months is shown in Fig. 3.22.

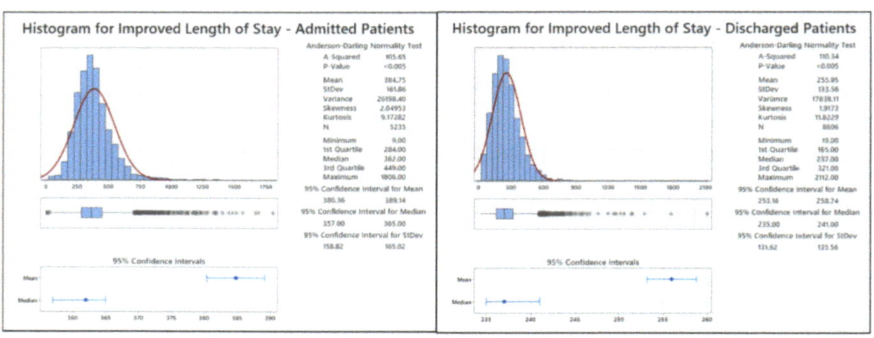

Fig. 3.19 Histograms of improved length of stay (minutes) admitted and discharged patients

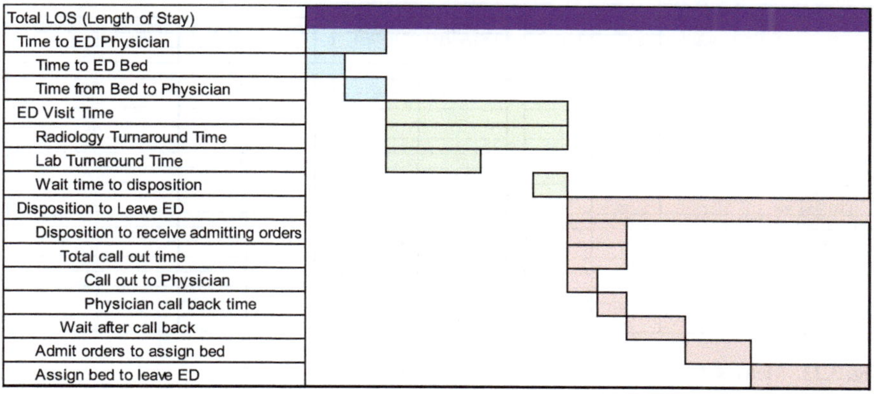

Fig. 3.20 Improved ED process timeline

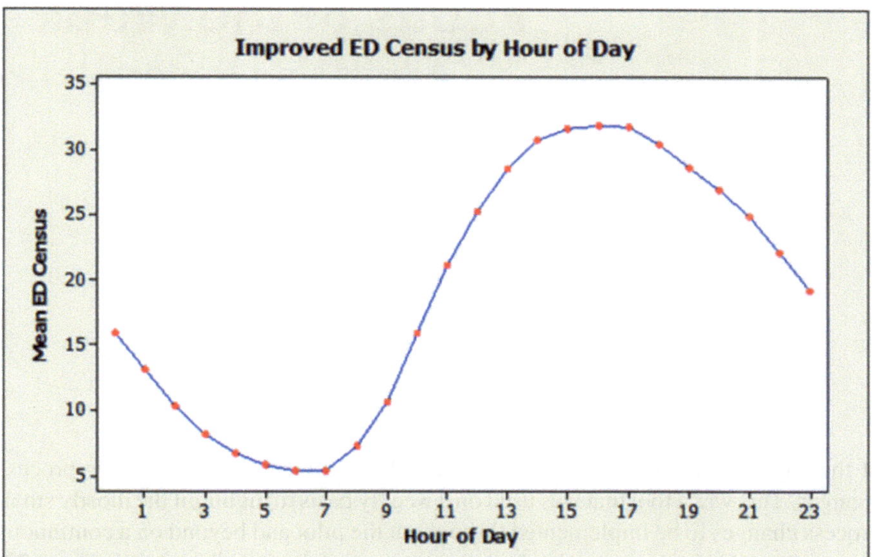

Fig. 3.21 Improved ED census by hour of day

Train and Execute

The team trained based on the needs of the improvement effort. To train the nurses in how to update the Hospital Bed Board, formal training sessions were held with the nurses and unit clerks. The Nurse Supervisors were formally trained in how to assign hospital beds and how to monitor the ED Bed Board to know when a bed was requested. They also received on-the-job training for their new positions. A "focus

Metrics	Baseline		Target		Improved		% Improvement	
	Admitted	Discharged	Admitted	Discharged	Admitted	Discharged	Admitted	Discharged
Total LOS:	8.7 hrs	5.8 hrs	5 hours (43%)	3 (48%)	6.1 hrs	4.1 hrs	30%	29%
% LWBS:	6.50%		3.50%		0.60%		91%	
Total Time to EDP:	93	109	35	35	32		66%	
Triage Time:	34	34	15	15	13		62%	
Wait time triage to EDP:	65	79	20	20	25		62%	
Time to ED Bed:	61	81	35	42	16		74%	
Bed to Physician:	34	55	20	20	16		53%	
Treat/Diagnose to Disposition Time:	192	175	109	91	166		14%	
Radiology Diagnostic Time:	138	270	154	154	170		-23%	
Lab Time:	119				57		109%	
EDP Time Visit:	192	175	109	91	175		9%	
Disposition to Leave ED	223	64	127	33	171	47	23%	27%
Disposition to get admit orders:	103		59		70		32%	
Total Call Out Time:	41		23		24			
Call Out to Physician	13		7		5			
Physician Call Back	29		17		19			
Wait After Call Back:	37		21		46			
Admit Orders to Assign Bed:	42		24		44		-5%	
Assign Bed to Leave ED:	112		64		67		40%	

Fig. 3.22 Metrics dashboard three-month improvement

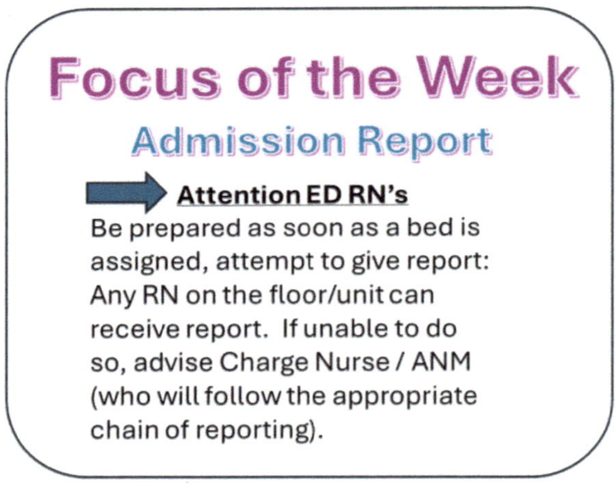

Fig. 3.23 RN focus of the week example

of the day" was used to train ED clinical and nonclinical staff in the new process changes. This was a tool that was used on a weekly basis to highlight the mostly small process changes to be implemented throughout the pilot and beyond on a continuous improvement basis. An example of a "clinical focus of the week" is given in Fig. 3.23.

Hypothesis Test

We performed a *t*-test to assess whether the improvement in LOS was statistically significant. We first performed a test to assess whether the variances were equal before and after improvements. The standard deviation for the LOS for each patient during the baseline period of January through April was 224.2, and 162.1 after implementing the improvements. The variance was significantly different (reduced) after the improvements, with a *p*-value of 0.000.

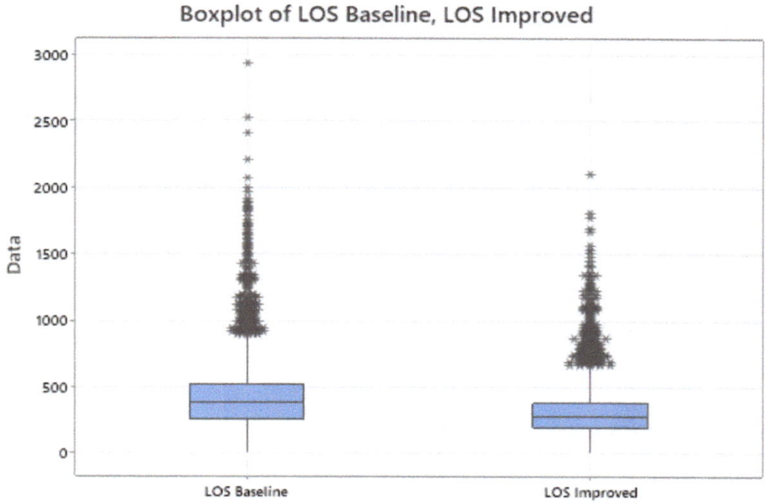

Fig. 3.24 Boxplots of average length of stay before and after

We then performed the two-sample T test to assess whether the LOS had shown a statistically significant improvement. The LOS before improvements was 411 min and after improvement was 303 min, showing a statistically significant improvement with a *p*-value of 0.000.

The box plots of the average LOS before and after improvement are shown in Fig. 3.24.

Patient Satisfaction

We also assessed whether the ED patient satisfaction increased. The patient satisfaction survey that is normally used to assess patient satisfaction was compared between the last two quarters. The patient satisfaction from April to June was 65.9%. The patient satisfaction for the third quarter, July through September, coinciding to the ED pilot, was 73.5%. This was a 12% improvement in patient satisfaction, which was statistically significant at the 95% confidence level. This patient satisfaction put the ED into the 90th percentile and second place compared to the other Regional Healthcare Centers in the system. For the previous year, October through September, the patient satisfaction was 68.9%. So, patient satisfaction had dipped prior to the pilot and increased significantly during the pilot. By the end of the project, the patient satisfaction had increased by 24 to 89.9%, putting the ED into the top 1% of hospitals nationally.

Control Phase

The purpose of the Control Phase is to implement control mechanisms and procedures to ensure that the process maintains the levels of improvements attained during the pilot activities.

The following activities were executed in the Control Phase:

1. Measure results and manage change.
2. Report scorecard data and create process control plan.
3. Identify replication opportunities.
4. Develop future plans.

Measure Results and Manage Change

The pilot activities started at the end of July. By the end of November, the LOS for all patients was 4.7 h, down from 6.9 h, having a 31% improvement of over 2 h for the LOS. The LWBS percentage was reduced by 6.2% from 6.5 to 0.34%, a 95% improvement.

The admitted patients' LOS was reduced from 8.7 to 6.0 h, a 2.7-h reduction and a 31% improvement. The discharged patients' LOS was reduced from 5.8 to 3.9 h, a 1.9-h reduction and a 33% improvement. The up-front times showed the greatest improvements, as follows:

- Triage time improved from 34 to 12 min, a 65% improvement.
- Time from Door to Doctor from 100 to 27 min, a 72% improvement.
- Time to ED Bed from 74 to 13 min, an 82% improvement.

The back-end times are not as dramatically improved as follows:

- Treat/Diagnose to EDP Disposition time from 184 to 165 min, a 10% improvement.
- Disposition to get admit orders from 103 to 77 min, a 25% improvement.
- Admit orders to assign bed from 42 to 34 min, a 19% improvement.
- Assign bed to leave ED from 112 to 70 min, a 38% improvement.

Change Management

To manage the change throughout the project, we used many change management techniques. The Define phase performing the project chartering and customer/stakeholder analysis helped with getting the team members and subject matter experts bought into the project. The CEO was at the initial project team kick-off meeting to speak with the team to describe the importance of the project and his expectations. A communication plan was developed as shown in Table 3.11.

Table 3.11 Communication Plan

Audience	Objectives/ Message	Media/ Mechanism	Frequency	Responsible
Project team	Weekly 2-h working meetings	Working meetings, email	Weekly	Team leader
Executive team	Weekly status, performance metrics update	Mission critical meetings, email	Weekly or as needed	Team leader
CEO	Project status	Meetings as needed, email	As needed	Team leader

To continue to manage the change after the initial improvement ideas were piloted, we kept the improvement team together, meeting weekly until the LOS targets were met, then we met every other week for just 1 h each week to discuss continuous improvement ideas and to ensure the gains were maintained. We continued to send out the Emergency Service Value Chain status and performance against the metric dashboard on a weekly basis to the executive team.

Report Scorecard Data and Create Process Control Plan

The control plan is a detailed guide for maintaining all of the positive changes that you have implemented. We developed the control plan defining the metric and control mechanisms for each step in the value chain. The extensive Emergency Department Bed Board information system and the ED Dashboard will be used to provide data and metrics to control the process. The control plan is shown in Table 3.12.

From a management perspective, the ED Director will monitor and control on a weekly basis to ensure that the process is in control and stable and determine when additional improvements are needed. A monthly ED process improvement meeting is held on an on-going basis to generate and implement continuous improvement.

Identify Replication Opportunities

The Lean Six Sigma problem solving approach can and has been replicated in other areas in the hospital. Many of the improvements can be replicated in other Emergency Departments.

Table 3.12 Control plan summary

Process Step	Control Mechanism	Measure/ Metric	Criticality (High/ Medium/Low)	Who	When
Triage	Dashboard: ED Bed Board	Triage time	Low	ED Director	Weekly
Place in ED Bed		Time to ED Bed	Medium	ED Director	
Nurse assesses patient		None	Low	ED Director	
Registration		Registration time	Medium	ED Director	
Enter orders		Time to enter orders	High	ED Director	
Draw specimens		Time to draw specimens	High	Lab Director	
Perform lab results		Time to perform lab results	Medium	Lab Director	
Perform radiology tests		Time to perform radiology tests	High	Radiology Director	
EDP visit		EDP visit time	High	ED Physician	
Call and obtain admit orders		Time to get admit orders	Medium	ED Physician	
Request and assign bed		Time to assign bed	Medium	Nurse Manager	
Report, move and admit to hospital		Time from assign bed to leave ED	High	ED Director	
Discharge patient		Time from disposition to leave ED	Medium	ED Director	

Develop Future Plans

The ED process improvement team will continue to meet to identify and implement additional improvements.

Conclusions

This project has been extremely challenging, due mainly to the large scope and the complexity of the processes. However, we achieved great success in improving the ED Value Chain, in all of the metrics that we measured, especially in overall LOS, the percent of patients leaving without being seen, the door-to-doctor time, and with patient satisfaction.

Project Critical Success Factors

The following key factors contributed to the success of this project:

- CEO's visible commitment; engagement; and gentle encouragement toward goals.
- Key team members selected to participate.
- Experienced facilitator with technical and change management skills.
- Information Technology involvement as key participant.
- ED Bed Board providing data, process metrics, and discipline.
- Admin team monitoring performance on a weekly basis.
- ED Director and Nurse Manager engaged and willing to change.
- ED Staff willing to pilot improvements.
- ED Physicians as key process innovators.

Even though the process changes will not be directly replicated on another project for the same hospital, many of the best practices can be shared with sister hospitals in the hospital system as well as other hospitals through this book and journal articles.

Reference

Crowding Resources Taskforce of the American College of Emergency Physicians. (2002). *Responding to emergency department crowding: A guidebook for chapters.* American College of Emergency Physicians.

Chapter 4
Operating Room Turnaround and On-Time Starts Improvement

Abstract The next two chapters describe the improvements made to a community-based acute care hospital's Operating Room environment with respect to the following areas: (1) In this Chapter Performing turnaround time between surgical cases, and improve on-time start of operating room cases. (2) Chap. 5. Improve OR location and organization of equipment, instruments, and supplies.

Keywords Turnaround time · Inventory control · Staff satisfaction · Secondary stakeholder · Sigma project

Lean, Six Sigma, Theory of Constraints, and Supply Chain Management tools and methods were applied to improve the OR processes.

The Value Chain for providing surgical services is one of the most complex for the hospital because it touches almost every function in the hospital. It starts with identifying the supplies, equipment, and instruments needed for the surgeries, customized for each procedure and surgeon. The supplies, equipment, and instruments are sourced, purchased, received, and managed. The supplies and instruments are prepared and picked for the surgeries. The operating rooms are prepared, including moving and setting the equipment, furniture, supplies, and instruments. The patient flow includes registration, testing and diagnosis, surgery scheduling, and preparation for surgery in Pre-op. The surgery is performed; the patient is recovered in Post-op and is either discharged or admitted/transferred to the hospital as appropriate. The OR is then turned around for the next surgical case. The Provide Surgical Services Value Chain is shown in Fig. 4.1.

Two projects were performed that focused on key elements of the surgical services value chain. One project focused on improving the turnaround of the OR and the preparation of the OR for the first cases of the day. The second project focused on the locations and organization of the equipment, furniture, supplies, and instruments in the OR area.

The remainder of this chapter will focus on the first project, improving turnaround of the OR and the preparation of the OR for the first cases of the day.

S. Furterer, *Lean Six Sigma for the Healthcare Enterprise*,
https://doi.org/10.1007/978-1-4471-7617-6_4

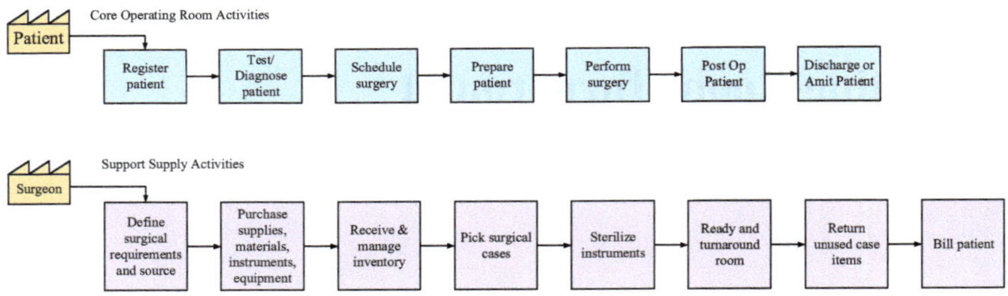

Fig. 4.1 Provide surgical services value chain

Lean Six Sigma Project for Operating Room (OR) Turnaround and On-Time Starts of Cases

The DMAIC (Define, Measure, Analyze, Improve, Control) problem-solving methodology was used to improve the OR turnaround and start the OR cases on time.

Define Phase

The following steps were performed in the Define Phase:

- Develop project charter.
- Perform stakeholder analysis.
- Identify critical to satisfaction (CTS).
- Select team and launch the project.
- Create project plan.

Develop Project Charter

The executive team has identified an opportunity to improve the time it takes to clean and set up the OR between cases, as well as start the first cases in each room on time in the morning. The corporate goal for turning the room around is 30 min or less; however, the average turnaround time in the OR is 40 min. The average time for starting cases on time is 23 min late.

Project Overview: The OR turnaround time is not meeting the deployed corporate targets of an average 30 min turnaround time. Turnaround time is defined as time from wheels out of the patient from the prior surgery to wheels in of the patient for the next surgery. Our current average turnaround time is 40 min.

Customer/Stakeholders The stakeholders for the project include the OR surgeons, nurses, anaesthesiologists, anaesthesiologist technicians, sterilizer processing, OR liaisons, OR case scheduling, OR director, charge nurse, inventory control, patients (in and out patients), and families.

The CTS criteria of the stakeholders include quick turnaround time of the OR, starting the OR cases on time, patient care and safety, patient satisfaction, physician and staff satisfaction, and staff productivity.

The goals of the project are to reduce the OR turnaround time between cases, while providing continued patient care and safety, to meet the average 30 min turnaround time.

The project includes the processes involved in cleaning the room after a case and preparing the room for the next case, and communication for making the turnaround happen. This time is the time from wheels out of the patient to wheels in of the following patient. The project also includes the activities related to starting cases on time in the morning for each OR.

We anticipate a reduction of time between cases and a decrease in overtime due to cases not starting on time.

The project charter is shown in Table 4.1, Project Charter.

Perform Stakeholder Analysis

The primary stakeholders are the OR surgeons and staff. The OR staff includes the nurses [including nurses, nurse managers (Circulator, Pre-op holding, Charge nurses, clinical coordinators)]. Other stakeholders are anaesthesiologists and anaesthesiologist techs who provide anaesthesiology care; sterile processing OR liaisons and sterilizer processors who are associates who provide communication between Sterile Processing and the OR, and Sterilize Processors who sterilize OR instruments. Patients and their families who have surgery in the OR are also primary stakeholders of the project. The primary stakeholder analysis is shown in Table 4.2.

There are several secondary stakeholders of the project. Inventory control manages OR materials and supplies. Ancillary services include the lab and radiology that provide diagnostics for surgery patients. Case scheduling schedules OR cases. Physician offices schedule surgeries for their patients. Vendors provide medical devices and perform diagnostic testing for OR patients. The last secondary stakeholder group is the recovery who cares for the patients when they complete the surgical case. The secondary stakeholders are shown in Table 4.3.

Suppliers-Inputs-Processes-Outputs-Customers

The Suppliers-Inputs-Processes-Outputs-Customers (SIPOC) for the case turnaround is shown in Table 4.4. There are five high-level process steps in the OR turnaround process. When the case is about 10 min from being complete, the OR nurse calls to the front desk to notify them that the technicians should be available to start cleaning the room. When the case is complete and the bandages

Table 4.1 OR turnaround and on-time case starts project charter

Operating Room (OR) Turnaround and On-Time Starts Project Overview, Problem Statement, & Goals
Project Statement: The operating room turn-around time is not meeting the deployed targets of an average 30 minute turnaround time. Turnaround time is defined as time from wheels (of patient bed) out to wheels in of the patient's bed. Our current average turn-around time is 40 minutes. Customers/Stakeholders: Surgeons, nurses, anesthesiologists, sterilizer processing, SPD/OR liaisons, case scheduling, administration, patients (in and out patients) and families, materials control. What is important to these customers: Quick turnaround time of operating rooms, patient care and safety, patient satisfaction, associate /employee satisfaction, productivity. Goal(s): To reduce the operating room turnaround time between cases while providing continue patient car and safety, to meet the desired targets. The targeted goals are: Average turnaround time of all cases = 30 minutes; general surgery & small cases = 15 minutes; ortho and large cases = 20 minutes; open heart and major spines = 30 minutes. We will define specific project and hospital goals after the Measure phase. Scope: The project includes processes involved in cleaning the room after a case and preparing the room for the next case, and communication for making the turnaround happen. This time is the time from wheels out of the patient to wheels in of the following patient. Projected financial benefits: Reduction of time between cases; enhanced OR utilization, revenue, efficiency, productivity, decrease in overtime.

are on, the OR team moves the patient from the OR table to a bed. The patient is moved to recovery. The team cleans the room, removing trash, used instruments, and equipment after it is wiped down. The materials for the next case have already been staged outside the room, and were prepared the day before the case. Once the room is cleaned from the prior case, the scrub nurse and circulating nurse set up the room for the next case.

The SIPOC for the processes for starting the OR cases on time in the morning is shown in Table 4.5. There are four high-level processes for starting a case on time in the morning. The focus for starting the cases on time in the morning for the first cases in each OR is important. If the first cases do not start on time in the morning, then the timeliness of the entire schedule is jeopardized for the entire day. The materials, equipment, and instruments are planned the prior day through the night shift based on the schedule for the surgical cases for the next day. The materials, equipment,

Table 4.2 Stakeholder analysis—primary

Stakeholder analysis					
Stakeholder	Type	Primary role	Potential impacts/ concerns	Initial receptivity	Future receptivity
Surgeons and physician assistants	**Primary**	Surgeons who perform surgery, and physician assistants who assist in surgery	Patient care and safety On-time surgery schedules OR turnaround times	Moderate Against	Strongly Support
Nurses and nurse managers (circulators, pre-op holding, charge nurses, clinical coordinators)	**Primary**	OR nurses and managers who care for patients in the OR	Patient care and safety On-time surgery schedules Efficiency OR turnaround times	Moderate Support	Strongly Support
Anesthesiologists	**Primary**	Anesthesiologists and techs who provide anesthesiology care	Patient care and safety On-time surgery schedules OR turnaround times	Strongly Against	Strongly Support
Sterile processing	**Primary**	Associates who provide communication between Sterile Processing and the OR; Sterile processors	Patient care and safety On-time surgery schedules Efficiency Turnaround times Patient satisfaction Staff satisfaction	Strongly Support	Strongly Support

(continued)

Table 4.2 (continued)

Stakeholder analysis					
Stakeholder	Type	Primary role	Potential impacts/ concerns	Initial receptivity	Future receptivity
Patients and families	**Primary**	Patients and their families who receive OR care	On-time surgery schedules Patient care and safety	Neutral	Neutral
Registration	**Primary**	Register the patient	Correct registration Accurate billing	Moderate Support	Strongly Support

and instruments are picked from the Core inventory room, that is stored within the center of the OR area. The OR team readies the room in the morning, ensuring that all of the supplies, equipment, and instruments are available and set up in the room. If the patient and OR team are ready, the case starts on time.

Identify Critical to Satisfaction

The CTS criteria for this project are:

- Quick turnaround time of OR: This includes cleaning the room from the prior case to readying the room for the new case.
- On-time case starts: This includes picking materials, equipment and instruments the day before for the case, and readying the room in the morning to start the first cases in each room on time.
- Patient care and safety: Ensuring and providing patient care and safety.
- Patient satisfaction: Ensuring the satisfaction of the patients.
- Associate satisfaction: Ensuring the satisfaction of the OR staff, balancing patient care, safety, and timeliness.
- Physician satisfaction: Ensuring the OR physicians' satisfaction.
- Productivity: Ensuring appropriate use of personnel resources in the OR to manage operating costs.

Table 4.3 Stakeholder analysis—secondary

Stakeholder analysis					
Stakeholder	Type	Primary role	Potential impacts/ concerns	Initial receptivity	Future receptivity
Materials control	**Secondary**	Provide sterile and non-sterile materials for surgery	OR turnaround times OR scheduling	Moderate Support	Strongly Support
Ancillary services	**Secondary**	Ancillary services such as lab, radiology who provide diagnostics for surgery patients	Patient care and safety Case scheduling Providing timely and accurate results	Neutral	Neutral
Case scheduling	**Secondary**	Schedule OR cases	Case scheduling On-time cases Accurate case schedule	Moderate Support	Strongly Support
Physician offices	**Secondary**	Physician offices who schedule patients for surgery	Patient care and safety On-time surgery schedules Getting patient scheduled Vendor notification Availability of devices	Strongly Support	Strongly Support

(continued)

Table 4.3 (continued)

Stakeholder analysis					
Stakeholder	Type	Primary role	Potential impacts/ concerns	Initial receptivity	Future receptivity
Vendors	**Secondary**	Vendors who provide medical devices and perform diagnostic testing	Patient care and safety On-time cases Use and availability of product	Strongly Support	Strongly Support
Recovery	**Secondary**	Staff who provide recovery services for patients	OR turnaround time On-time surgery schedules Patient care and safety	Strongly Support	Strongly Support

Table 4.4 Case turnaround SIPOC

SIPOC				
Suppliers	Inputs	Process	Outputs	Customers
Medical staff Patient Materials control Ancillary services Case scheduling Vendors Physician offices	Materials, equipment, supplies, devices Patient information OR schedule Anesthesia, meds	Complete surgery case	Patient information Materials Call of completion	Associates Patients Physicians
Physicians Associates Patients	Patient Patient information	Move patient to recovery	Patient information	Associates Patients Physicians Recovery
Associates Material control Anesthesia	Correct supplies	Clean room	Room status	Associates Patients Physicians
Case scheduling Vendors Material control Associates	Case schedule Pick list	Prepare materials	Equipment, etc	Associates Patients Physicians Anesthesia
Associates Physicians Patients Anesthesia	Materials, equipment, supplies, devices OR Schedule Anesthesia, meds	Prepare room	OR status	Associates Patients Physicians

Table 4.5 On-time start SIPOC

SIPOC				
Suppliers	Inputs	Process	Outputs	Customers
Case scheduling Inventory control	Case preference cards OR case schedule	Prepare and plan	Needs for materials Needs for equipment Needs for instruments	Inventory control Sterile processing
Inventory control Sterile processing	Printed preference cards	Pick materials	Materials Equipment Instruments Preference cards	OR team
OR team Diagnostics Pre-op	Materials Equipment Instruments Preference cards Results / patient information	Ready case in morning	Room status Patient status	OR team Patient
OR team Patient	Patient information Room status	Start case on time	Room status	OR team Administration

Select Team and Launch the Project

The team was formed, consisting of the OR Executive Director, the OR surgical coordinators, inventory control, instrument sterile processing, Pre-op, Post-op, Chief of Anaesthesia, Operational Excellence Master Black Belt, and Black Belts.

Create Project Plan

The project was started in October. The Define phase was completed by October 31. The Measure phase started in November and was completed on January 8 of the next year. The Analyze phase ended on January 22nd. The Improve phase finished on December 31st. The Control Phase was completed in April of year 3. The project milestones are shown in Fig. 4.2.

Rules of Engagement:

The following rules of engagement were embraced by the project team:

- Be present and be committed at the meetings.
- Look at issues, not people; not to be defensive.
- Be open to new ideas.

	Estimated Finish Date	Activities
D	10/30	Define project charter and scope, stakeholders and initial Critical to Satisfaction (CTS); document value chain, identify constraint processes
M	1/8	Measure and document current As-Is process; baseline throughput time; collect data on current process; identify constraints & bottlenecks
A	1/22	Analyze problems and their root causes; identify waste and non-value added activities.
I	12/31	Identify improvement ideas; pilot improvements; assess improvement to throughput time
C	4/2	Identify and implement control mechanisms; create procedures and training for new improvements

Fig. 4.2 Project milestones

- Think outside the box.
- Help each other work together as a team.
- Be respectful for people to finish their thoughts.
- Focused structure: agendas, outcomes, parking lot issues.
- Limit our time; keep our meetings to the time; finish meetings.
- Fix the process, do not fix blame.
- Everyone's point of view is appreciated and important.
- It's your meeting and project, feel comfortable to contribute.

The rules of engagement are critical to ensure the smooth running of the team meetings, so that the team can stay on topic, and change management can begin.

The Define phase is important in defining a clear project scope, identifying the project stakeholders, and planning for the rest of the project. The team moved next into the Measure phase.

Measure Phase

The Measure Phase included the following activities:

- Define the current process.
- Define detailed voice of customer (VOC).
- Define the voice of process (VOP) and current performance.
- Validate measurement system.

Define the Current Process

The first step was to map the current process by creating detailed process maps of the processes to turnaround the OR between surgical cases, shown in Figs. 4.3 and 4.4, the processes for readying the rooms for on-time starts of cases in the morning, shown in Fig. 4.5.

We also filmed the turnarounds of the rooms, including the cleaning of the room, and the set up of the supplies and instruments by the OR team.

Data Collection Plan

To measure the CTS we developed the data collection plan shown in Table 4.6. We measured and baselined the time to turnaround the OR, as well as the time to start the first cases in each room for the day. We measured the delays to starting the cases on time for the first cases and turning the cases during the day. We assessed staff and physician satisfaction using surveys developed for this project. We used the existing productivity measures that assess the productive staff hours compared to the number of procedures performed.

We operationally defined each measure to ensure a consistent measurement system.

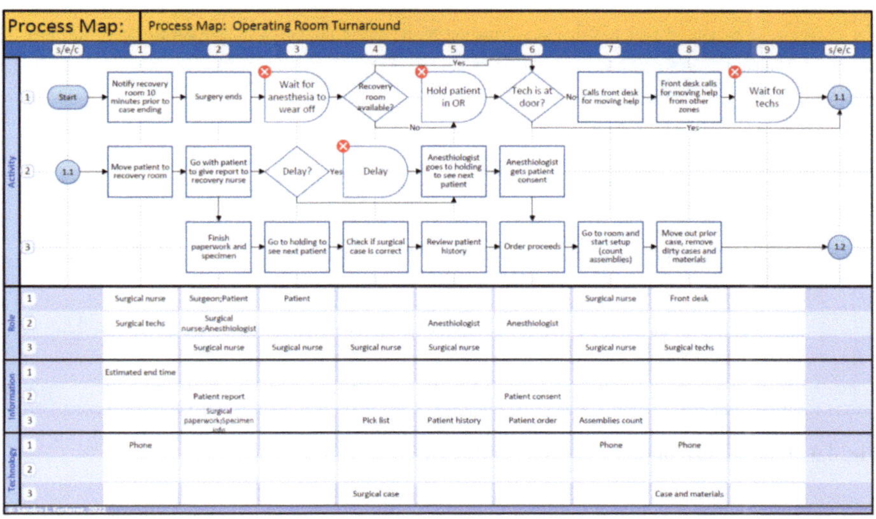

Fig. 4.3 Process map of OR turnaround

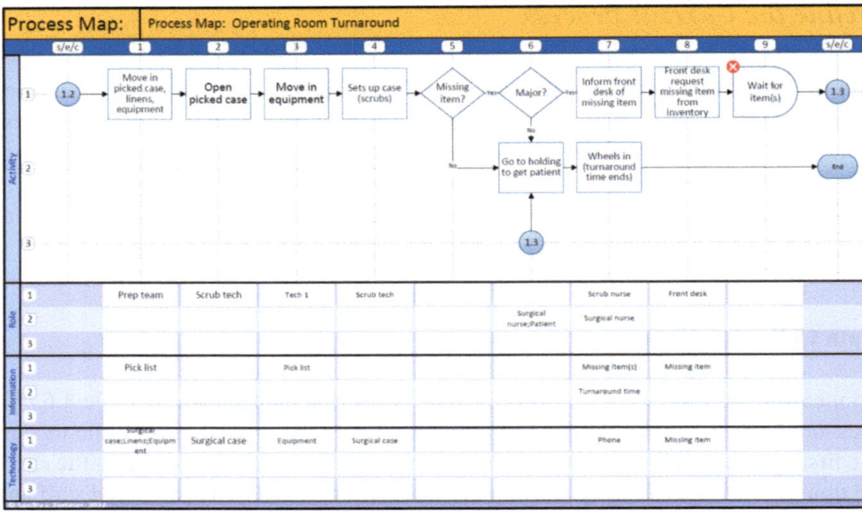

Fig. 4.4 Process map of OR turnaround

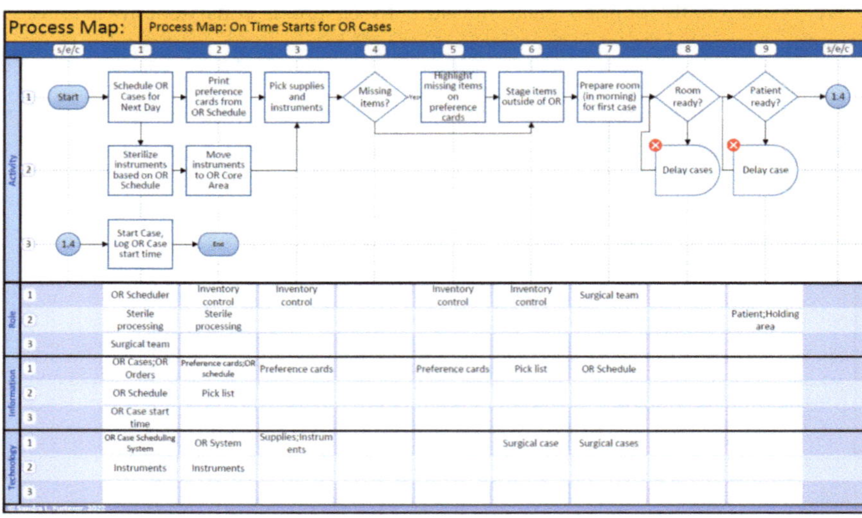

Fig. 4.5 Process map of on-time starts of OR cases

OR Turnaround Time

OR turnaround time is the time from when the patient is wheeled out after surgery is complete to when the next patient is wheeled into the OR. The purpose is to understand the time it takes to turnaround the OR from one case to the following

Table 4.6 Data collection plan

Data collection plan							
Critical to satisfaction (CTS)	Metric (short title)	Operational definition (metric description)	Data collection source	Analysis mechanism	Sampling plan (size, frequency)	Process to collect and report	Baseline
Quick turnaround time of operating rooms	OR turnaround time	Time from wheels out to wheels in of following patient; by surgeons, and by room	ORM (OR Management Systems	Statistics	Jan–Sept	Verify and validate data with team and subject matter experts	Mean = 40 min; Stdev = 21 N = 1179
Staring cases on time	On-time starts	First cases for a surgeon (1st 2 cases in 1st 2 rooms of day; 7:30 and 8:00 am start times)	ORM (OR Management Systems	Statistics	Jan–Sept		Mean = -22.90 min Stdev = 40.10 Median = -13
Starting on time cases (by time buckets)	Starting on-time cases	Time from when the patient is seen by the ED Physician to when they complete their disposition	ORM (OR Management Systems	Statistics	Jan–Sept		Time Buckets: Early or on-time: 11% (201/1824) Late 10 min or less: 36% Late 11 min < = 20 min: 26% Late 21 min < = 30 min: 12% Late > 30 min: 14%
Patient care and safety delays	Delays impacting turnaround time, patient care, and safety	Counts within occurring categories	Purple sheet, and put into Excel spreadsheet	Counts and frequencies	Jan–Sept		See Pareto chart: Largest delay is MD Late

(continued)

Table 4.6 (continued)

Data collection plan

Critical to satisfaction (CTS)	Metric (short title)	Operational definition (metric description)	Data collection source	Analysis mechanism	Sampling plan (size, frequency)	Process to collect and report	Baseline
Associate satisfaction	Associate satisfaction	Survey of satisfaction	Survey results	Survey results	Most recent		See results
Productivity	Total productive minutes	Current standard 7.85 h of staffing for 1 h of surgery	ORM (OR Management Systems	Percentage	YTD		YTD: 994%

case within the same room, excluding surgeons who have two rooms. We extracted the data from the Operating Room Management system and calculated the moving range of the difference between the actual start time of the next case minus the actual end time of the prior case, for the same room and same surgeon. We removed values greater than 90 min as cases not intended or scheduled to follow directly after the prior case in the same room. We also removed cases with the same surgeon at the same time in different rooms since one of the rooms is purposely held for the specific surgeon to reduce surgeon downtime.

On-Time Starts (of Cases)

The On-Time Start measures the timeliness of the OR cases starting when scheduled. The purpose is to assess the timeliness of the start of the OR cases based on their scheduled start times. For the first case for a surgeon, we measured the first case of a surgeon that started by either 7:30 or 8:00 a.m. start times. For surgeons with two rooms used sequentially, we removed those times since the range of time would appear as negative across the surgeon. We also measured the on-time starts of all cases.

Delays Impacting Turnaround Time and On-Time Starts of Cases

This measure assesses delays that impact starting cases on time, and turning around the room in a timely manner. The purpose is to understand the root causes of delays that impact turnaround time. The delays are currently recorded in the OR and entered into an Excel spreadsheet for tracking. We created a bar chart of the frequency by delay reason.

Associate Satisfaction

The associate satisfaction survey assesses the job satisfaction of the staff that works in the OR. The purpose is to understand the perception of staff satisfaction with their jobs.

This assesses whether the staff believes that they have the tools to do their jobs, whether management supports them and enables them, and removes barriers to provide great patient care. We created a staff satisfaction survey that was distributed to all OR staff to measure a baseline. We analyzed the data through statistical Chi-Square analysis by question to determine whether the associates are satisfied with their jobs in the OR.

Productivity

The productivity metric measures the total productive time of the staff in the OR. The purpose is to assess the efficiency of the staff that provides services in the OR. We used the existing productivity measures as a baseline for the fourth quarter of year one.

Define Detailed Voice of Customer

The voice of the customer surveys were used to assess patient, staff, and physician's satisfaction with the OR processes.

All but one of the questions' ratings was statistically significant. All but one of the questions were rated the highest percent of ratings in the 5-Strongly Agree category. The question, "5. My case started on time today." Had 7 ratings of strongly disagree and agree, and five rated as strongly agree, with no ratings of disagree, and only one rating of neither agree nor disagree. This indicates that some of the cases start on time, while others do not.

The following questions all received a high percentage of strongly agree ratings: front desk communication; the turnaround time; patient safety; equipment and supplies available; people worked together; it bothers staff when cases do not start on time; patients were ready in holding. The ratings are shown in Table 4.7.

The results show overall fairly high satisfaction with the processes around OR case readiness and turnaround of rooms, as well as communication and management. The only area where there was a mix of responses was in "My case started on time today." This supports the need for this project.

Define the Voice of Process and Current Performance

The Pareto Chart in Fig. 4.6 shows the percentage of cases by surgery procedure category. Orthopaedic cases consist of 31% of the cases in the OR, followed by general surgery cases consisting of 25% of the cases.

The graphic in Fig. 4.7 illustrates the major activities and average timeline for the cardiac cases. There is an average of 21 min to clean and set up the room, and then another 23 min to set up anaesthesia equipment for the surgery, before wheels in. There is an average 34 min of patient prep and anaesthesia patient time before incision. The surgeries averaged 84 min, with a 10-min patient wake up after the close.

Table 4.7 Staff satisfaction survey

Question	% Rating strongly disagree	% Rating somewhat disagree	% Rating neutral	% Rating somewhat agree	% Rating strongly agree	Chi-Square p-value/ significant (Yes/No)
1. The front desk/ management team communicated effectively with me today regarding my cases	0	2	4	3	12	.008 (Yes)
2. The turnaround time was appropriate for this procedure	2	1	2	3	12	.000 (Yes)
3. Patient safety took precedence over starting on time today	0	0	1	3	17	.000 (Yes)
4. I had the equipment and supplies I needed to do my cases	0	2	1	5	14	.000 (Yes)
5. My case started on time today	7	0	1	7	5	.187 (Yes)
6. People worked together in the department today to get the job done	1	1	0	5	16	.000 (Yes)
7. It bothers me when cases do not get started on time	0	0	2	4	14	.002 (Yes)
8. My patients were ready in holding today, leaving me time to attend to room preparation	2	2	4	3	11	0.011 (Yes)

Turnaround Times

The baseline for the turnaround time is an average of 40 min with a standard deviation of 21 min. The histogram is shown in Fig. 4.8.

The average turnaround time by surgical procedure category is shown in Table 4.8. Neurology cases have the longest average turnaround time of 50 min, followed by vascular cases with an average of 47 min. The lowest average turnaround case types are ICD with 24 min, Ear nose throat (ENT) with 25 min, and Cardiac cases average 31 min.

The average case turnaround times also differed by surgeon. We stratified the average turnaround time for each surgeon.

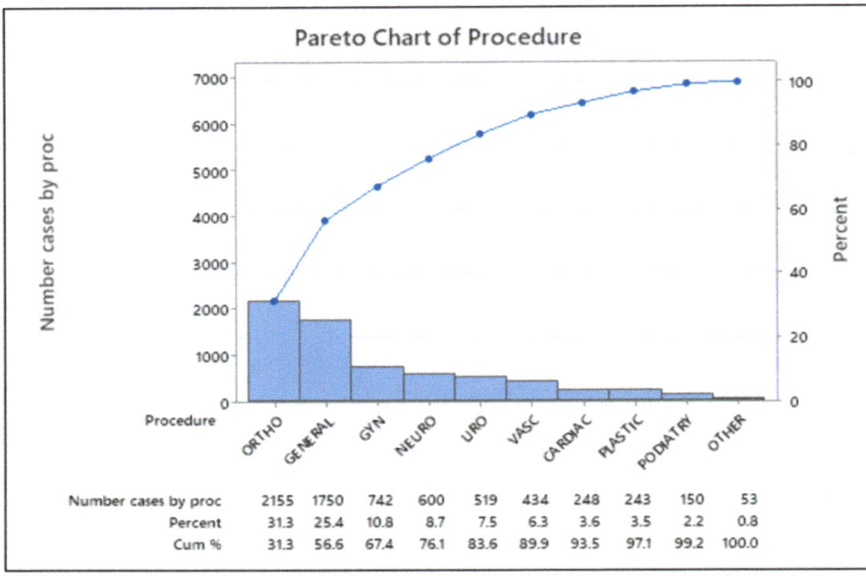

Fig. 4.6 Pareto chart of surgical cases by procedure category

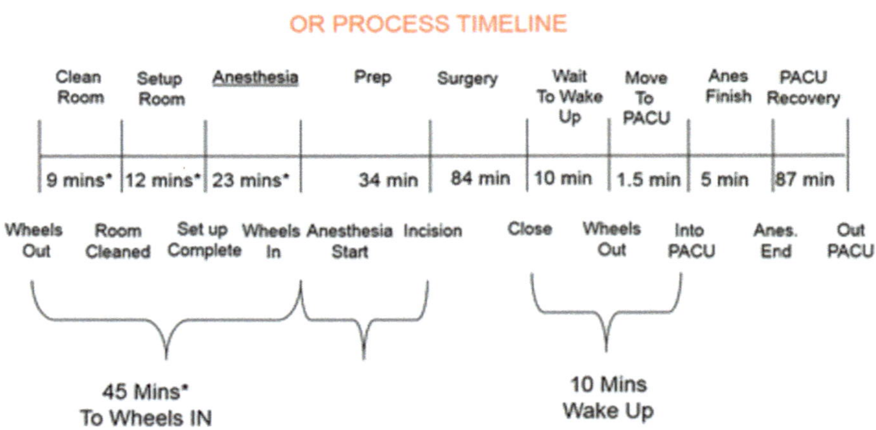

Fig. 4.7 OR cardiac process timeline

On-Time Starts

The on-time start average time is minus 19 min, representing start times that are 19 min late, with a standard deviation of 22 min. The average on-time start for those cases that start on time or early averages 10 min, with a standard deviation of 31 min.

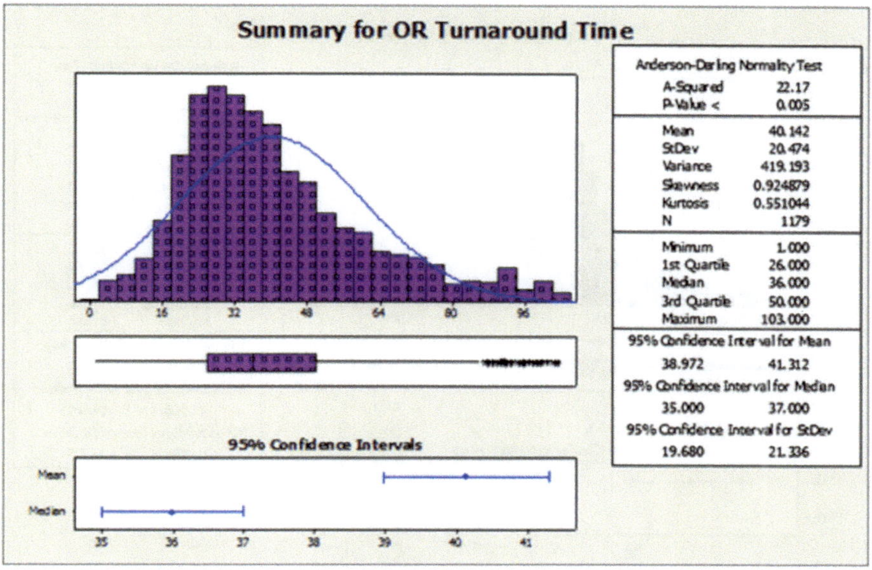

Fig. 4.8 Baseline of turnaround time

Table 4.8 Average turnaround time by procedure category

Procedure type	Mean turnaround time (minutes)	Standard deviation	Median turnaround time (minutes)
Bariatric	36	19	34
Cardiac and Pacer	34	20	30
ENT	25	6	27
General	40	20	36
ICD	24	28	24.5
OBGYN	42	21	37
Neuro	50	17	47
Oral	35	14	33
Ortho	35	19	31
Plastic	35	18	31
Podiatry	40	24	33
Urology	32	19	27
Vascular	47	25	44.5
Average turnaround time	40		

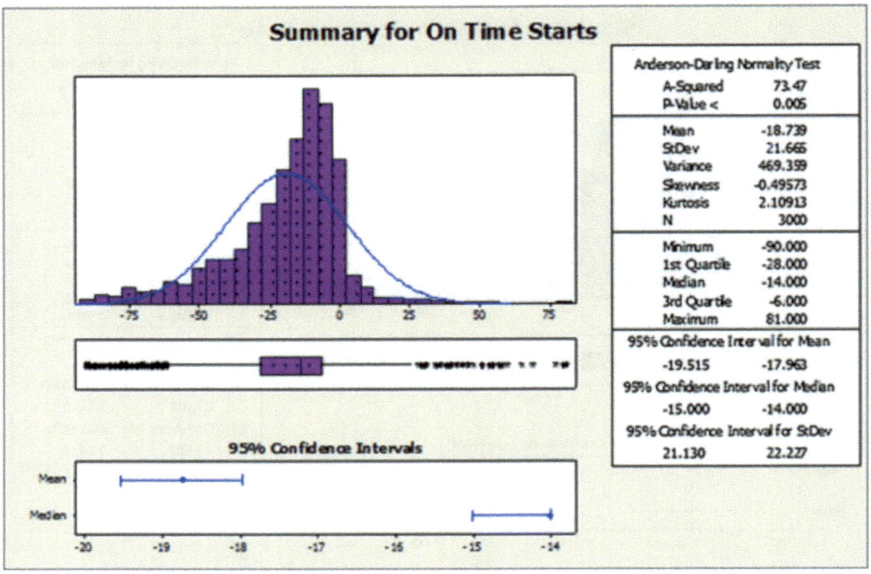

Fig. 4.9 On-time starts (late, on time, early)

The average on-time start for those cases that start late is −23 min, with a standard deviation of 19 min.

Eleven percent (11%) of the cases start early or on time. Thirty-six percent (36%) of the cases start 10 min late or less. Twenty-six percent (36%) of the cases start between 11 and 20 min late. Twelve percent (12%) of the cases start between 21 and 30 min late. Fourteen percent (14%) of the cases start more than 30 min late. The on-time start histograms are shown in Figs. 4.9, 4.10, and 4.11. The percentage of on-time starts for the first case of the day and for all cases by percent in the early/on-time and late categories are shown in Table 4.9. We also stratified the on-time start data by surgeon and procedure category to determine if there were particular surgeons or procedure types that had longer on-time start times. The on-time start times by procedure category are displayed in Table 4.10.

Delay Reasons

The most frequently occurring delay reason for starting cases on time or turning them around in a timely manner is the MD (surgeon) being late. The Pareto chart is shown in Fig. 4.12.

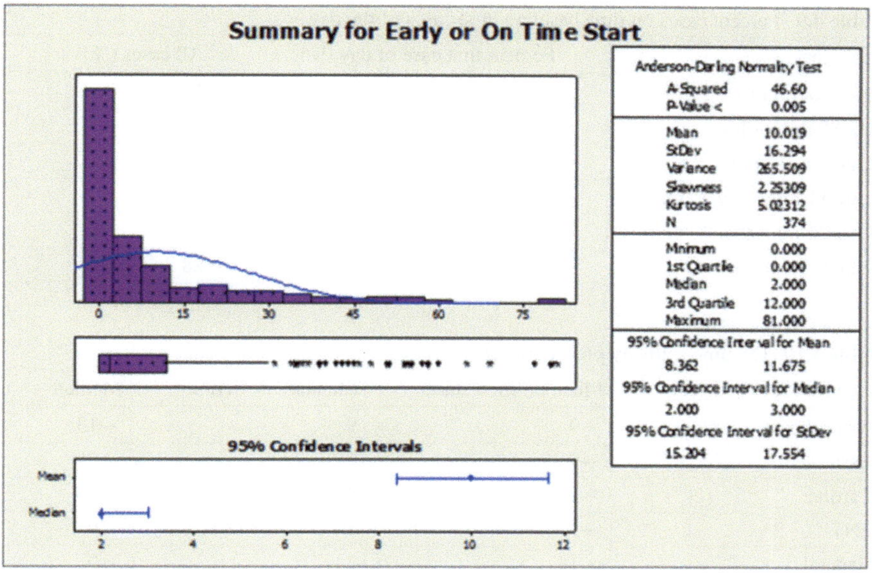

Fig. 4.10 On-time starts (early, on-time starts)

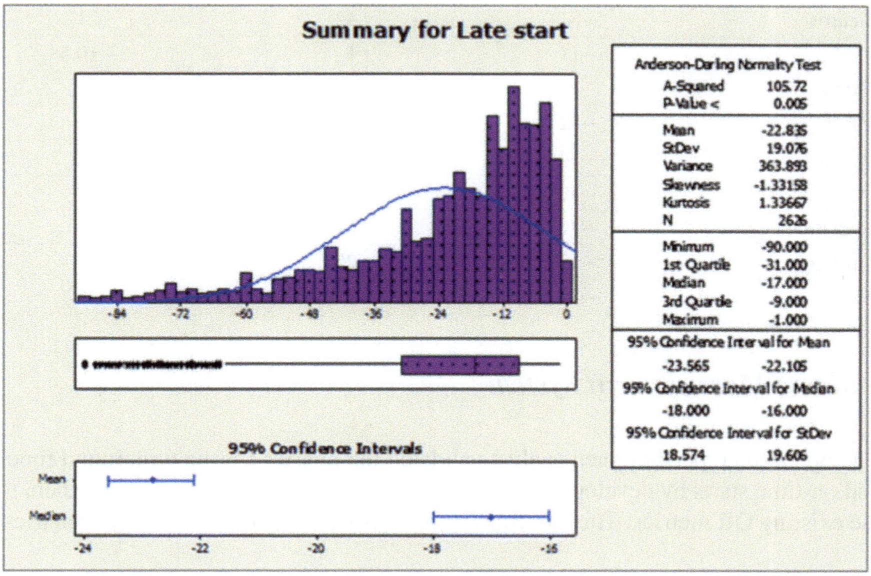

Fig. 4.11 On-time starts (late starts)

Table 4.9 Percent cases on time baseline: first cases of the day

Category	Percent first case of day (%)	All cases (%)
Early or on time	11	11
Late 5 min or less	16	10
Late 6 to 10 min	20	14
Late 11 min < = 20 min	26	22
Late 21 min < = 30 min	12	15
Late > 30 min	14	28

Table 4.10 On-time starts by procedure category

By procedure category	Mean on-time starts	Standard deviation	Median
Anaesthesia	−9	9	−13
Bariatric	−9	9	−7
Cardiac	−26	42	−17
ENT	−11	10	−9
General	−15	34	−9
Gyn	−20	40	−12
ICD	−13	3	−14
Neuros	−31	42	−21
Oral	−11	1	−10.5
Ortho	−17	42	−8
Pacer	−42	57	−19
Plastic	−14	15	−9.5
Podiatry	−10	9	−9.5
Radiation	−7	6	−6
Urology	−12	25	−9
Vasc	−34	46	−22

Validate Measurement System

The operational performance analyst validated the metrics for the turnaround times and on-time starts by developing the baseline measurements and comparing them to the existing OR metrics. The baselines were consistent with the established metrics.

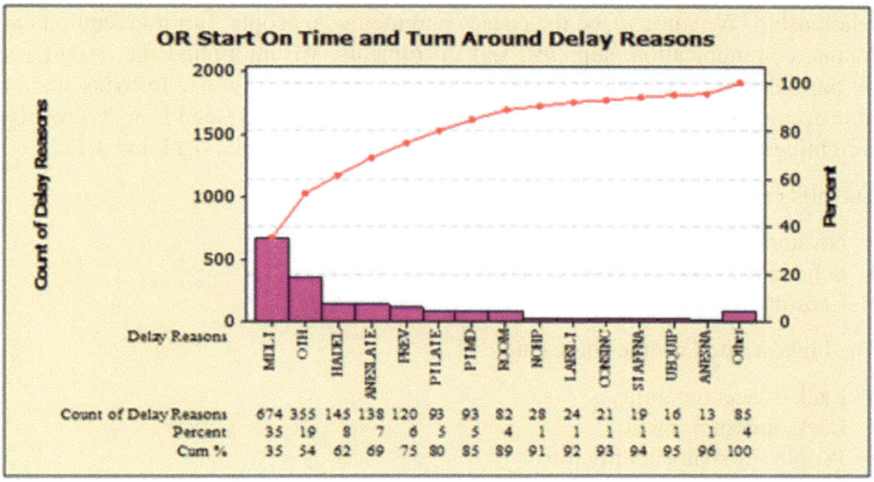

Fig. 4.12 Delay reasons

Analyze Phase

The following activities were performed for the Analyze Phase to identify and validate the root causes and to assess whether the process is currently capable of meeting the desired specifications.

1. Develop cause-and-effect relationships.
2. Determine and validate root causes.
3. Develop process capability.

Develop Cause-and-Effect Relationships

We filmed the OR turnarounds, both the cleaning activities as well as the set up of the next surgical case. Filming the activities was incredibly helpful in having the team understand the activities and the barriers that caused inefficiencies in the process. Many of the people would never have believed the variability in the process unless they actually saw the recorded turnaround processes. The team viewed several turnarounds in both general surgery and orthopaedic rooms. They made observations regarding what was done well and what was not consistent. The team also brainstormed potential causes of inefficiencies regarding starting OR cases on time. We then created a cause-and-effect matrix to understand the root causes for both the timely turnaround of the OR cases and the on-time starts of the first cases of the day in each room. We rated the importance of the effects of the OR turnaround as high (10) and the on-time starts as medium (5). We then rated the level of each cause related to each effect on a scale of 9-high, 3-medium, 1-low, and blank as no relationship. We categorized the causes into methods, people, furniture/equipment/rooms, communication, supplies, and instruments. We multiplied the importance by the rating for each cause to get a relative weighting of where to focus first on improvements. We grouped them into High (A), Medium (B), and Low (C) relative weightings. The cause-and-effect matrices are shown in Tables 4.11 and 4.12.

The highest rated methods causes are:

- No standard process and order of activities.
- Schedule changing as physicians see fit.
- Labs not ready.

The highest rated people causes are:

- Lack of accountability
- Lack of commitment
- People say that's not my job
- Surgeon not here on time.

The highest rated Furniture/Equipment/Rooms cause is:

- Equipment not ready in room.

The highest rated Communication cause is:

- Miscommunication with vendors, not being notified of devices.

The highest rated Supplies and Instruments causes are:

- Case cart lost.
- Wrong rep for their surgery.
- Instruments not ready.
- Case not picked correctly.
- Accuracy of preference cards.
- Rep trays not here.

Determine and Validate Root Causes

To ensure that we are getting to the root causes, we performed a Why-Why analysis and created a Why-Why diagram.

Develop Process Capability

The goal of the turnaround time is 30 min or less. We set the upper specification to 30 min, and the lower specification to 15 min. The goal of the on-time start times is to be between 30 min late and 30 min early.

Table 4.11 Cause-and-effect matrix: methods

	Effects			
CTS	Turnaround barriers	Delays in on-time starts	Total	Relative weighting
Importance	10	5		
Causes for methods				
No responsibility for organizing sub-sterile rooms	3	3	45	C
Cases finishing at same time, people not available to clean at same time	9		90	B
Assumption that you need more than 1 tech person	1		10	C
Circulating nurse leaves and never comes back	1		10	C
No standard process and order of activities	9	9	135	A
Schedule changing as physicians see fit	9	9	135	A
Anaesthesia clinicians not coming promptly to rooms	3	3	45	C
Bed not available when patient needs to come out	9		90	B
No hospital room available	1		10	C
Holding area nurses have too many patients		9	45	C
All cases start at same time		9	45	C
Patient still being processed to admitting	9	9	135	A
Admitting batching and not available		3	15	C
Cases run late at night, instruments go down late, low staffing at night in Sterile Processing	3	3	45	C

(continued)

Table 4.11 (continued)

	Effects			
CTS	Turnaround barriers	Delays in on-time starts	Total	Relative weighting
Importance	10	5		
Patient not ready in holding	3	9	75	B
Physician offices not consistently getting paperwork that is needed	9	9	135	A
Tech not available to clean room, busy, break	9		90	B
Anaesthesia not seeing patient in holding area or night before (pre-screening)	9	9	135	A
Patient not told correct time to be here		9	45	C
Labs not ready	9	9	135	A
Changing priorities when in middle of cleaning room; no set priorities	9	3	105	B
Need to be able to communicate what is done in a room	9		90	B

The goal of the on-time starts are as follows:

1. Average of all cases: $\leq$ 30 min.
2. General and small cases: 15 min.
3. Orthopaedic and large cases: 20 min.
4. Heart and neuro (major spine) cases: 35 min.

To develop the process capability, we first had to determine if the distributions for the metrics were Normal distributions. The null hypothesis is that the data distributions are Normal, and if we fail to reject the null hypothesis then we can conclude that the data follows a Normal Distribution. However, the OR turnaround time and the On-time Start times are not Normal Distributions (p-values less than 0.005). We fit non-normal statistical distributions for the OR turnaround times, and found the Largest Extreme Value distribution to fit the best, shown in Fig. 4.13. For the on-time start times, since there were negative values, and most of the distributions (Weibull, Exponential) required positive values only, we used the Normal distribution assumption. We developed the process capability for the turnaround time, and found that

Table 4.12 Cause-and-effect matrix: other causes

	Effects			
CTS	Turnaround barriers	Delays in on-time starts	Total	Relative weighting
Importance	10	5		
Causes for people				
Have perception that they don't have time to do their work	1		10	C
Lack of accountability	9	9	135	A
Lack of commitment	9	9	135	A
People say that's not my job	9	9	135	A
Physician perception of turnaround time	3		30	C
Surgeon not here on time	9	9	135	A
People are coming in late to morning report and don't hear everything	1	1	15	C
OR team not being trained on the new process	9	9	135	A
Furniture/Equipment Causes				
Having nothing to transport trash on (heavy bags)	1		10	C
Not able to find furniture	3	3	45	C
Reps have stuff laid out on every surface	1		10	C
Equipment not ready in room (depends on night before)	9	9	135	A
Communication				
Lack of communication	3	3	45	C
Don't get communication from front desk	3		30	C
Morning report runs late because it's full of fluff (not patient related)		3	15	C

(continued)

Table 4.12 (continued)

	Effects			
CTS	Turnaround barriers	Delays in on-time starts	Total	Relative weighting
Importance	10	5		
Miscommunication with vendors, not being notified for devices	9	9	135	A
Supplies and instrument				
No sterile batteries available	1	1	15	C
Case cart lost	9	9	135	A
Missing items for case	9	9	135	A
Need to take back unused items to Core	1		10	C
Wrong rep there for surgery	9	9	135	A
Instruments not ready	9	9	135	A
Cleaning supplies not available	1		10	C
Cases not picked correctly	9	9	135	A
Too much checking of supplies and instruments	3	3	45	C
Accuracy of preference cards	9	9	135	A
No room for extra supplies not used in surgery	3		30	C
Rep trays not here	9	9	135	A

currently the process is not capable of meeting the specification of 15 to 30 min, shown in Fig. 4.14. The Pp = 0.14 and the Ppk = −0.06. The process is not capable of meeting our upper and lower specifications for on-time starts of −30 min late to 30 min early. The Cp = 0.27 and the Cpk = 0.09, shown in Fig. 4.15. This is much lower than our goal of a Cp of 1.33 or greater. We then moved on to the improvement phase to identify improvements to the current process.

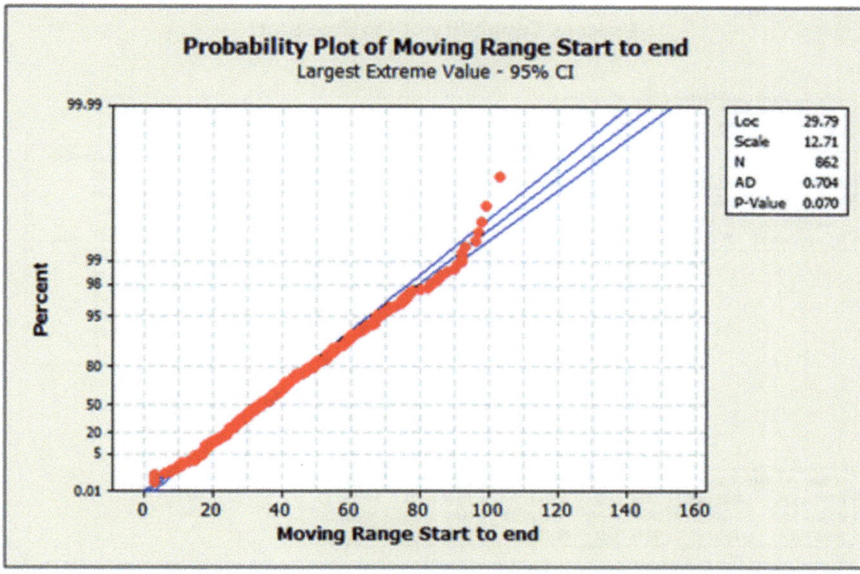

Fig. 4.13 Distribution fit for process capability: OR turnaround

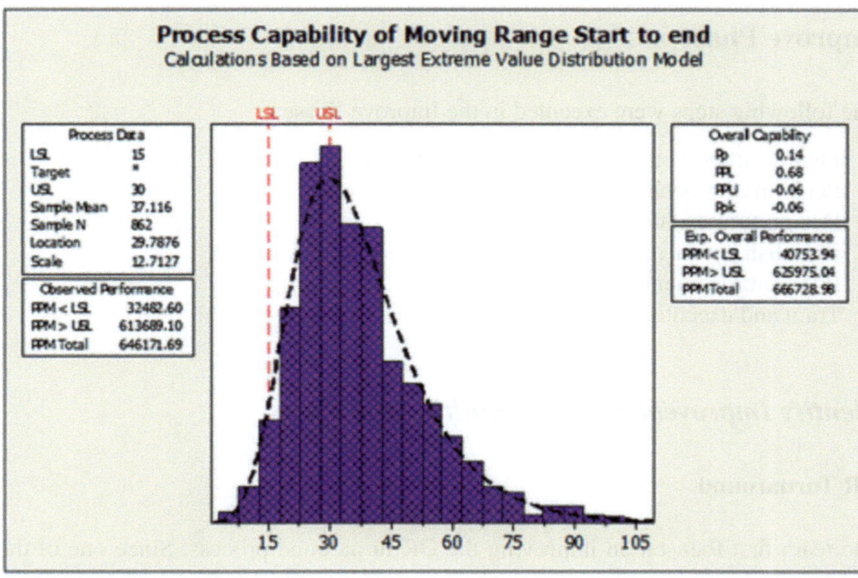

Fig. 4.14 Process capability: OR turnaround

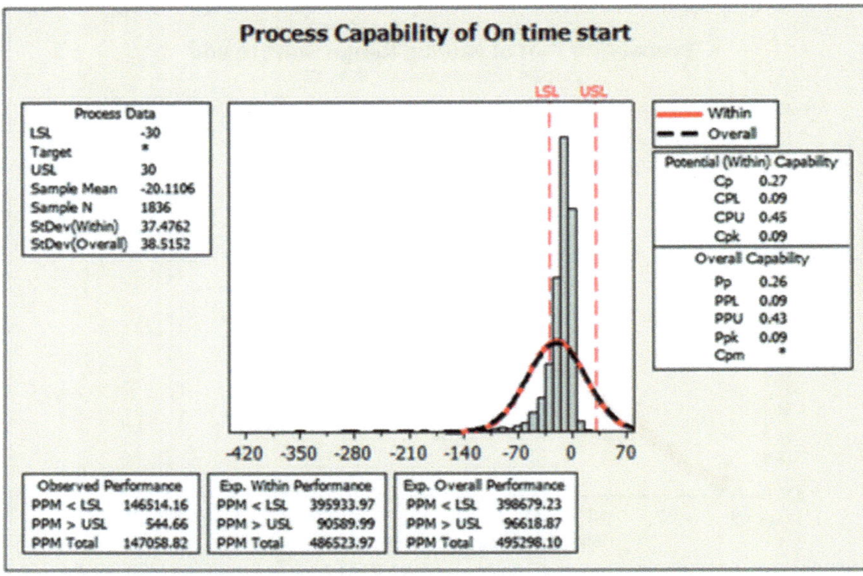

Fig. 4.15 Process capability: on-time case starts

Improve Phase

The following steps were executed in the Improve Phase:

1. Identify improvement opportunities and plans.
2. Perform cost/benefit analysis.
3. Design future state.
4. Establish performance targets and project scorecard.
5. Pilot improvements.
6. Train and execute.

Identify Improvement Opportunities and Plans

OR Turnaround

The team first focused on improving the OR turnaround process. Since one of the highest rated root causes was the lack of a standard process, the team developed a standard turnaround process, focusing on eliminating the inconsistencies and lack of alignment with the AORN (American OR Nurses) requirements, and the policies and procedures. The team gathered some of the staff that perform the OR turnarounds to develop the new process. We clearly identified the order of the tasks, ensuring first that the trash is removed and that the dirty room is cleaned, communicating that the cleaning is complete, and then setting up the room for the next case.

The team decided to practice the new process that was designed on paper to see if it worked well, and to determine if improvements could be made by aligning some of the activities to be performed by different people in parallel. We then revised and standardized the process. We then filmed the new process, and our media personnel put together a very professional training film of the new OR turnaround process. We introduced the new process at one of the OR staff meetings, with the CEO stressing the importance of improving the OR turnarounds to increase OR capacity, as well as patient, staff, and surgeon satisfaction.

Improvement Plan OR Turnaround and On-Time Starts:

There were barriers that were identified that had to be removed before we could implement the new OR turnaround process, and that would also help to improve on-time start of cases. We developed an improvement plan that addressed and removed these barriers and root causes.

Perform Cost/Benefit Analysis

For implementing the new OR turnaround process, and the 5S organization of the OR, there was very little cost, and none that was not covered by the existing OR budget.

Design Future State

We designed the future state by developing the OR turnaround film and the implementation plans. The process maps did not change, since the OR process changes were at a more detailed work instruction level, rather than the higher process map level.

Establish Performance Targets and Project Scorecard

Our performance targets for turnaround time are as follows:

1. Average of all Cases: $\leq$ 30 min.
2. General and small cases: 15 min.
3. Orthopaedic and large cases: 20 min.
4. Heart and neuro (major spine) cases: 35 min.

Pilot Improvements

Our formal piloting of the improvements began in June, when we kicked off the new OR turnaround process. The CEO and Chief Nursing Officer, as well as the OR management and the Master Black Belt, spoke to the OR staff to describe the importance of standardizing the OR processes and improving efficiency throughout the OR. The CEO said that this project had the complete backing of the administrative council, his direct reports, and was the number one priority in the organization.

There was quite a gap of time between when we completed the initial design of the new process and when we rolled it out. The OR staff and OR management were extremely comfortable with the current process and were resistant to change. The CEO, CNO, VP of Human Resources, the OR Director, and the Master Black Belt met to map out an approach to implementing the change. We decided to bring in an outside consultant that would be able to focus on a daily basis on implementing the detailed improvement plans, help to train the OR staff on the new process, implement accountability, process discipline, and incorporate the new daily OR turnaround and On-time start metrics.

The OR turnaround and on-time start improvements are shown in Tables 4.13, 4.14, 4.15, and 4.16. A summary of the improvements is as follows:

Process Standardization and Improvement

- Implement a standard turnaround process.
- Implement team huddles to plan and debrief at the end of the day.
- Bring patients from admitting to Pre-op earlier in the morning, and do not batch.

Organize the OR (5S)

1. Implement shift handoffs to verify equipment and cases are ready.
2. Remove unneeded equipment, supplies, and beds.
3. Develop check sheets to ensure sub-steriles, hallway supplies, and OR cabinets are restocked.
4. Organize and identify all equipment and bed locations, "a place for everything and everything in its place."

People, Skills, Competencies, and Evaluation

- Implement turnaround and cleaning competencies, incorporate into performance evaluation.
- Ensure all appropriate staff are following the process.

Instruments

- Implement instrument tracking to reduce missing and incomplete trays.
- Replace missing instruments to complete trays.

Metrics and Flow Control

1. Implement process metrics and dashboards that track improvement in on-time starts, turnaround, and manage flow.

Table 4.13 OR turnaround improvement plan: high-priority items

Improvement idea	Causes	Owner	Implementation date
Develop and implement new process. Assign techs with hall person with 2 rooms in a team and have each team huddle in the morning to discuss the day (cases, equipment, supplies, instruments needed, etc.) Debrief at end of day with a few of the teams	No standard process or order of activities; assuming need more than 1 tech; circulating nurse leaving and not coming back; have perception don't have time; lack of organization; people coming in late for morning report	OR staff	4/3 for the huddle meetings
Look at scheduling process in more detail, scheduled and planned time for case, and average for cases	Schedule changes	Enterprise Performance Excellence team	June
Talk with admitting to quantify time to admit. Have dedicated admission person for surgery from 7 to 10 am	Patient still being processed in admitting; admitting batched	Patient Access	April
Physician offices, stat labs, added labs, look at pre-screening prep patient Six Sigma project	Labs not ready; miscommunication with vendor (Dr. office)	Green Belt Lean Six Sigma process team	July
Update preference cards; standardize preference cards, implement pref-cards accuracy metric. Implement audit	Case not picked correctly, Accuracy of preference cards, taking items back to Core	Coordinators, Sterile Processing	On-going
Standardize picking case carts. Track accuracy of case carts	Checking of supplies for cases; no room for extra supplies in surgery	OR Scheduler and Materials Management	4/9

2. Implement a flow control sheet to manage flow and ensure on-time starts.
3. Implement surgeon and anaesthesia dashboards to assess and control physician lateness.

Flow Control Board: A Major Breakthrough

The OR pilot progressed at a snail's pace for the first six months. We saw some improvement, mostly in fits and starts. We really struggled with buy-in and engagement at all levels and with all functions throughout the OR, including the OR staff, the anaesthesiologists, and the surgeons. The general consensus in the OR was that we were trying to get everyone to work harder at the potential risk of patient safety.

Table 4.14 OR turnaround improvement plan: high-priority items (continued)

Improvement idea	Causes	Owner	Implementation date
Implement 5S organization and standardization Pref-cards are being updated; Board outside on door for required equipment for next case. Verify boards are on doors. Night people should write on board what is missing from the room, for each room for first case. (checklist of night shift accountability	Equipment not ready	Circulators to implement putting required equipment on board on door	4/2
Talk with Materials Management about getting appropriate linens—specific to what OR needs	Linen not available	Enterprise Performance Excellence (EPE)	4/30
Implement dashboard (manual calculation—need to automate) and incorporate process into performance evaluation and skill competencies; getting inline with synergy model	Lack of accountability and commitment	OR Director	4/30
Implement surgeon dashboards; require surgeons by 10 min early and send letters to communicate how much it costs when late or how it reduces delays. Recognize on-time physicians and nominate for physician of quarter	Surgeon late; physician getting lost	OR staff	4/30 for huddle meetings
List of vendor trays by procedure and surgeon	Vendor trays	Material management	4/30
Implement 5S organization and standardization	Case cart lost, missing items	5S	5/15
Rationalize most used instruments, and 5S organization	Instruments not ready, and checking of instruments	5S	5/15

Whether this was a very gross misunderstanding perpetuated out of poor communication, lack of understanding of how Lean Six Sigma can help simultaneously with quality and service delivery, or whether it was just a smoke screen to divert the focus away from change, we weren't sure. We constantly had to reinforce and explain the value of the approach and the importance of the improvements. What finally enabled us to turn the corner and gain buy-in from the OR staff, the anaesthesiologists and the surgeons was implementing the Flow Control Board.

It became painfully obvious that functional silos were alive and thriving in the OR. Each function, scrub tech, surgical nurse, inventory control, sterile processing,

Table 4.15 OR turnaround improvement plan: medium priority items

Improvement idea	Causes	Owner	Implementation date
Implement dashboard measuring OR teams for cleanliness and organization; implement checklists; Nurse's job to keep it clean, monitor trash on floor, keep up with it during surgery. Put in audit to measure cleanliness of room when techs start cleaning	Lack of organization	Circulator; scrub tech	4/30
How to coordinate and pull techs from entire OR, know when surgeon is out of room, so front desk better coordinates and anticipates multiple rooms coming out at same time. Long term: electronic status reporting, with electronic charting	Cases finishing at same time; techs not available	Charge front desk coordinator	4/30
Stretcher should stay outside of room until bed is available	Bed not available when patient comes out		4/30
Six Sigma project for holding	Patient not ready in holding	Six Sigma project	7/8
Will assign techs on night shift; assign runners to floor; and others have assigned rooms	No set priorities, changing priorities of rooms	Evening OR supervisor	4/12
Part of OR process to communicate what is done in rooms	Need to be able to communicate what is done in rooms		4/30

anaesthesia, and surgeon, would perform their "job" but had no way to communicate the status of their outcome. For example, identifying when the room is ready, the patient is ready, anaesthesia has seen the patient, the surgeon has seen the patient, and the patient is ready to be wheeled in. There was also incredible fear of being blamed for the case starting late, so very few problems were recorded or surfaced. More often than not, even if there were a series of issues that could cause the case to be delayed, only the last delay would be recorded. This last delay was typically "MD late," since all of the other issues had been resolved as they waited for the surgeon. Why was the surgeon late? Because there was no predictability or communication that the OR team was close to starting on time. It was an impossible situation that was almost impossible to fix.

The Master Black Belt and the Black Belts came up with a solution that was seeded by the CEO. He had made the comment that what the OR needs is a "traffic cop" to direct traffic. When he had been in the OR trying to fix the problems, he

Table 4.16 OR turnaround improvement plan: low priority items

Improvement idea	Causes	Owner	Implementation date
5S organization, assign responsibilities for sub-sterile room organization and stocking	No one responsible for sub-sterile rooms	5S	5/15
Do a Six Sigma project with Anaesthesiologists	Process	New Six Sigma Project	7/8
Part of Six Sigma project for holding	Holding area nurses have too many patients	Green Belt LSS project	7/8
Nursing doing a lessons learned from busy season	No hospital room available to assign	Nursing	5/30
Included in board on door improvement	All cases start at same time		
Additional Sterile Processing staff. Down several positions. Six Sigma project	Cases run late at night, instruments not ready	Green Belt LSS project	7/8
Look on tracker and call admitting to process patient right away. Send letter to physicians' offices to have their patients here 2 h prior to surgery time	Patient not told correct time. Late for other reasons	Physician office, patient late	4/30
Dashboard showing turnaround times, weekly or monthly	Physician perception of turnaround		5/15
Post fluff notices on board, instead of covering in morning meeting	Morning meetings, people coming in late	OR Front Desk	3/29
Use back table for set up and then clean	Nothing to transport trash on	OR staff	4/30
5S organization	Not being able to find furniture. Cleaning supplies not available	5S	5/15
Reps should be instructed to only bring what they need for the case	Reps have laid out stuff on all surfaces	Coordinators	4/30
Front desk more proactively pulling and coordinating staff	Lack of communication, not communicating with front desk	OR Front Desk	4/30

could apply pressure and direct traffic, and even show improvement, but as soon as he left, it went back to the same chaos.

We had been waiting for a large Information Technology (IT) project to progress that would implement an OR tracking board that would enable the following process steps to be tracked:

1. Room ready.
2. Patient ready.
3. Anaesthesia arrived.
4. Surgeon arrived.

When it became obvious that the IT solution was going to be very laborious and not very user friendly, we realized that all we really needed was to have a simple shared Excel spreadsheet that allowed different people in the OR to notify everyone else that their "job" was complete. The Chief of Anaesthesia saw the value and required the anaesthesiologists to notify everyone that they had arrived and saw the patient via the Flow Control Board. Prior to this milestone, there was no way to easily know if anaesthesia had arrived or had seen the patient in Pre-op. This was a huge success, since anaesthesia had been a very huge bottleneck up to this point in moving the patient to the ready state. The Flow Control Board is shown in Fig. 4.16. From the beginning of the project to about 1 year later, we had improved the on-time starts for first cases by 8%, from 11 to 19%. Within one week of implementing the Flow Control Board, we further improved the on-time starts for first cases by another 16 to 35%.

OR FLOW BOARD

					Readiness								
OR Room	Scheduled Start Time	Procedure	Surgeon	Preop Room	Patient	Preop	Room	Anesthesia	Surgeon	Wheels In	Close	Wheels Out	Room Clean
1	8:00	Total Knee	PC	4	Y	Y	Y	Y	Y	N	N	N	N
2	7:30	Total Knee	JT	3	Y	N	N	Y	Y	N	N	N	N
3	7:30	Prostate	KF	6	Y	I	N	Y	Y	N	N	N	N
4	7:30	Excision	DF	5	Y	Y	Y	Y	Y	Y	N	N	N
5	7:00	Endoscope sinus	CC	2	Y	Y	Y	N	N	N	N	N	N
6	8:00	Repair hernia	AE	1	Y	Y	Y	N	N	N	N	N	N
7	8:00	Hysterectomy	DC	11	Y	Y	Y	Y	Y	Y	N	N	N
8	8:00	Excision	DW	7	N	I	Y	Y	Y	N	N	N	N
9	7:30	Replace Aortic	DJ	9	N	Y	Y	I	Y	N	N	N	N
10	7:30	Total Knee	EB	10	I	Y	Y	Y	Y	N	N	N	N
11	8:00	Total Hip	EB	12	Y	Y	Y	Y	Y	N	N	N	N
12	8:30	Total Knee	RL	8	Y	Y	Y	Y	Y	N	N	N	N
14	8:00	Total Hip	KB	14	Y	Y	Y	Y	Y	N	N	N	N
15	7:00	Hysterectomy	KD	15	Y	Y	Y	Y	Y	N	N	N	N
16	7:30	Hernia	KP	16	Y	I	Y	Y	Y	Y	N	N	N
17	8:00	Cerv Lami	RK	17	Y	N	N	Y	Y	N	N	N	N

Fig. 4.16 Flow control board

Train and Execute

When we kicked off the new OR turnaround process, we showed the new OR film and made process checklists and the film available to all of the OR staff. The OR Manager ensured that the surgical coordinators understood the new process and assessed their ability to perform the new process by direct observation. The surgical coordinators then assessed the OR staff. Each responsible person was assessed through direct observation of them performing the OR turnaround process, and the competency assessment was added to their performance evaluation. Disciplinary action was also implemented if the new process and the 5S organization were not adhered to by each OR staff associate.

Control Phase

The purpose of the Control Phase is to implement control mechanisms and procedures to ensure that the process maintains the levels of improvement attained during the pilot activities.

The following activities were executed in the Control Phase:

1. Measure results and manage change.
2. Report scorecard data and create process control plan.
3. Identify replication opportunities.
4. Develop future plans.

Measure Results and Manage Change

On-Time Starts

By the beginning of January 21st in year 3, we achieved improvement in on-time starts, as follows:

- First Cases of Day On-time starts from 11 to 35% (24%).

The p-value for the two-proportion test is 0.000, showing a significant improvement of 24% in the percent of on-time starts for the first case of the day. The Minitab output for the two-proportion test to assess whether there was a statistically significant improvement in the on-time starts from 11 to 35% follows:

Test and CI for Two Proportions

Method

p_1: proportion Sample 1 = Event	where
p_2: proportion Sample 2 = Event	where
Difference: $p_1 - p_2$	

Descriptive Statistics

Sample	N	Event	Sample p
Sample 1	1824	201	0.110197
Sample 2	54	19	0.351852

Estimation for Difference

Difference	95% CI for difference
−0.241654	(−0.369833, −0.113476)

CI based on normal approximation

Test

Null hypothesis	H_0: $p_1 - p_2 = 0$
Alternative hypothesis	H_1: $p_1 - p_2 \neq 0$

Method	**Z-Value**	**P-Value**
Normal approximation	−3.70	0.000
Fisher's exact		0.000

The percent of on-time starts, starts within 5 and 10 min, is shown for weeks 44 of year 2 through week 4 of year 3, in Fig. 4.17.

Turnaround Time

By the end of year 3, the Turnaround Time improved as follows:

- Mean: 40 to 34 min (15% improvement).
- Standard Deviation: 20.4 to 14 min (34% improvement).

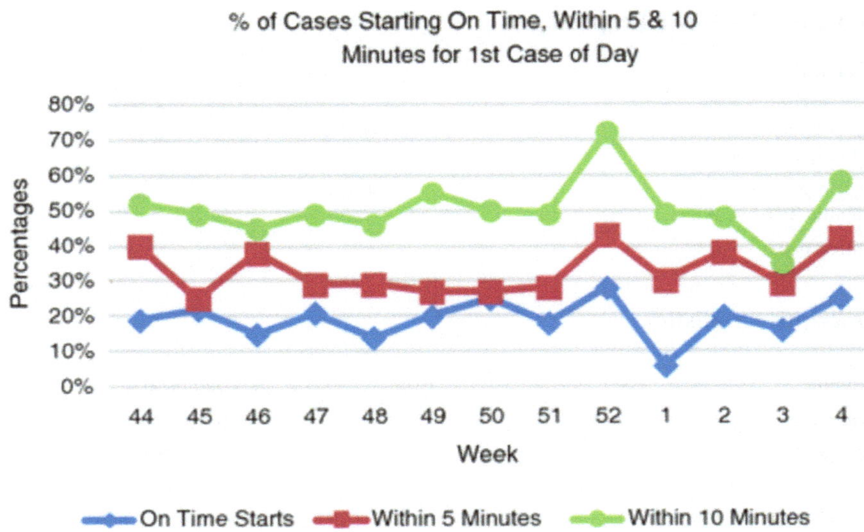

Fig. 4.17 Graph of on-time starts, starts within 5 and 10 min

Both the mean and variance showed a statistically significant improvement, with a *p*-value of 0.000 for the mean and a value of 0.001 for the variance. The Minitab output for the two-sample *t*-test and the test of two variances follows:

Two-Sample T-Test and CI

Method

μ_1: population mean of Sample 1
μ_2: population mean of Sample 2
Difference: $\mu_1 - \mu_2$

Equal variances are not assumed for this analysis.

Descriptive Statistics

Sample	N	Mean	St Dev	SE mean
Sample 1	1179	40.0	20.4	0.59
Sample 2	219	33.9	14.0	0.95

Estimation for Difference

Difference	95% CI for difference
6.10	(3.90, 8.30)

Test

Null hypothesis	H_0: $\mu_1 - \mu_2 = 0$
Alternative hypothesis	H_1: $\mu_1 - \mu_2 \neq 0$

T-Value	**DF**	**P-Value**
5.46	412	0.000

Test and CI for Two Variances

Method

σ_1: standard deviation of Sample 1
σ_2: standard deviation of Sample 2
Ratio: σ_1/σ_2
F method was used. This method is accurate for normal data only

Descriptive Statistics

Sample	N	StDev	Variance	95% CI for σ
Sample 1	1179	4.517	20.400	(4.341, 4.707)
Sample 2	219	3.742	14.000	(3.421, 4.129)

Ratio of Standard Deviations

Estimated Ratio	95% CI for Ratio using F
1.20712	(1.086, 1.332)

Test

Null hypothesis	H_0: $\sigma_1/\sigma_2 = 1$
Alternative hypothesis	H_1: $\sigma_1/\sigma_2 \neq 1$
Significance level	$\alpha = 0.05$

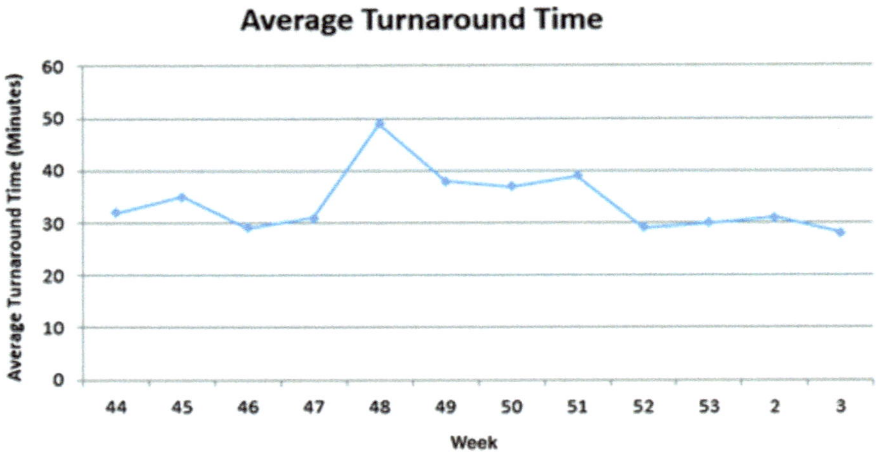

Fig. 4.18 Improvement of turnaround time

Method	Test statistic	DF1	DF2	P-Value
F	1.46	1178	218	0.001

The turnaround time from October 25th year 2 to January 11th year 3 is shown in Fig. 4.18.

The turnaround time for each procedure category before and after improvement is shown in Table 4.17. Plastic surgery had a 38% improvement in Turnaround Time, OB/GYN had a 32% improvement, General and Urology Surgery each had a 21% improvement. The other specialty areas did not yet see a significant improvement in the turnaround time. Although the difference in turnaround time for Vascular was 13 min with a 28% difference, because of the small sample size and the large standard deviation, the difference was not statistically different. This demonstrates the importance of performing the hypothesis tests to assess whether the difference was statistically significant.

Change Management

The OR staff, surgeons, and anaesthesiologists were at first extremely resistant to change, so we brought in an external consultant that was experienced in implementing process improvements and change. We also used the Dr. John Kotter change model (Kotter, 2011) to implement the improvements.

(1) Acting with Urgency:

   a. We gained top management support, and they visibly communicated this support for the project and emphasized the importance of the project to the organization.

Table 4.17 Improved turnaround time by procedure category

Procedure type	Mean turnaround time	Standard deviation	Improved mean turnaround time	Improved standard deviation	% Improvement (%)	Statistically different? P-value
Cardiac	34	20	35	6.5	−4	No, .86
General	40	20	32	13.4	21	Yes, 0.0
OBGYN	42	21	29	8.1	32	Yes, 0.0
Neuro	50	17	45	14.4	9	No, .294
Ortho	35	19	38	15.1	−8	No, .132
Plastic	35	18	22	9.1	38	Yes, .013
Urology	32	19	25	4.8	21	Yes, 004
Vascular	47	25	34	15.6	28	No, .304
Average Turnaround Time	40		34		16	

(2) Develop the Guiding Coalition:

a. We formed the project team and developed the scope of the project in the Define phase. Throughout the project we brought in subject matter experts who helped develop the new process and generate improvement ideas.

(3) Developing a Change Vision:

a. We developed the new OR Turnaround process, creating a film, process work instructions, and checklist.
b. We developed the implementation plan and a detailed work plan to stage the improvements.

(4) Communicate the Vision Buy-in:

a. We communicated the changes at the new process rollout meeting, attended by the CEO, CNO, Master Black Belt, and OR management. We communicate the changes and status of the improvement at the OR staff morning meetings and huddles.
b. The OR surgical coordinators model the behavior of the new process, being assessed on the new process, and assessing the OR staff on the new process. The circulating nurses on each team ensure that the new process is followed.

(5) Empowering Broad-Based Action:

a. We identified and removed obstacles to change. We implemented a suggestion system to get ideas from any and all of the OR staff. The project team evaluated these ideas and incorporated many of the ideas into the new process.
b. We incorporated accountability to adhere to the new process.

(6) Generating Short-Term Wins:

a. We implemented a reward system to provide rewards (meal tickets, cakes, etc.).
b. We communicated the wins and significant improvements to the OR staff and hospital administration.

(7) Do not Let Up:

a. We incorporated the new metrics dashboard into the daily huddles and morning meetings.
b. We incorporated the process competencies into the performance evaluation system.

(8) Make Change Stick:

a. We provided change management and Lean Six Sigma training to OR staff and management to ensure the succession of the change in thinking to a fact-based decision process and one of continuous improvement.

Report Scorecard Data and Create Process Control Plan

The control plan is a detailed guide for maintaining all of the positive changes that you have implemented, shown in Table 4.18. We developed a dashboard that will help us to ensure control. Additionally, the 5S sustainability will be discussed in the next chapter, that will help to maintain the gains for the OR organization of equipment, furniture, supplies, and OR turnaround cleaning and setup. The dashboard that we implemented is shown in Table 4.19.

Identify Replication Opportunities

The concepts and ideas of the 5S organization could be implemented throughout the hospital, and we will use the 5S changes as a guide for other departments.

We can also use the learnings from our change management approach to better implement change in other Lean Six Sigma projects and change initiatives within the hospital and within the hospital system.

We can also implement the concepts of standardized and repeatable processes documented at a work instruction level in almost every hospital process. We will communicate the success of this process to share the value of Lean Six Sigma in healthcare.

Table 4.18 Improvement control plan

Improvement control plan

The control plan is a detailed assessment and guide for maintaining all of the positive changes that you have implemented

Process steps	Control mechanism	Measure/ Metric	Criticality (H M L)	Action taken if problems occur	Owner
5S: OR organization					
Sub-sterile stocking	Sub-sterile checklist	Percent stocked	M	Assign cause	Surgical Coordinator
Equipment rooms	Equipment room checklist	Percent in place	M	Assign cause	Front Desk Coordinator
Verify hallways, furniture, equipment, OR beds, sub-sterile cabinets, leftover supplies in OR room returned	Shift handoff setup checklist	Percent in place	M	Assign cause	Exec. Director
5S Audit	5S Audit checklist	5S ratings	M	Assign cause	Exec. Director
Surgery execution					
Coordinate first case of day; room ready, patient ready, Anesthesia here, Surgeon here; time wheels in	Flow Control Board	On-time starts first cases	H	Assign cause	Nurse Manager
On-time starts first case of day	OR Metrics	On-time starts first cases	H	Assign cause	Exec. Director
Turnaround time	OR Metrics	Turnaround time	H	Assign cause	Exec. Director
Number cases by service	Number cases	Number cases	L	Assign cause	Exec. Director
Surgeon and anesthesia late analysis	Surgeon and Anesthesia late analysis	Surgeon and Anesthesia percent late	H	Assign cause	Exec. Director
Sterile processing					
Tray average cycle time	IMPRESS report	Tray average cycle time	H	Assign cause	Exec. Director
Number wet loads	IMPRESS report	Number wet load	M	Assign cause	Exec. Director

(continued)

Table 4.18 (continued)

Improvement control plan					
Percent missing trays	IMPRESS report	Percent missing trays	M	Assign cause	Exec. Director
Percent incomplete trays	IMPRESS report	Percent incomplete trays	M	Assign cause	Exec. Director
Tray errors	IMPRESS report	Tray errors	M	Assign cause	Exec. Director

Table 4.19 OR turnaround and on-time start scorecard

Metric	Baseline	Improvement	% Improvement
% of On-Time Starts	11%	35%	24%
% Starts within 10 min	47%	51%	4%
Average Turnaround Time	Mean = 40 min Stdev = 20.4	Mean = 34 min Stdev = 14	Mean = 15% Stdev = 34%

Develop Future Plans

The future plans are to implement the more complex changes in the OR related to the following areas, which are shown in Fig. 4.19.

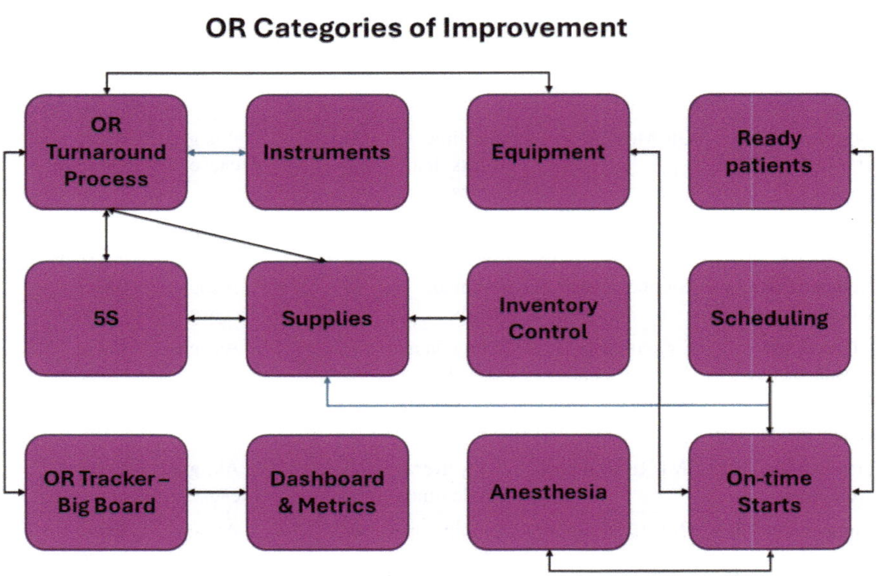

Fig. 4.19 Future OR plan categories

- *Outpatient Patient Readiness Six Sigma Project*—Improve the patient pre-admission process to ensure charts are ready the day before.
- *OR Tracker*—Implement an OR tracker, big board, to identify where the patient is in each step of the surgical process (Pre-op, surgery, recovery).
- *Supplies Inventory Control System*—Implement inventory control and supply process improvements.
- *Instrument Tracking Six Sigma Project*—Implement instrument tray tracking.
  - *Pre-op Inpatient Readiness*—Improve inpatient readiness and handoff to surgery.
  - *Flow Control*—Implement a tool to ensure on-time starts to assess the following:
  - Patient Ready
  - Room Ready
  - Anaesthesia Ready
  - Surgeon Ready
  - *Pre-op Readiness*
  - Analyzing coordinating who does what, in what order (Pre-op nurse, circulating nurse, surgeon, and anaesthesia).
  - Visual control to assess status of each patient.
  - Reducing labs and tests to be done day of surgery.
  - *Same Day Case and Resource Planning*—Reduce case time and room changes, and improve resource planning.
  - *Preference Cards*—Improving the accuracy and efficiency of the preference cards.

Summary and Project Critical Success Factors

This project was extremely successful, yet very difficult from a change management perspective. The OR staff was somewhat addicted to the natural chaos of their jobs, and believed that constant schedule changes, searching for instruments, equipment, and supplies, putting out constant "fires," or issues with room turnarounds were just part of the job. It wasn't until we standardized the processes, put metrics in place, held people accountable to the metrics and change, and demonstrated the value of the new process and improvements that we achieved improvement.

The following key factors contributed to the success of this project:

- CEO commitment and support, visible communication of the organization-wide priority of this project.
- Following the Lean Six Sigma methodology to identify root causes, and identify improvements that engaged the OR staff to make suggestions and provide their input and feedback on the piloted improvements.

- Operationally defining the metrics, validating the measurement system and electronic data, and implementing the metrics dashboard that holds people accountable to these metrics.
- Using a formal change process, partially imbedded in the Lean Six Sigma methodology to make change happen.

Reference

Kotter, J. P. (2011). *8 steps for leading change.* Kotter International. Retrieved from January 09, 2011 from http://www.kotterinternational.com/KotterPrinciples/ChangeSteps.aspx

Chapter 5
Applying 5S to Improve OR Organization

Abstract This chapter describes the project that was initiated with the OR Turnaround and On-time Case Starts project from Chap. 4. The need for this project became painfully obvious early in the OR turnaround project. When we first toured the OR to better understand the layout and processes, we saw the challenges the OR was facing with storage and management of the OR supplies, instruments, and equipment. The hallways were cluttered with supplies, both in storage from the vendor (vendor owned and managed), as well as storage that did not fit into the Core inventory storage area, and cases that were staged for the cases for the same day. There were also beds waiting for patients to come out of surgeries, and trash bags being staged while rooms were cleaned. These challenges resulted in the need for improving the OR location and organization of equipment, furniture, instruments, and supplies. The Lean 5S approach was used to improve organization within the OR suite.

Keywords Operating room · Preventive maintenance · Storage area · Storage room · Sterile processing

Introduction

The Lean 5S approach was used to improve organization within the OR suite. Before pictures of the OR, prior to the start of the project, are shown in Figs. 5.1 and 5.2. In Fig. 5.1, the OR case cart staging area is shown. One of the OR hallways is shown in Fig. 5.2. As is evident from the pictures, the areas lack organization, labeling, and visual control, so it is difficult to find desired supplies and equipment quickly.

The main Lean tool that was used in this project was 5S, which is a tool that helps to organize a workplace. The 5S' are (Anything Lean, 2013):

- *Sort* Clearly distinguishes between what is necessary and what is unnecessary, disposing of the unnecessary. A red tag is used to identify items that should be reviewed for disposal.

S. Furterer, *Lean Six Sigma for the Healthcare Enterprise*,
https://doi.org/10.1007/978-1-4471-7617-6_5

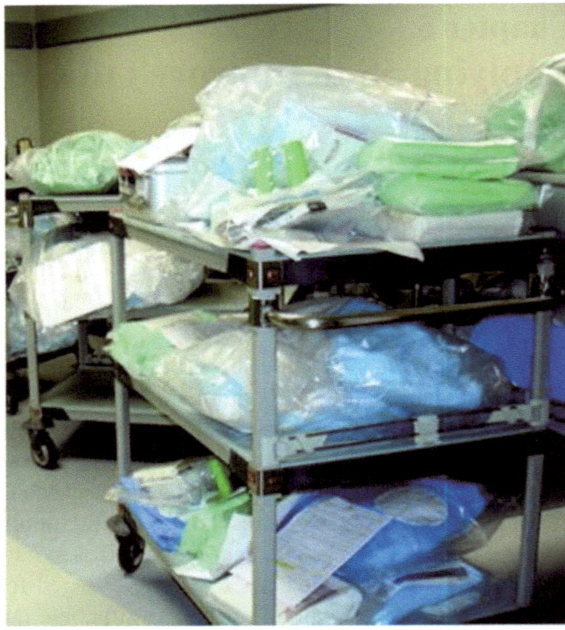

Fig. 5.1 Before picture of OR case staging area

- *Straighten(Set in Order)* Organizes the necessary items so they can be used and returned easily.
- *Shine* Fixes the root cause of the dirt or disorganization.
- *Standardize* Maintains and improves the standards of the first three S'.
- *Sustain* Achieving the discipline or habit of properly maintaining the organization.

The Lean Six Sigma DMAIC (Define, Measure, Analyze, Improve, Control) problem-solving methodology was used to guide the 5S activities.

Define Phase

This project was initiated based on the need to organize the Operating Room (OR) to enable the OR staff to locate, store, and manage equipment, furniture, supplies, and instruments that are used on a daily basis for surgical cases.

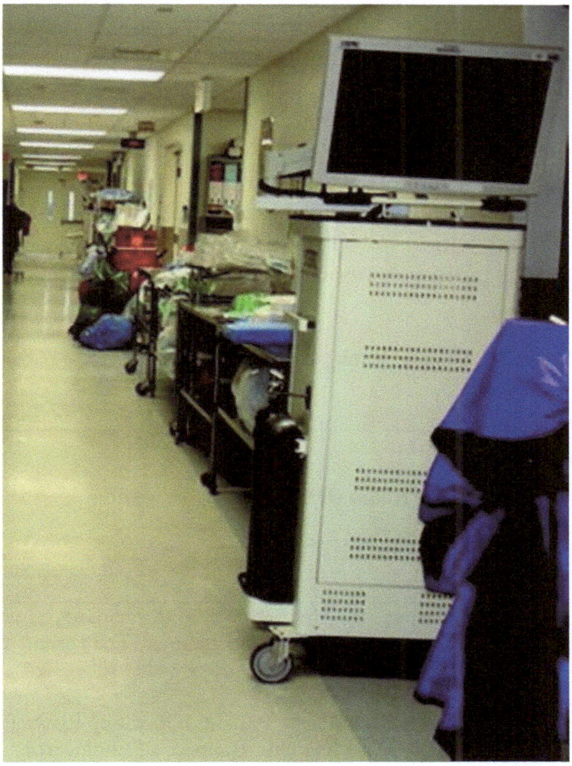

Fig. 5.2 Before picture of OR hallway

Project Charter

Problem Statement: There is no current standard organization being used in the OR area to store equipment, furniture, supplies, and instruments. The current state of the OR areas lack of organization is causing inefficiencies within the OR processes.

Customer/Stakeholders: The stakeholders of this project are the OR staff including nurses and technicians who provide OR patient care; OR physicians including surgeons, anaesthesiologists, and physician assistants who provide surgical care; Sterile Processing who clean, sterilize, sort, and store surgical instruments; Inventory and Materials Control personnel who store, pick, and manage supplies and instruments; Biomedical Engineering who maintains clinical equipment; Pre-op staff who care for patients prior to surgery; and Recovery room staff who care for patients after their surgery.

Goal of the Project: Improved visibility of problem conditions, improved safety, reduced waste, improved morale, an increased sense of ownership of the workspace, improved productivity, improved quality, improve maintenance and shorter OR turnaround times. A new sense of discipline and order that carries over into all activities is a goal of the project.

Scope Statement: The scope of the project includes implementing 5S in the OR Central Core, Operating Rooms, OR storage areas, Pre-op areas, Recovery Room, Sterile Processing area, and remote storage areas on the second floor.

Critical to Satisfaction: Available space; Time to find equipment, supplies, etc.; OR turnaround time; Reduce Preventive Maintenance (PM) time for OR Bio-med equipment.

The project champions are the Chief Executive Officer (CEO) and the Chief Nursing Officer (CNO).

The project charter is shown in Table 5.1, Project Charter.

Table 5.1 Project charter

Operating Room (OR) 5S Implementation
Project Statement: There is no current standard or methodology being used in the OR area to store equipment or supplies. The current state of the operating rooms is causing inefficiencies and errors within processes. Customers/Stakeholders: Patients, Operating Room Physicians, Nurses, Sterile Processing, OR technicians, Biomedical Engineering, Materials, Pre-op Staff, Recovery Room staff. Critical to Satisfaction: Available space; time to find equipment, supplies; OR turnaround time; reduced preventive maintenance time for OR biomedical equipment. Goal(s): Improved visibility of problem conditions, improved safety, reduced waste, improved morale, an increased sense of ownerships of the workspace, improved productivity, improved quality, improved maintenance, shorter OR turnaround times. A new sense of discipline and order carries over into all activities. Scope: The scope of the project includes implementing 5S in the operating room, central Core storage area, operating rooms, OR storage areas, Pre-op areas, recovery room, and remote storage areas. Projected financial benefits: Reduction of time between cases; enhanced OR utilization, revenue, efficiency, productivity, and decrease in overtime.

Stakeholder Analysis

The stakeholders of this project are the OR staff including nurses and technicians who provide OR patient care. They clean and set up the OR for surgical procedures. This stakeholder group also includes the Pre-operative and Recovery Room staff. The better organized the Pre-op, Recovery, and OR areas are, the more efficient the processes can be to provide care to the OR patients. The OR physicians include surgeons, anaesthesiologists, and physician assistants who provide surgical care. They have requirements for having all supplies, instruments, equipment, and furniture available when needed. Sterile Processing clean, sterilize, sort, and store surgical instruments. The Inventory and Materials Control personnel store, pick, replenish, and manage supplies. Biomedical Engineering maintains clinical equipment.

The stakeholder analysis is shown in Table 5.2.

Suppliers-Inputs-Processes-Outputs-Customers

The Suppliers-Inputs-Processes-Outputs-Customers (SIPOC) for the 5S is shown in Table 5.3. The five high-level processes for the 5S activity are Sort, Straighten, Shine, Standardize, and Sustain. The sort activity includes sorting the various areas in the OR to remove unneeded furniture, equipment, instruments, and supplies. The Straighten activity encompasses defining standard setups for the OR, Core, and storage areas, so that each item has a home. Shine is the activity of cleaning the equipment and areas for storing the item while not being used. Standardize is the labeling and color coding of all items, so that it is visually easy to control and find items. Sustain is the activity that helps to maintain the organization and implement the discipline of organization. It includes identifying roles and responsibilities and auditing the level of cleanliness and organization in the OR.

The Critical to Satisfaction criteria for this project are:

- Available space—the space that is available and freed up due to better organization and the reduction of unneeded items.
- Time to find equipment, supplies, instruments, and furniture.
- OR turnaround time—time to clean and set up the operating room for the next case.
- Reduced Preventive Maintenance time for OR Bio-med equipment, due to reducing the quantity of equipment, by eliminating unneeded equipment.

Project Kick off

The project was started in November 2009. We held a team kick-off meeting to explain the project and identify the initial sorting activities as follows.

Table 5.2 Stakeholder analysis

Stakeholder	Type	Primary role	Potential impacts/ concerns	Initial receptivity	Future receptivity
OR staff	Primary	Nurses and techs that work in the Pre-op, OR, and Recovery areas	Readily available instruments, supplies, and equipment; adequate workspace	Moderate support	Strongly support
OR physicians	Primary	OR physicians, anesthesiologists, physician assistants, who provide care to the patients in the OR	On-time starts; fast turnarounds	Neutral	Neutral
Inventory and material control	Primary	Associates who are in charge of the central Core, OR supply inventory	Adequate workspace; visibility of materials and inventory levels; standard location for materials	Moderate support	Strongly support
Sterile processing	Secondary	Associates who are in charge of OR instruments that need sterilizing	Receive instruments in a timely manner; standard location for instruments; sterilization turnaround time	Strongly support	Strongly support
Biomedical engineering	Primary	Associates that are in charge of preventive and corrective maintenance of biomedical equipment	Ease of finding equipment to perform preventive and corrective maintenance	Strongly support	Strongly support

We mapped the DMAIC phases to the following 5S activities, as shown in Table 5.4.

The define phase is important in defining a clear project scope, identifying the project stakeholders, and planning for the rest of the project. The Measure phase will be discussed in Measure Phase.

Table 5.3 SIPOC for 5S Project

SIPOC				
Suppliers	Inputs	Process	Outputs	Customers
OR Staff biomed inv. control sterile proc physicians	Usage items	Sort	Standard setups Disposition	OR staff biomed inv. control sterile proc physicians
OR Staff biomed inv. control sterile proc physicians	Items requirements	Straighten	Standards Baseline	OR staff biomed inv. control sterile proc physicians
OR staff biomed inv. control sterile proc physicians	Cleaning supplies items	Shine	Cleaned items	OR staff biomed inv. control sterile proc physicians
OR staff biomed inv. control sterile proc physicians	Items	Standardize	Labeling Visual Controls	OR staff biomed inv. control sterile proc physicians
OR staff biomed inv. control sterile proc physicians	Controls Audits	Sustain	Audit results	OR staff biomed inv. control sterile proc physicians

Measure Phase

The Measure Phase included the following Sort activities:

1. Identify rooms and areas to be sorted.
2. Identify roles and responsibilities of team members.
3. Define time frames of sorting activities.
4. Set up sorting area.
5. Create OR room standard furniture and equipment set ups.
6. Establish criteria for separating needed and unneeded items.
7. Separate needed from unneeded items.
8. Red tag unneeded hold items.
9. Remove excess items from working areas.

We held a two-day sorting event of furniture and equipment. The team members used one of the OR rooms that was not in use for the two-day sorting event. They removed any furniture or equipment that was not a part of the standard OR set up from each OR room. They also moved the equipment out of the equipment room and staged it in the sorting area. The surgical coordinators who are responsible for equipment and furniture went through the furniture and equipment in the staging area to identify the frequency of usage (daily, weekly, monthly, annually, or less than annually) and disposition of the furniture and equipment. If the usage was daily or weekly, the equipment or furniture was stored in the immediate OR area either in an OR as the standard set up, in the equipment room, or in another designated storage area. If the usage was monthly a storage area away from the OR was used. If the usage was

Table 5.4 DMAIC and 5S activities mapping

DMAIC phase	5S phase	5S activities
Define	Define 5S need	Define 5S need
Measure	Sort	(1) Identify rooms and areas to be sorted (2) Identify roles and responsibilities of team members (3) Define time frames of sorting activities (4) Set up sorting area (5) Create OR room standard furniture and equipment set ups (6) Establish criteria for separating needed and unneeded items (7) Red tag unneeded hold items (8) Remove excess items from hold area
Analyze	Straighten	(1) Decide where things belong (2) Decide how they will be put away (3) Make it as easy as possible to obey the put away rules (4) Color coding areas for supplies, type of equipment (5) Same category of equipment/supplies in close areas (unless posed safety issue) (6) All items that are used in conjunction placed close together (7) Outline aisles, places where items are kept (8) Standardize height, size of shelves if necessary
Improve	Standardize	(1) Create visual management so that abnormal state can be easily identified (2) Find abnormalities (3) Engage Environmental Services and Engineering to repair root causes
Control	Sustain	(1) Create 5S committee (2) Create audit plan and schedule (3) Periodic spring-cleaning schedule (4) Assess improvement

annually or rarely, but the item was to be kept, it was stored in another area outside of the immediate OR area. The rest of the equipment and furniture was removed to a secondary staging area in the basement, so that all appropriate people could review whether they needed any of the items. Some of the outdated, unnecessary, broken, and not reparable equipment and furniture were discarded. The discarded equipment was removed from the equipment tracking system by Biomedical Engineering.

A similar sorting activity was performed in the Pre-Op and Post-Op areas, as well as the Sterile Processing areas.

The team assessed the baseline 5S audit. The audit criteria are shown in Table 5.5. The baseline audit was 1.6. From a Sort perspective, the team has sorted necessary and unnecessary items. From a Straighten perspective, items are still randomly located throughout the workplace. From a Shine perspective, the work and break areas are clean, and key items to check have been identified. From a Standardize perspective, methods are being improved and practiced continually but have not been documented. From a Sustain perspective, workplace checks are randomly performed, and/or there is no visual measurement of 5S performance.

Table 5.5 Baseline 5S audit

	Sort	Straighten	Shine	Standardize	Sustain
Level 5 Focused on prevention measures	Employees are continually seeking improvement opportunities.	A dependable, documented method has been developed to provide continual evaluation and a process is in place to implement improvements.	Area employees have devised a dependable, documented method of preventive cleaning and maintenance.	Every team member is continually seeking the elimination of waste with changes being documented and information being shared.	A general appearance of a confident understanding and adherence to the 5S principles is maintained. A culture of cleanliness and orderliness is expected of all.
Level 4 Focused on consistency	A dependable, documented method has been established to keep the work area free of unnecessary items. "Red tag" area is being utilized.	A dependable, documented method has been established to recognize in a visual sweep if items are out of place or exceed quantity limits.	5S individual assignments are agreed upon or 5S is practiced by all consistently.	Standard Work is consistently followed by all team members. Areas consistently follow 5S standards.	All team members are actively engaged in driving continuous improvement in 5S scores. Scorecards are posted.
Level 3 Making it visual	Unnecessary items have been removed from the workplace. "Red tag" area is well identified.	Designated locations are marked to make organization visible. Visuals (tape, paint) are in good shape.	Work and break areas have been cleaned on a daily basis, and visual controls have been established for key items. Floors are clean.	Visual controls and Standard Work is in place and proven out.	Weekly 5S reviews are being conducted by the Manager and /or others. Feedback is being acted upon.

Cumulative improvements

(continued)

Table 5.5 (continued)

Level 2 Focus on the basics	Necessary and unnecessary items have been separated.	Yes	Designated locations have been established for items.		Work and break areas are clean, and key items to check have been identified.	Yes	Methods are being improved and practiced continually, but haven't been documented.	Yes	A recognizable effort has been made to improve the condition of the workplace.	
Level 1 Just beginning	Necessary and unnecessary items are mixed throughout the workplace.	Yes	Items are randomly located throughout the workplace.	Yes	The workplace is dirty, disorganized, and key items are not marked or are not identified.	Yes	Workplace improvement methods are not consistently followed and/or are undocumented.	Yes	Workplace checks are randomly performed and/or there is no visual measurement of 5S performance.	Yes
Score	2 Partially complete		1 Incomplete		2 Partially complete		2 Partially complete		1 Incomplete	

Supervisor comments:
Opportunities:

Analyze Phase

In the Analyze phase, we performed the following Straighten and Shine activities:

Straighten:

- Decide where things belong.
- Decide how they will be put away.
- Make it as easy as possible to obey the put away rules.
- Color coding areas for supplies, type of equipment.
- Same category of equipment/supplies in close areas (unless poses safety issue).
- All items that are used in conjunction are placed close together.
- Outline aisles, places where items are kept.
- Standardize height, size of shelves, if necessary.

Shine:

1. Find root causes for things to get disorganized.
2. Find abnormalities.
3. Engage Environmental Services and Engineering to repair root causes.

Straighten:

We identified locations for patient transport beds, case carts, equipment, furniture, supplies, and instrument trays in the OR space. We worked with the vendors to remove unneeded items from the OR space. We painted the equipment and furniture to be color coded for each OR room, and put a color-coded label on each OR door, so it is obvious if the wrong furniture and equipment are in a room. We worked to remove all unnecessary equipment, supplies, furniture, and instrument trays from the OR space, and only store what is used on a daily or weekly basis in the OR Core and surrounding areas. We standardized the OR case cart as much as possible, identifying what types of items are placed in which location on the cart, as well as where the cart will be stored. We also better labeled the OR case carts, so they could be more easily found when needed.

Shine:

We identified root causes of disorganization in the OR space, as shown in Fig. 5.3.

1. No standard process for identifying the frequency of use of equipment, supplies, instruments, and furniture.
2. No methodology for storing items, identifying where they belong, or assigning responsibility for ownership and organization.
3. No process for identifying what is needed in the OR space, and what is not needed.
4. Lack of buy-in and commitment from the OR staff and management for implementing the 5S.
5. No defined locations for the patient transport beds outside of the OR.

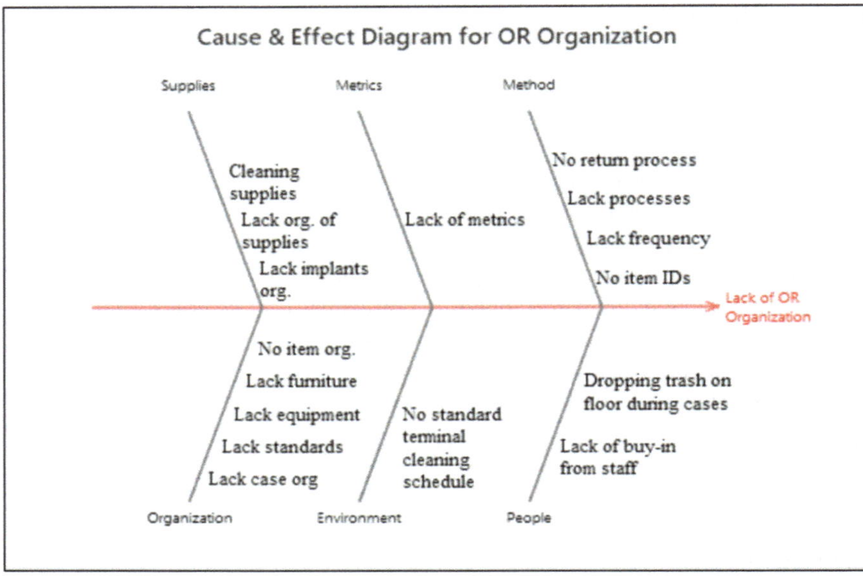

Fig. 5.3 Cause-and-effect diagram

6. No process for vendors to only bring what they need for a case each day, allowing vendors to store several days (weeks) worth of supplies in the OR space.
7. Some consignment implants were recently purchased by the hospital, but no usage and frequency analysis of the supplies was performed.
8. No standardized terminal cleaning schedule.
9. Not placing trash in receptacles during surgical cases.
10. Equipment and furniture not returned to rooms where they belong and are used.
11. No process for returning unused supplies to the Core and restocking them.
12. Cleaning supplies not available.
13. No set case cart organization.
14. No standard set up for OR back tables.
15. Cleaning supplies are not available because no set place is identified, and not enough cleaning supplies are available for each room.
16. No 5S metrics have been implemented.

We identified any areas that needed to be repaired or cleaned.

We then moved into the Improve Phase.

Improve Phase

We performed the following Standardize 5S activities during the Improve phase:

1. Create visual management so that the abnormal state can be easily identified.
2. Color coding and labeling, shadow boards, indicators of where things should be put.

We created ideas for improving the organization of equipment, instrument trays, supplies, and furniture as follows:

5S Ideas for Improvement:

- Create schedule for cleaning patient bathrooms during busy times. (10 to 11 am)—Holding area.
- Have Environmental Services garbage and linen bins in PACU, during the day. (3 pm).
- Keep same bear warmer blanket used in Pre-Op (not soiled).
- Get folders for staff—get rid of big cabinet (PACU).
- Find a home for the X-ray machine.
- Get rid of the forms cabinet in Pre-op area, put up on wall.
- Get IVs in a dispensing unit.
- Fix isolation room.
- Get rid of cabinets, sink in clean utility room.
- Get pedestal sink in Pre-op holding, get rid of cabinets.
- Get new, standardized, rolling chairs for PACU and holding, need to be able to be moved higher or lower (like golden ones in Phase 2).
- Visitor chairs that do not get stained, no wheels, not arm chairs.
- Skinny garbage cans at Pre-op at bedside.
- IV Poles too tall, need to be mobile.
- Pre-op needs IVs for all rooms.
- Holding area has no med room.
- Go through the soiled utility closet and get rid of things that are not needed. (split up the room)—only need sink and hopper (make med room for Pre-op).
- Ask to have an environmental person dedicated.
- Do we need the suction machine?—Can respiratory maintain them?
- Breathing bags identification.
- Pre-op to have wipes to wipe off at the end of the day. In the basket, wipe down cables.
- Standardize tables, with wheels that roll, uniform Pre-op to Post-op.
- Clean out PED room area in Post-op.
- Put pull out tables on the walls in Pre-op, get rid of rolling tables.
- Convert storage area in Pre-op to nurses' station and nurses' station into med room—put blankets inside.
- Create cubbies/get shelving for SCD machines.
- Get rid of sink in isolation room.

- Need a patient going home chair.
- Reduce number of soiled linen and garbage cans.
- Identify standardized patient transport beds; bring beds closer to when the case is complete.

Another organization activity was done in the sub-sterile storage rooms. These are very small rooms right between two OR suites, where sub-sterile items are stored. These are the rooms where people enter after they are scrubbed to go into the OR. Since they are very small spaces and must be kept clear for the surgical teams to constantly walk through, these areas are important to organize. Before we organized these rooms, no one was responsible for restocking the rooms. The analyst created checklists for each sub-sterile room, with par levels for restocking. He then redesigned the rooms with consistent bins and created markings for par levels.

Another activity was to organize the cabinets in the OR. The team plans to organize and assign ownership for restocking the OR cabinets on a daily basis.

Control Phase

We performed the following Sustain 5S activities in the Control phase:

1. Create 5S committee.
2. Create Audit, Control Plan, and Schedule.
3. Periodic spring-cleaning schedule.
4. Assess improvement.

Create 5S Committee

We used the same OR Turnaround Process Improvement team as the 5S committee. This included team members from the OR, Sterile Processing, Pre-op, PACU, Anaesthesia, and Enterprise Performance Excellence.

Create Audit Control Plan and Schedule

We developed the improvement control plan, shown in Table 5.6. The 5S improvement control plan includes a sub-sterile checklist. The sub-sterile instructions and checklist are shown in Figs. 5.4 and 5.5, respectively. The sub-sterile checklist is to be completed by the Tech 1 s on a daily basis after they have restocked their assigned sub-sterile rooms. The equipment room checklist is shown in Fig. 5.6. The equipment room should be reviewed, and the checklist should be completed by the equipment tech on a weekly basis. The shift handoff setup checklist is shown in Fig. 5.7. It is

Table 5.6 Improvement control plan

Improvement control plan					
The control plan is a detailed assessment and guide for maintaining all of the positive changes that you have implemented					
Process steps	Control mechanism	Measure/ metric	Criticality (H M L)	Action taken if problems occur	Owner
Sub-sterile stocking	Sub-sterile checklist	Percent sub-sterile stocked	H	Assign cause	Exec. director
Equipment rooms	Equipment room checklist	% equipment stocked	H	Assign cause	Exec. director
Verify hallways, furniture, equipment, OR bed, sub-sterile, cabinets, leftover supplies in OR rooms for surgery or returned	Shift handoff setup checklist	% handoffs complete	M	Assign cause	Exec. director
5S audit	5S audit checklist	5S audit ratings	M	Assign cause	Exec. director

reviewed and completed at shift handoff by both surgical coordinators between each of the shifts. It is used to verify the organization of the hallways, furniture, equipment, beds, subseries, and the leftover supplies are restocked. The 5S audit checklist is shown in Table 5.7. It is used to assess the overall progress of the 5S program.

Periodic Spring-Cleaning Schedule

The next step in the Sustain phase is to set up a periodic spring-cleaning schedule, to continue the removal of unneeded equipment and furniture.

Assess Improvement

We used the 5S audit checklist to assess improvement of the 5S program. The original or baseline 5S Audit checklist is shown in Table 5.5. There were three 2 s and two 1 s. For the improved checklist, shown in Table 5.7, there are now three 3 s and two 2 s. Even though this showed improvement, there is still additional discipline and control that need to be enhanced.

We eliminated nearly 400 square feet of outdated and unneeded equipment. We organized the OR hallways, so it looks better, and it is easier to navigate the patients

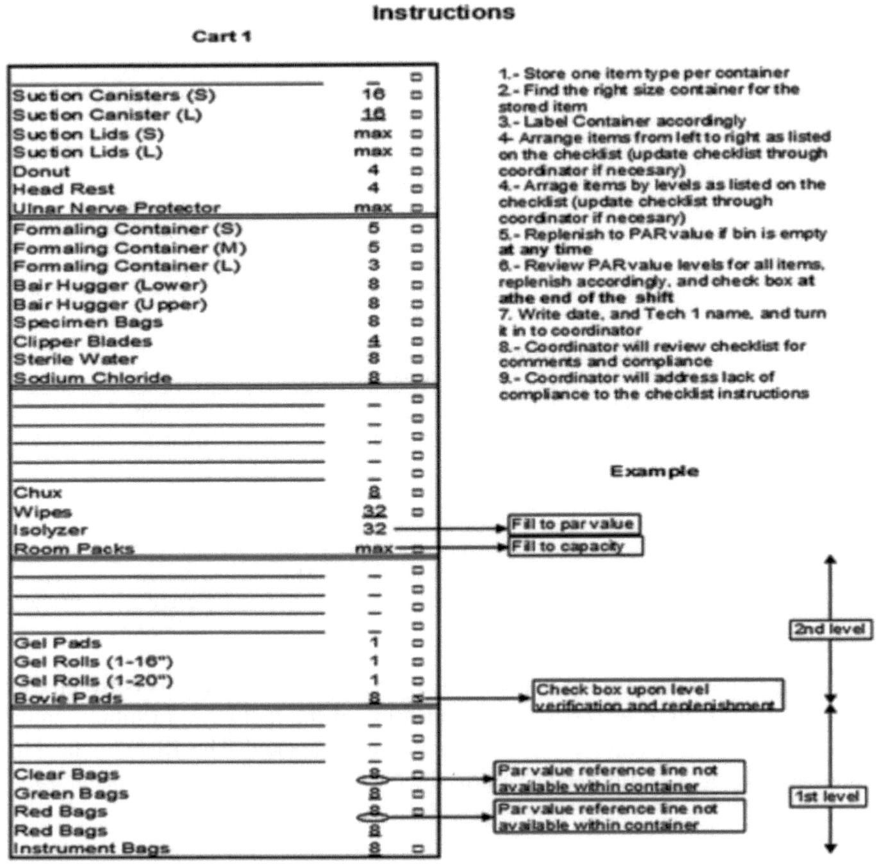

Fig. 5.4 Sub-sterile checklist instructions

on beds to the OR. Figure 5.8 shows one of the OR hallways after improvement. Table 5.8 shows the OR color schemes for each room. We painted the equipment and furniture, so that it is easy to return back to the correct OR when it is moved. Figure 5.9 shows the equipment room before the 5S, where it was nearly impossible to maneuvre, and equipment had to be removed to get to other equipment. Figure 5.10 shows the equipment room after removing the unneeded equipment and defining a place for each piece of equipment. There is now space to walk between equipment and retrieve the needed equipment. The OR case cart storage area, before and after, is shown in Figs. 5.11 and 5.12. There is also another equipment storage room further from the OR that stores equipment that is used less frequently. This room was packed wall to wall, as shown in Fig. 5.13before, but improved immensely as shown in Fig. 5.14. Figures 5.15 and 5.16 show the sub-sterile room before and after it was reorganized. The 5S Audit ratings before and after the 5S are shown in Table 5.7.

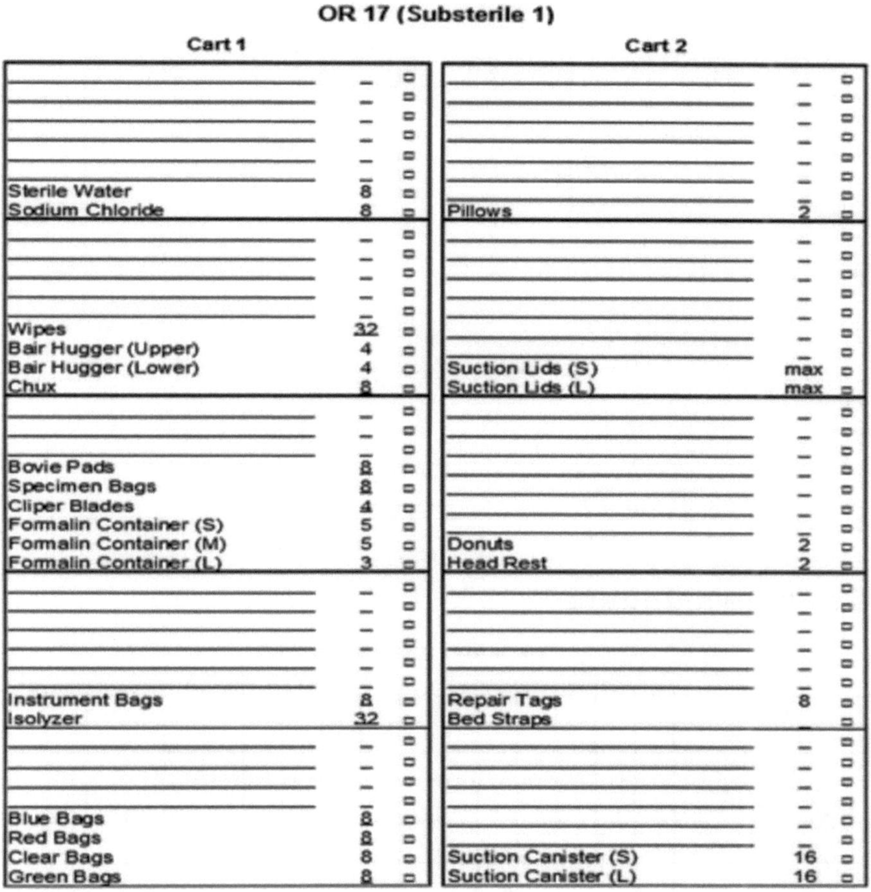

Fig. 5.5 Sub-sterile checklist

The 5S scorecard ratings compared to the desired standard are shown in Fig. 5.17. The 5S scorecard ratings before and after improvement are shown in Fig. 5.18.

Conclusions

The 5S project was very successful in organizing the OR. It was critical to remove equipment and furniture that was not used from the OR hallways and storage rooms. This enables the OR to be able to function at a higher level of efficiency and orderliness that is expected in an OR environment. It is now easier to clean and keep the OR clean, which is vital in the OR.

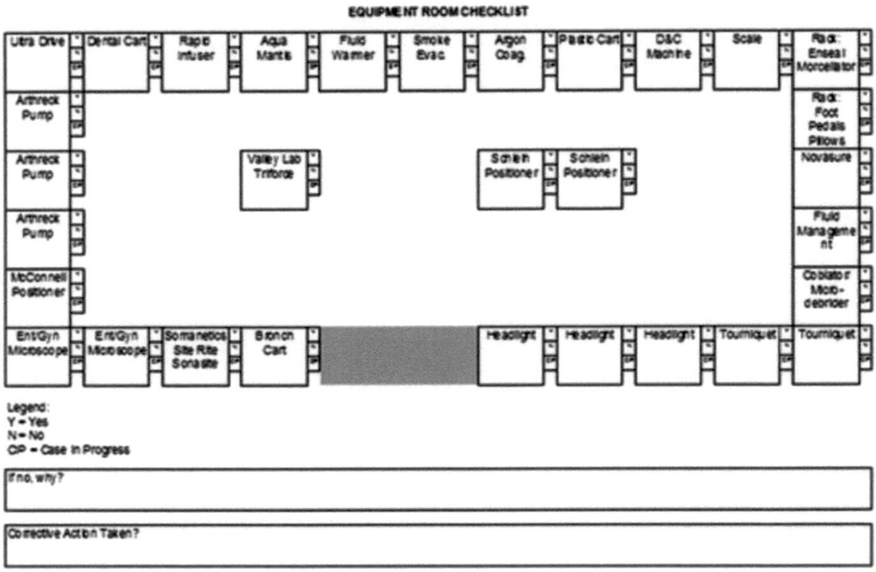

Fig. 5.6 Equipment room checklist

SHIFT HANDOFF SETUP CHECKLIST

		OR 1	OR 2	OR 3	OR 4	OR 5	OR 6	OR 7	OR 8	OR 9	OR 10	OR 11	OR 12	OR 14	OR 15	OR 16	OR 17	OR 18
1	Is all furniture present?																	
2	Is all furniture in its place?																	
3	Is the bed in the correct location?																	
4	Is there equipment in the room that needs returned to its location?																	
5	If yes to 4 above, did you move the equipment to its proper location?																	
6	Are there left over supplies in the room?																	
7	If yes to 6, did you return left over supplies to the CORE?																	
8	Are cabinets stocked?																	
9	Is the substerile stocked, and blanket warmers and fluid warmers stocked?																	
10	Is all equipment in its hallway location?																	

Legend: Y = Yes N = No CIP = Case in Process

If no, why?

Corrective Action Taken?

Fig. 5.7 Shift handoff setup checklist

Table 5.7 Audit checklist improved

	Sort		Straighten		Shine		Standardize		Sustain	
Level 5 Focused on prevention measures	Employees are continually seeking improvement opportunities.		A dependable, documented method has been developed to provide continual evaluation, and a process is in place to implement improvements.		Area employees have devised a dependable, documented method of preventive cleaning and maintenance.		Every team member is continually seeking the elimination of waste with changes being documented and information being shared.		A general appearance of a confident understanding and adherence to the 5S principles is maintained. A culture of cleanliness and orderliness is expected of all.	
Level 4 Focused on consistency	A dependable, documented method has been established to keep the work area free of unnecessary items. "Red tag" area is being utilized.		A dependable, documented method has been established to recognize in a visual sweep if items are out of place or exceed quantity limits.		5S individual assignments are agreed upon or 5S is practiced by all consistently.		Standard Work is consistently followed by all team members. Areas consistently follow 5S standards.		All team members are actively engaged in driving continuous improvement in 5S scores. Scorecards are posted.	
Level 3 Making it visual	Unnecessary items have been removed from the workplace. "Red tag" area is well identified.	Yes	Designated locations are marked to make the organization visible. Visuals (tape, paint) are in good shape.	Yes	Work and break areas have been cleaned on a daily basis, and visual controls have been established for key items. Floors are clean.		Visual controls and Standard Work are in place and proven out.	Yes	Weekly 5S reviews are being conducted by the Manager and or others. Feedback is being acted upon.	
Level 2 Focus on the basics	Necessary and unnecessary items have been separated.	Yes	Designated locations have been established for items.	Yes	Work and break areas are clean, and key items to check have been identified.	Yes	Methods are being improved and practiced continually, but haven't been documented.	Yes	A recognizable effort has been made to improve the condition of the workplace.	Yes
Level 1 Just beginning	Necessary and unnecessary items are mixed throughout the workplace.	Yes	Items are randomly located throughout the workplace.	Yes	The workplace is dirty, disorganized, and key items are not marked or are not identified.	Yes	Workplace improvement methods are not consistently followed and/or are undocumented.	Yes	Workplace checks are randomly performed, and /or there is no visual measurement of 5S performance.	Yes
Score	3 Partially complete		3 Partially complete		2 Partially complete		3 Partially complete		2 Partially complete	

Cumulative improvements

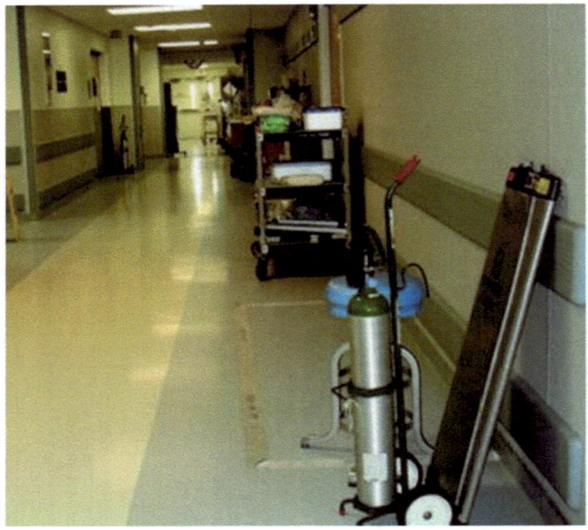

Fig. 5.8 After pictures—OR hallway

Table 5.8 Operating rooms color schemes

Operating room	Color scheme	
1	Green	Orange
2	Orange	
3	Orange	Blue
4	Green	
5	Yellow	
6	Green	Blue
7	Blue	
8	Blue	Pink
9	Blue	White
10	Pink	
11	Purple	White
12	Red	
14	Red	White
15	White	
16	Purple	Orange
17	Purple	Green
18	Purple	

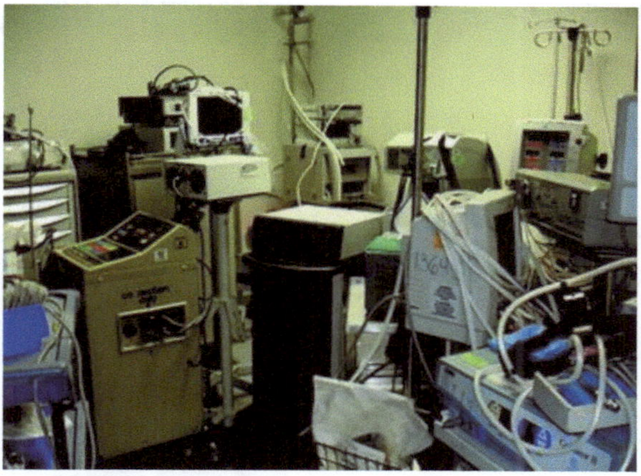

Fig. 5.9 Equipment room before

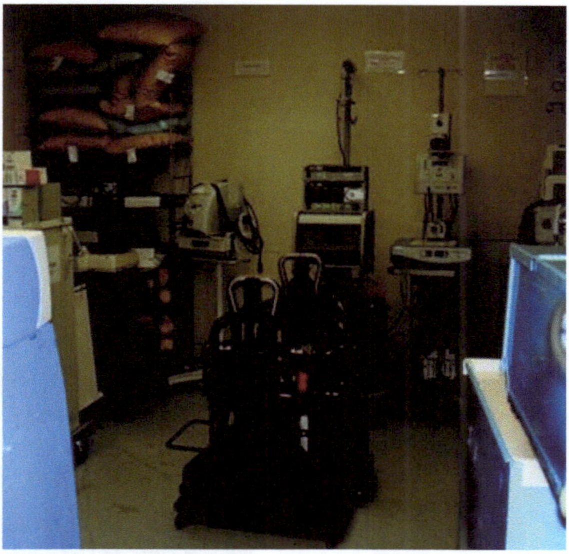

Fig. 5.10 Equipment room after

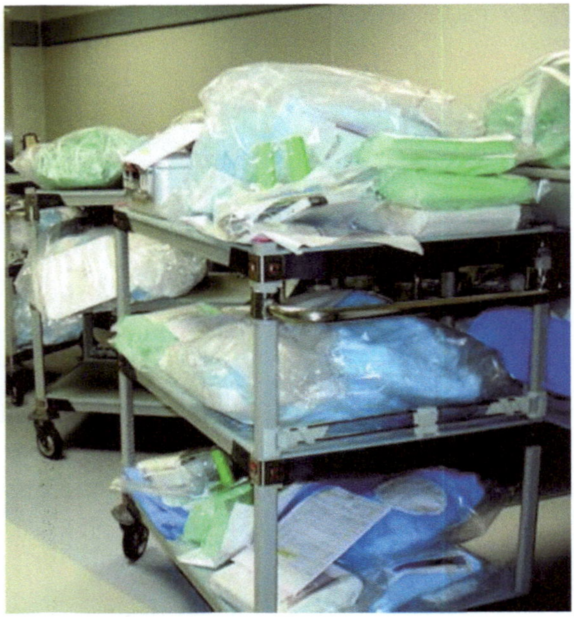

Fig. 5.11 Case cart store before

Fig. 5.12 Case cart storage after

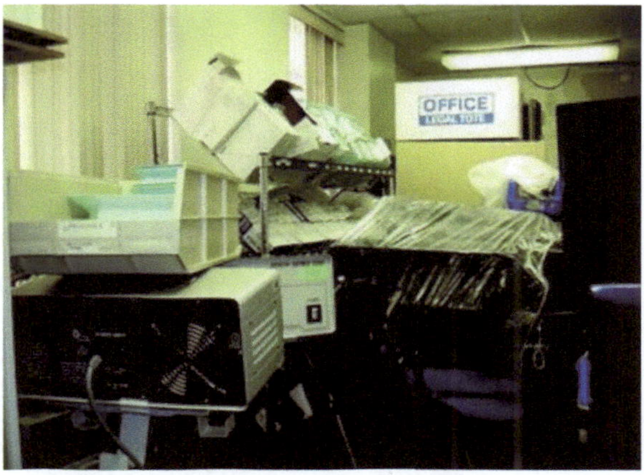

Fig. 5.13 Additional storage before

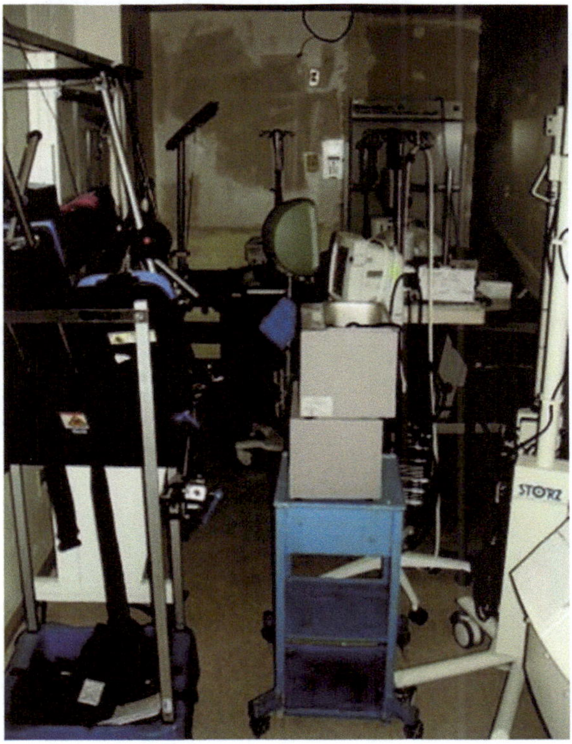

Fig. 5.14 Additional storage after

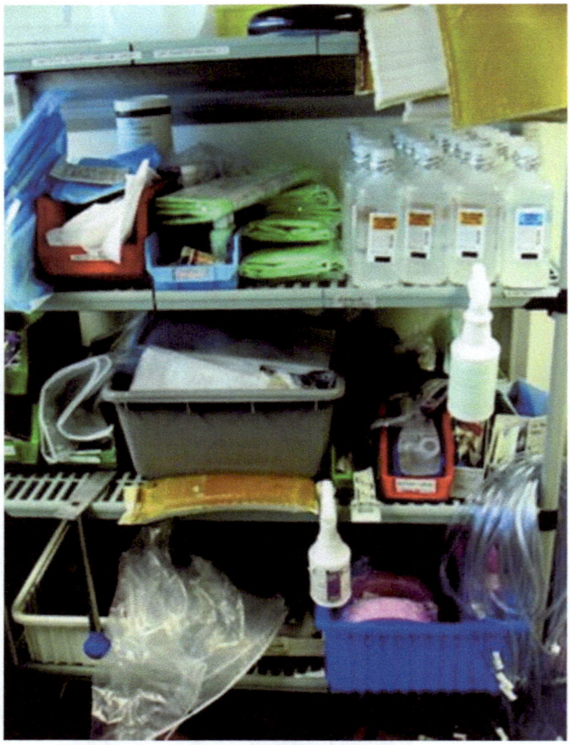

Fig. 5.15 Sub-sterilize before

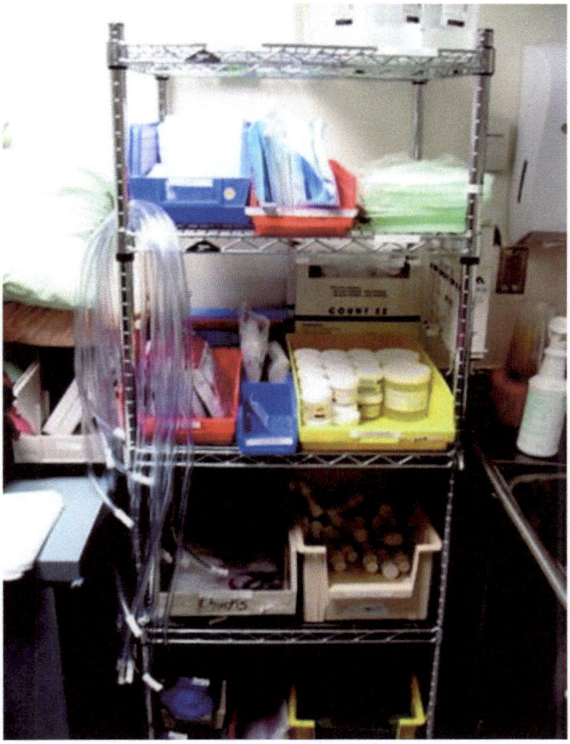

Fig. 5.16 Sub-sterile after

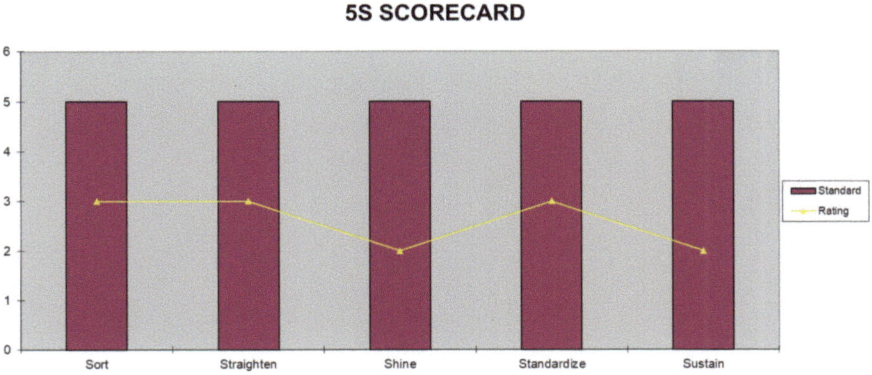

Fig. 5.17 5S audit rating after improvement compared to the standard

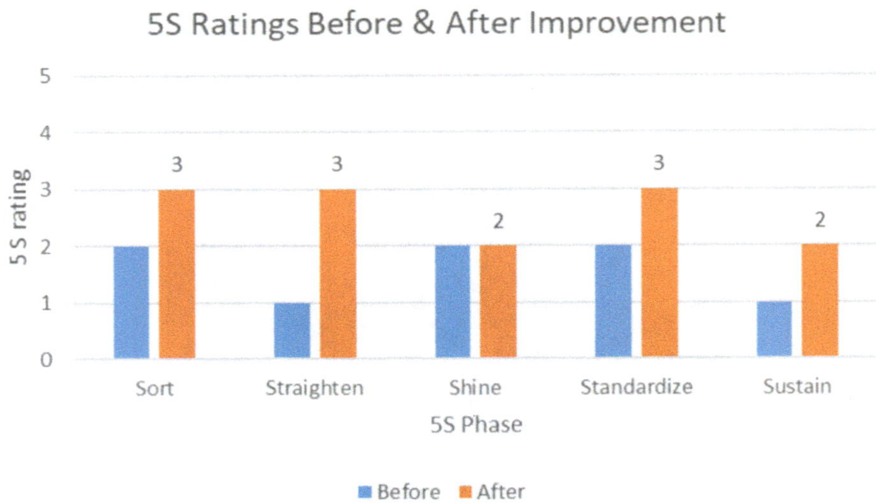

Fig. 5.18 5S audit rating before and after improvement

Reference

Anything Lean. (2013). 5S housekeeping. http://www.5sleanmanufacturing.com/5s_Housekeeping_s/232.htm

Chapter 6
Leaning the Surgical Outpatient Screening Process in an Acute Care Hospital

Abstract Elective surgery is a surgery that is a planned, nonemergency surgery, that can be scheduled at the convenience of the patient or the physician. The surgical procedure may extend life or improve the quality of life. In the year 2000, there were over 40 million inpatient surgical procedures performed in the United States, and 31.5 ambulatory or outpatient surgeries in 1996, based on the National Center for Health Statistics of the U.S. Centers for Disease Control CDC (http://www.hea lthtools.com, 2012). Typically, a complete health history, physical examination, lab and diagnostic tests may be part of the preparation of the surgical patient prior to the elective surgery. Additionally, typing and screening of the patient's blood is also performed prior to surgery (http://www.healthtools.com, 2012). A percentage of the patients who receive the elective surgery, may already be a patient in the hospital, and this preparation for the surgery is performed under the guidance of the inpatient nurse. A patient that is not already in the hospital but has scheduled their elective surgery is considered an outpatient from the perspective of the surgical preparation, even though the patient may be admitted after the surgery. This case will describe how the Lean Six Sigma team at an acute care hospital improved the pre-screening process for surgical outpatients applying Lean tools, with the Six Sigma DMAIC problem solving approach.

Keywords Chart preparation · Pre-screening process · Design future state · Takt time · Entire value stream

Define Phase

The project was initiated as part of the operating room project to improve the efficiency while ensuring patient care and safety of the operating room processes. The focus of this project was on streamlining the surgical outpatient preparation activities prior to their scheduled elective surgery. It is actually a consolidation of several similar case studies across several hospitals.

S. Furterer, *Lean Six Sigma for the Healthcare Enterprise*,
https://doi.org/10.1007/978-1-4471-7617-6_6

The surgical patient pre-screening process was not providing the charts to the Operating Room in a timely, or high-quality manner. The complete patient pre-screening information is often not available when the charts are sent to the Operating Room by 5 pm the day before the scheduled case. The staff is often working on the charts the same day that they are due to the OR but are not able to complete all of them prior to the deadline. The patient presurgical documentation is usually faxed to the pre-screening department, and comes in as separate documents, sometimes without the patient's name and date of birth on each page of the document, as required. The pre-screening process requires extensive rework, including calling the surgeons' offices multiple times to get the required information. Since the primary care physicians provide the medical clearance and the history and physical, the surgeon's office must make multiple phone calls to the primary care physicians' offices. The pre-screening department makes appointments with the patients for them to be interviewed to collect the pre-screening information, such as lists of medications that the patient is taking, health and physical information, etc. The patients sometimes come to the hospital on an average of three times to complete all appropriate diagnostic tests, pre-screening interviews and educational seminars to prepare for the surgery. The percent of charts that are complete by 5 pm for the first cases of the day is 23%. The percent of charts that are complete by 5 pm for all cases that are scheduled for the next day is not currently measured.

The following steps were executed in the Define Phase:

1. Develop project charter
2. Perform stakeholder analysis
3. Perform initial Voice of Customer (VOC) and Identify Critical to Satisfaction (CTS)
4. Select team and launch the project
5. Create project plan.

Develop Project Charter

The team generated the initial draft of the project charter in the first team meeting. The project charter is shown in Table 6.1, Project Charter.

Project Charter

This project is focused on streamlining the outpatient surgery pre-screening processes prior to the day of surgery. This includes gathering all of the appropriate documentation, interviewing the patient to collect additional information by a pre-screening nurse, entering the data in an electronic medical record, and organizing the documentation into a chart.

The goal of the project is to assess the current outpatient pre-screening area processes to identify inefficiencies and improvement recommendations to lead to throughput improvement from the pre-screening area to the surgical suite for the day of surgery.

Table 6.1 Project charter for pre-operative screening process improvement project

Project Charter

Project Overview, Problem Statement, & Goals

Provides a description of the problem to be solved. Be as descriptive as possible. Include improvement goals that are SMART (Specific, Measurable, Attainable, Realistic, and Time-Based).

Project Name: Pre-operative Screening Process Improvement
Project Overview: This project is focused on streamlining the outpatient surgery pre-screening processes prior to the day of surgery.
Problem Statement: The project includes gathering all of the appropriate documentation, interviewing the patient to collect additional information by a pre-screening nurse, entering the data into an electronic medical record and organizing the documentation into a chart. The existing process can cause delays for the day of surgery.
Goal(s): To assess the current Outpatient Pre-screening area processes to identify inefficiencies and improvement recommendations to lead to throughput improvement from the Pre-op area to the surgical suite for the day of surgery.
Scope Statement: The scope of the project includes an assessment of the Outpatient Pre-screening area, and a process design with improvement recommendations.
Projected Financial Benefits: Reduced delays in pre-screening processes, reduced delays in surgery due to missing labs, diagnostics, and results or other misses in the pre-screening process.

The scope of the project includes an assessment of the outpatient pre-screening area, and a process design with improvement recommendations. The project benefits of this project are to identify process inefficiencies and improvement recommendations that can reduce the delays in the pre-screening area on the day of surgery, enhance the throughput of the Operating Room pre-screening area, and reduce unplanned cancellations of surgeries, due to the patient's pre-screening not being ready. The CTS criteria include patient chart readiness the day prior to surgery, reduced number of labs and diagnostic tests on the day of the surgery, and reduced number of surgery cancellations due to the patient's chart not being complete.

Perform Stakeholder Analysis

The stakeholders are the patients, pre-screening nurses, Director of Surgery, OR pre-screening nurses, surgeons, anesthesiologists, lab, and radiology. The patients who have surgery are concerned that they are prepared for surgery, and do not have to come to the hospital several times to perform assessments and tests. The pre-screening nurses review the patients' charts, physician orders, perform assessments and document information in the patient charts, and update the medical information system. They are concerned with not receiving the physician orders and information from the physicians' offices timely to ensure the pre-screening of the chart and patient is complete. They are also concerned about having to work overtime to complete the charts and negatively impacting their productivity metrics. The Director of Surgery who manages the surgical department is concerned that the patients are ready for surgery, that the department meets their productivity goals, that there are reduced labs and diagnostic tests that need to be performed on the day of the surgery, and that the number of cancellations are reduced due to the patient not being properly prepared and screened for their surgery. The OR pre-screening nurses prepare the patients for the surgery on the day of the surgery. They are concerned with having the patient charts ready when the patients arrive for pre-screening on the day of the surgery. They also want to readily know what tests and assessments still need to be performed on the day of the surgery. The surgeons who perform the surgeries want to ensure that the scheduled surgery starts on time with no delays. The anesthesiologists assess the patient's pre-surgery and are concerned about ensuring that the patient and the chart have been screened by anesthesia prior to surgery. The Lab provides lab specimens and tests for pre-surgery readiness. They are concerned about needing to provide staff to collect and perform labs on the day of the surgery, instead of ensuring that the labs are complete. The Radiology department provides radiology diagnostic tests readiness and is concerned about needing to provide staff to perform tests on the day of the surgery. The stakeholder analysis is shown in Table 6.2.

SIPOC (Suppliers-Inputs-Processes-Outputs-Customers)

The SIPOC is used to ensure the scope of the project is well understood as shown in Table 6.3. It is also a check and balance for the customer and stakeholder analysis. Any customers and suppliers should also be identified as a project stakeholder. The pre-screening secretary reviews the surgery schedule in the OR information system and provides it to the pre-screening department. The pre-screening secretary puts the standard blank forms into patient charts for the patients scheduled for surgery and provides them to the pre-screening nurses. The pre-screening nurses assign the charts to pre-screening nurses. The pre-screening nurses match the required documents according to the surgery type. The pre-screening nurses perform the pre-screening activities with the patient, documenting the health and physical, reviewing lab and test results, and complete the patient chart. The pre-screening nurses place the patient ID stickers on each page of the patient's chart. The pre-screening nurse copies the

Table 6.2 Stakeholder analysis

Stakeholder	Type	Primary role	Potential impacts/concerns	Initial receptivity	Future receptivity
Patients	Primary	Patients who have surgery	Prepared for surgery. Reducing number of times called and having to come to hospital for pre-screening activities	Neutral	Neutral
Pre-screening nurses	Primary	Nurses who pre-screen the patients prior to surgery	Not receiving orders and information from surgeons' offices timely. Needing to work overtime to complete charts that negatively impacts productivity	Strongly support	Strongly support
Director of surgery	Secondary	Manage the surgical department	Patients ready for surgery. Meeting productivity goals. Reduced labs and diagnostics on the day of the surgery. Reduce surgery cancellations due to pre-screening not complete	Strongly support	Strongly support
OR pre-op nurses	Primary	Prepare the patients for surgery on the day of the surgery	Having charts ready when the patients arrive in pre-op on the day of the surgery. Knowing what tests and assessments still need to be performed on the day of the surgery	Neutral	Strongly support
Surgeons	Primary	Perform surgery	On time starts, no delays in surgery	Moderate support	Strongly support

(continued)

Table 6.2 (continued)

Stakeholder	Type	Primary role	Potential impacts/concerns	Initial receptivity	Future receptivity
Anesthesiologists	Primary	Assess the patient pre-surgery	Ensuring the patient and/or chart has been screened by anesthesia prior to surgery	Neutral	Strongly support
Lab	Primary	Provide lab specimens and tests for pre-surgery readiness	Needing to provide staff to collect and perform labs on the day of the surgery, instead of ensuring labs are complete	Neutral	Strongly support
Radiology	Secondary	Provide radiology diagnostics tests for pre-surgery readiness	Needing to provide staff to perform radiology tests on the day of the surgery, instead of ensuring tests are complete	Neutral	Moderate support

worksheet with the key patient information and provides the charts to the OR surgical holding area by 5 pm the day prior to the scheduled surgeries.

Perform Initial Voice of Customer and Identify Critical to Satisfaction

The CTS Criteria for this project are:

1. Patient chart readiness the day prior to surgery.
2. Reduced number of labs and diagnostics tests on the day of surgery.
3. Reduction in number of cancellations due to the patient's pre-screening not being complete.

Select Team and Launch the Project

The cross-functional team was identified and formed. The following areas were included as team members:

Table 6.3 SIPOC

Suppliers	Inputs	Process	Outputs	Customers
Pre-screen secretary	OR systems	Review surgery schedule	Surgery schedule	Pre-screening department
Pre-screen secretary	Standard blank forms	Ready pre-chart	Standard blank forms in chart	Pre-screening nurses
Pre-screening nurses	Surgery schedule Charts	Assign charts to nurse	Patient chart with stickers	Pre-screening nurses
Pre-screening nurses	Surgery schedule	Match required documents	Standard forms	Pre-screening nurses
Pre-screening nurses	Database Health and physical Orders Medical clearances	Ready patient chart	Completed patient record/ chart	Pre-screening nurses
Pre-screening nurses	Patient ID stickers	Add patient stickers	Patient chart with stickers	Pre-screening nurses
Pre-screening nurses	Original worksheet	Copy worksheet	Copied worksheet	Pre-screening department
Pre-screening nurses	Patient record	Transfer charts to holding	Completed patient record	Surgery holding area

- Director of Surgery
- Director of Anesthesia
- Manager of pre-screening department
- Pre-screening nurses
- Holding area nurses
- Lab
- Radiology.

Create Project Plan

The project was started on April 14 and the Define phase was completed by May 15. The Measure phase was completed on June 16. The Analyze phase was completed on July 30. The Improve phase was completed in October and the Control phase was completed on July 17.

Measure Phase

The Measure Phase included the following activities:

1. Define the current process
2. Define detailed Voice of Customer (VOC)
3. Define the Voice of Process (VOP) and current performance
4. Validate measurement system

Define the Current Process

Value Stream Map

The value stream map provides a systems view of the departments and functions involved in preparing the patient for surgery. It shows the requirements for each function, the delays, wastes, and opportunities for improvement in the value stream. It helps us to understand where the departmental silos exist and where we optimize processes in some departments at the price of suboptimizing the flow of the entire value stream. Admitting/registration recently made some process changes that optimized the use of the admissions staff but negatively impacted the processes in the pre-admitting area. By understanding the process from the start to the end and the hand-offs between each function, we can better streamline the entire process that the patient is experiencing.

We have also provided a rough estimate of the lead time in the process compared to the value-added processing time. The lead time includes the wait time and delays between the functions. The processing time is the value-added time that is achieving the goals of the value stream, to ready the patient for surgery. In most value streams, there is typically only 2–5% of the time spent in value added processing time, versus 95–98% of the time spent in delays, wasteful, non-value-added activities. In our estimated value stream lead time compared to process time, we found a 2% value added time. For planned outpatient elective surgical cases, the lead time is roughly 16 days (11,513 min) from when the case is scheduled to when the surgery is performed. The value-added processing time is approximately 227 min, which is 2% of the total lead time. Therefore, we have great opportunity to reduce the time to prepare the patient's chart.

The value stream map is shown in Fig. 6.1.

The yellow kaizen blitz symbols identify opportunities for improvement in the value stream as follows:

- Scheduling pre-screening visits, instead of trying to catch the patient on the phone.
- Verifying insurance and preregistering the patients that are not currently meeting the needs of timely data entry of the nursing assessments.
- The charts are touched, worked on, and reviewed multiple times throughout the process. The goal should be to touch the chart once, or as few times as possible.

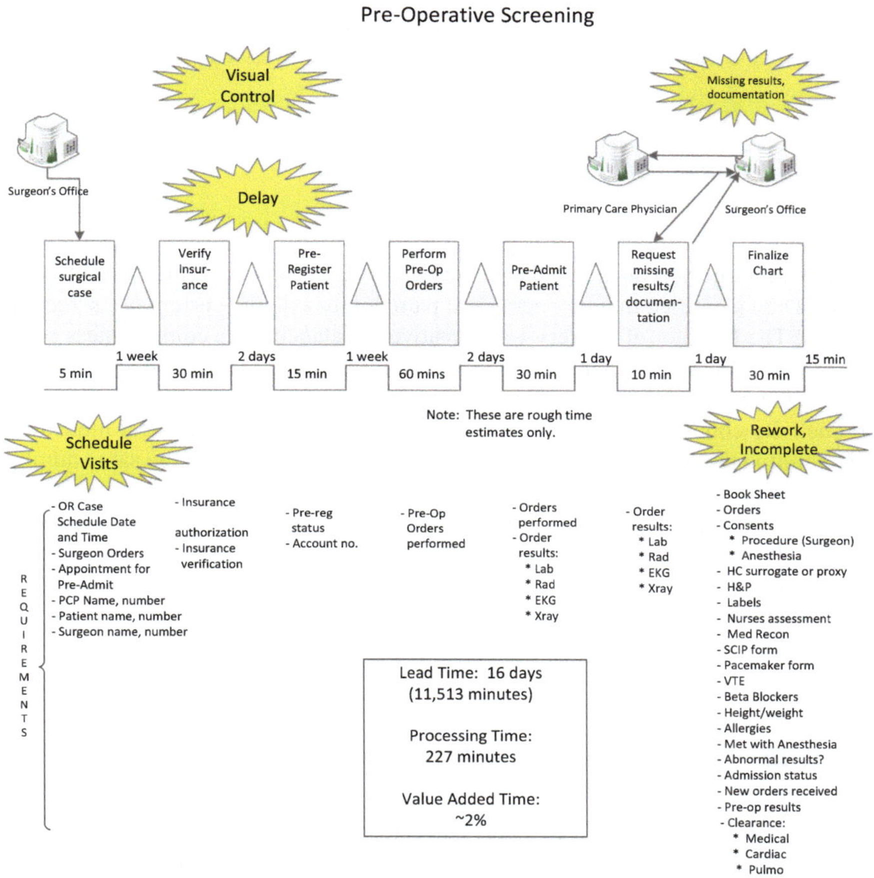

Fig. 6.1 Value stream: preoperative screening

- The charts move to pre-screening holding without being completed consistently and therefore require stat orders and additional review and completion of the charts.
- There are no visual controls that provide the status of the charts moving through the process.

Process Maps

The current state process maps show the more detailed activities that are performed in the pre-screening area. The process maps illustrate that the process is lengthy, with much redundancy and "touching" the chart by many people multiple times throughout the process. The surgical stakeholders are concerned that many activities are done on the day of the surgery in the pre-screening holding area that could be done prior to the day of the.

Additionally, there is lack of consistency in how the presurgical assessments are performed and how the documentation is gathered and reviewed prior to the surgery.

Define Detailed Voice of Customer (VOC)

The internal Voice of the Customer (VOC) was gathered in interviews with the Director of Surgery, Manager of the pre-screening department, and the Chief of Anesthesia. They identified their concerns and pain points with the current pre-screening process. The Director of Surgery was sensitive to balancing the completeness of the charts with meeting the department's productivity goals. If the productivity goals were not met, the Director of Surgery would need to explain the reasons at the Productivity council weekly, until they met the targets. There was great motivation to streamline the processes, so that more work could be done with the same number of people. The Manager of the pre-screening department voiced frustration with not obtaining the physician's orders, medical clearances, and other patient information in a timely manner without having to call and "harass" the offices. The Chief of Anesthesia wanted a more streamlined way to know which patients are needed to be seen in person by an anesthesiologist and which charts needed to be reviewed and when.

Define the Voice of Process (VOP) and Current Performance

The Voice of the Process (VOP) and current performance was an extensive data collection experience. The data collection plan was developed by the team and shown in Table 6.4. The goal is to measure the CTS criteria identified in the Define phase, including patient chart readiness the day prior to the surgery, reduced number of labs and diagnostics tests on the day of the surgery, and reduction in the number of cancellations due to the patients' pre-screening not being complete.

The operational definitions were developed for each metric.

Patient chart readiness on the day of the surgery

The pre-screening Manager began collecting the number of scheduled first surgical case charts daily that are ready by 5 pm the day prior to the surgery. The data for January to April are shown in Fig. 6.2. The overall average is 35% charts complete for the first cases of the day. This shows a large opportunity for improvement. The target is 100% of the first case charts should be ready by 5 pm the prior day.

Reduced number of labs and diagnostic tests on the day of surgery

The pre-screening area collected data on missing patient data in the charts over a period of time. The different data types include Health and Physical assessment,

Table 6.4 Data collection plan

Critical to satisfaction (CTS)	Metric (short title)	Operational definition (metric description)	Data collection source	Analysis mechanism	Sampling plan	Process to collect and report
Patient chart readiness the day prior to surgery	Percent charts ready	Percent charts ready day prior to surgery	Patient readiness check sheet	Statistics	2 months	Verify and validate data with team and subject matter experts
Reduced number of labs and diagnostics tests on the day of surgery	Missing patient data	Patient readiness check sheet	Patient readiness check sheet		13 months	
Reduction in number of cancellations due to the patient's pre-screening not being complete	Number surgery cancellations	Number of cancellations day of surgery due to pre-screening not complete	OR scheduler		2 week manual study (4/12–4/27)	

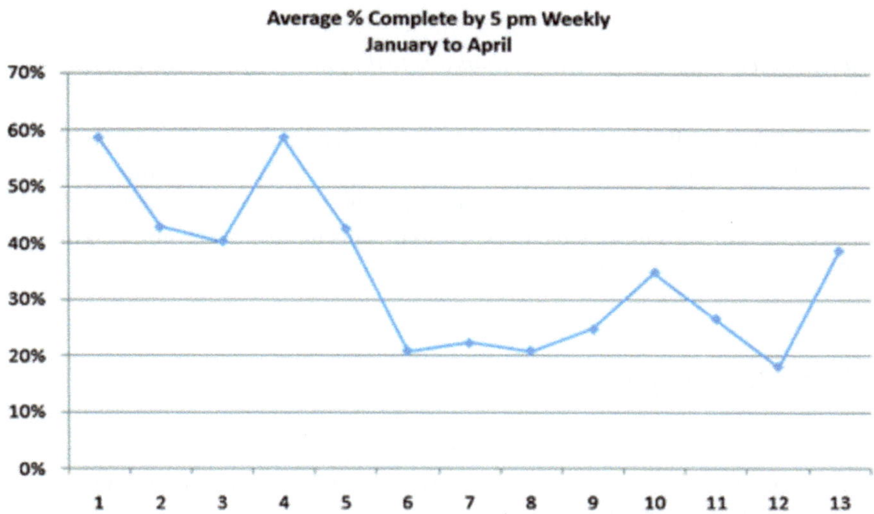

Fig. 6.2 Average percent of first case charts completed by 5 pm

orders, new orders, tests, database, surgical content, chart label, and labs. The missing patient data is shown in Fig. 6.3.

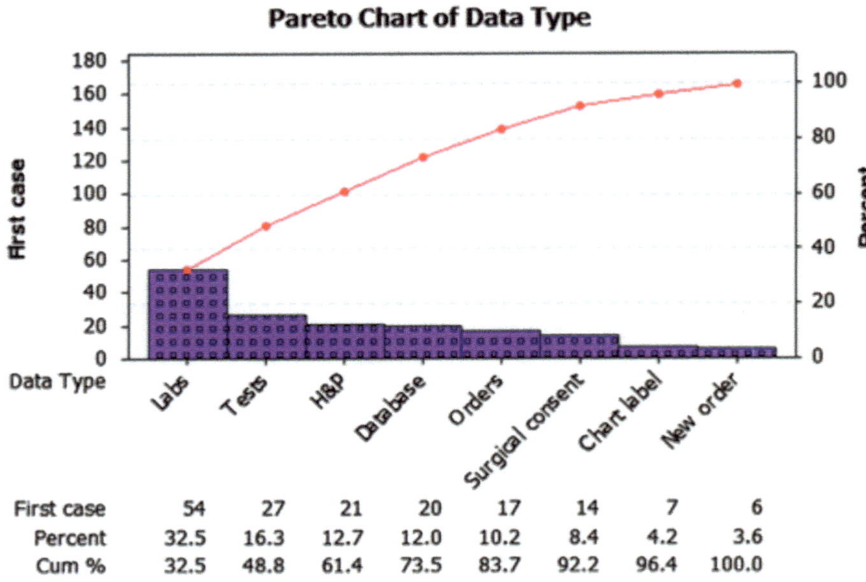

Fig. 6.3 Pareto chart of missing data

Reduction in the number of cancellations due to the patient's pre-screening not being complete

This would be a collection of the number of cancellations on the day of the surgery due to pre-screening not complete. However, this was not collected, due to it being difficult to track and determine the reason for the cancellation. Additionally, the Director of Surgery anecdotally believed that cancellations were not a significant problem, so we decided not to track this metric.

Validate Measurement System

The pre-screening Manager validated the data collection process for assessing whether the cases were complete and the types of data missing and found the data collection to be appropriate.

Analyze Phase

The following activities were performed for the Analyze Phase to identify and validate the root causes and to assess whether the process is currently capable of meeting the desired specifications for the Length of Stay:

1. Develop cause and effect relationships
2. Determine and validate root causes
3. Develop process capability.

Develop Cause and Effect Relationships and Determine and Validate Root Causes

A cause-and-effect analysis was performed to identify the root causes of the delays and inefficiencies in the process. The wait time between steps is quite significant, as can be seen in the Value Stream Map shown earlier in Fig. 6.4. The cause-and-effect diagram is shown in Fig. 6.4. Some of the root causes included the following:

Materials

- Missing Health and Physical
- Missing labs

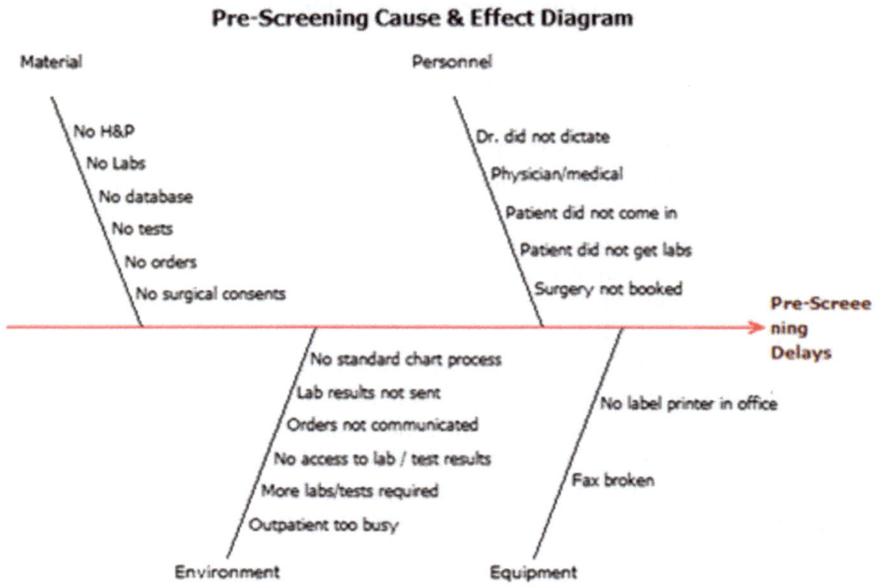

Fig. 6.4 Cause and effect diagram

- Missing database
- Missing tests
- Missing orders
- Missing surgical consents.

Personnel

- Dr. did not dictate the results
- Physician or medical office delays
- Patient did not come in for the pre-screening assessment
- Patient did not get labs run
- Surgery was not booked.

Environment

- No standard chart process
- Lab results not sent from provider
- Orders not communicated to patient
- No access to lab or test results
- More lab or tests required
- Outpatient is too busy to print labels.

Equipment

- No label printer in pre-screening office
- Fax broken, can't receive orders/results.

Takt Time, Identifying Constraints/Bottlenecks and System Capacity

We calculated the takt time for the pre-screening process to assess whether we have the pre-screening nurse capacity to complete the required demand (patient charts). With the complexity of the process that we are trying to improve, it is important to identify the key bottlenecks so that it focuses our improvement efforts. The takt time describes the time that is required to complete each unit to be able to meet the customer's demand. Note that these are hypothetical numbers, not actual case numbers to protect the confidentiality of the healthcare provider.

Takt Time = Total Available Time/Number of required charts

Total Available Time = 5 pre-screening nurses * 435 min per day = 2,175

Number of charts per day = 52

Takt time = 2,715/52 = 41.8 min per day = 0.7 h per day

We then calculated the patient chart volume capacity.

Patient chart volume capacity = Total Time Available/System cycle time

Patient chart volume capacity = 2,175/48 = 45 patient charts per day

Since the average number of patient charts per day is 52, and the department can only complete 45, the department is required to work at 115% of capacity (52/45). This supports the finding that the charts are not complete by 5 pm, with only 33% of the charts complete by 5 pm. This is due mainly to the nonvalue added delays in the process of waiting for orders, waiting for the patient to come to the hospital for their interviews and tests, waiting on results, etc.

Why–Why Analysis

We performed a why-why analysis to better understand the root causes. Sometimes asking why five times can better identify the root causes.

The why–why diagrams are shown in Figs. 6.5 and 6.6.

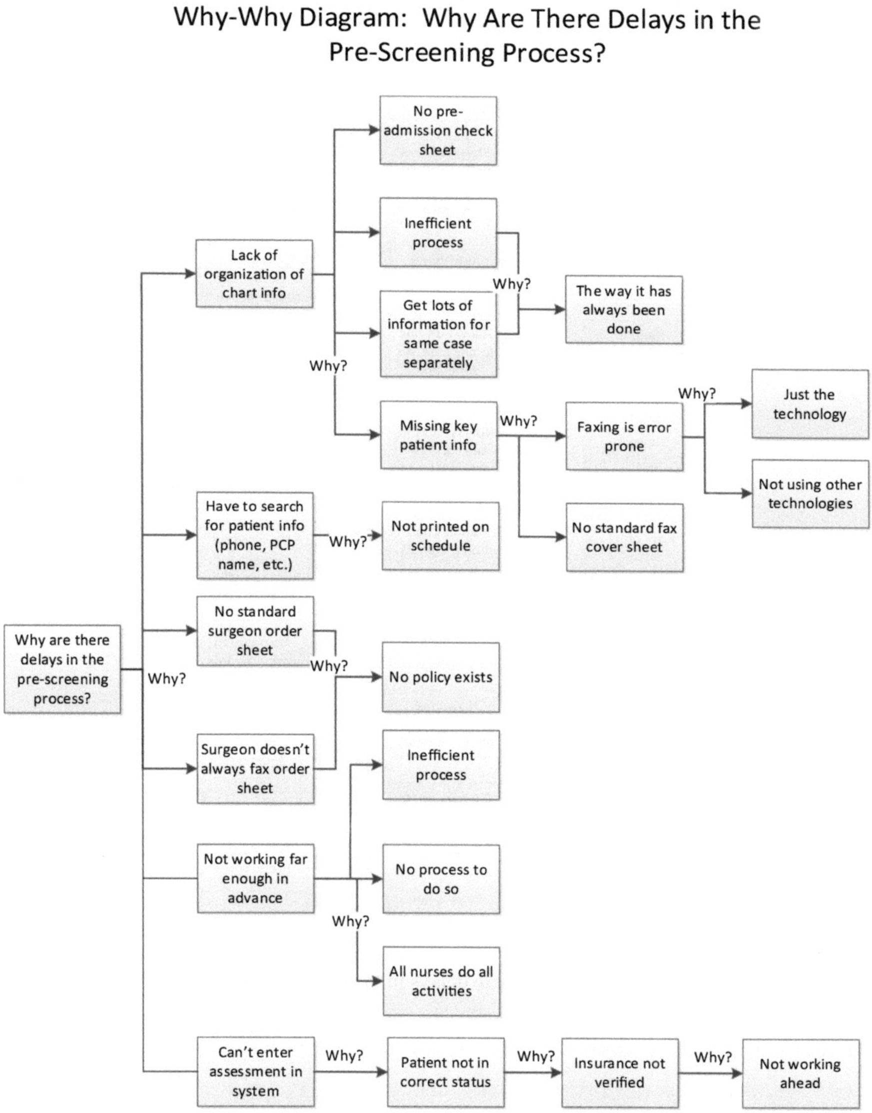

Fig. 6.5 Why–why diagram—Why are there delays in the pre-screening process?

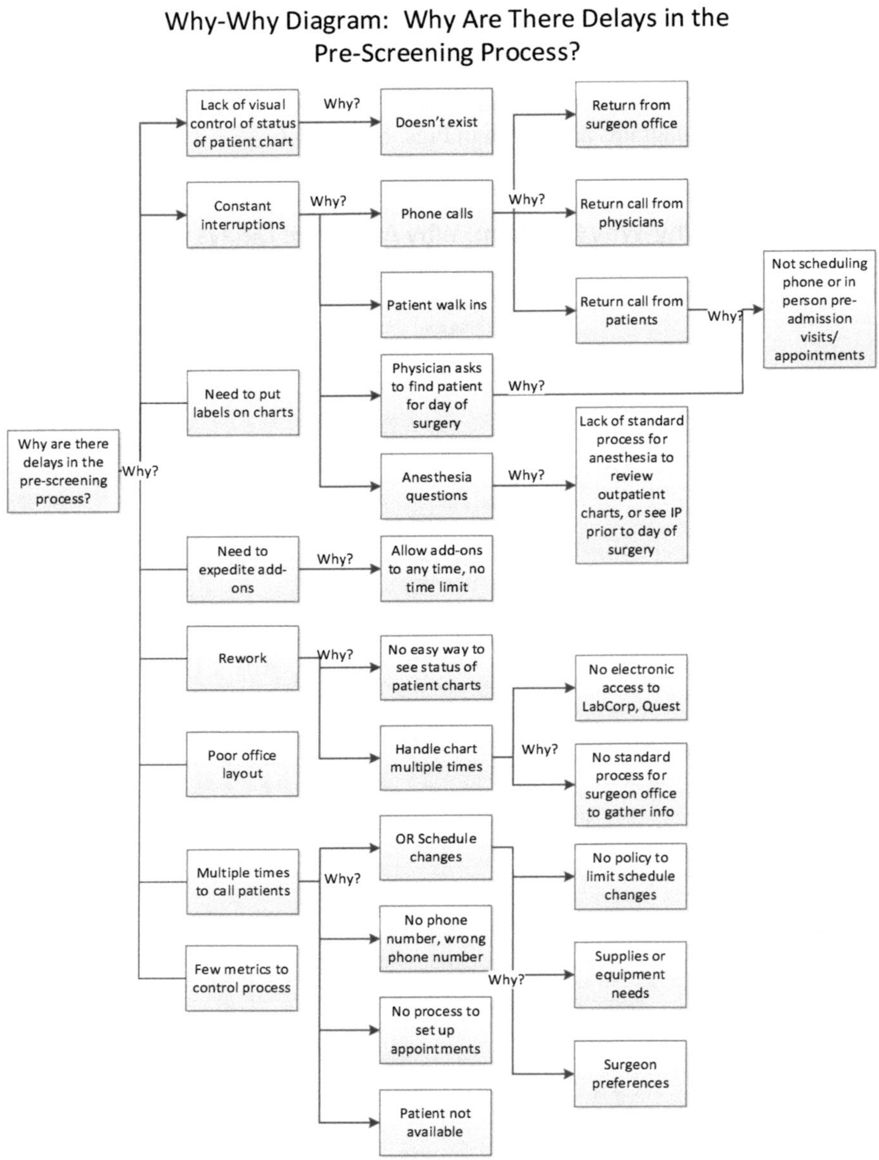

Fig. 6.6 Why–why diagram—Why are there delays in the pre-screening process? (continued)

Eight Wastes Analysis

The Eight Wastes Analysis is used to identify process inefficiencies and redundancies. The wastes are shown in Table 6.5.

Issues and Root Cause Analysis

Table 6.6 is a summary of the issues and root causes for delays in the surgical pre-screening process.

Develop Process Capability

We performed a process capability analysis for the percent of charts complete for first cases for January to April. The process is not stable, as is shown in Fig. 6.7. However, since the points that are initially out of control represent a higher percent of charts complete, we will calculate the process capability with those data points left in the study since a higher percentage of complete charts is desirable. The process capability becomes the center line of the chart, which is 33%.

The process is not capable of achieving our target of having 100% of the charts complete for first cases by 5 pm the day prior to the scheduled surgery. The binomial process capability analysis is shown in Fig. 6.8. The Process Z value is 0.45, which shows that there is a great deal of room for improvement, which also is shown by the center line of 33% of the charts being complete, compared to our target of 100%.

Improve Phase

The following steps were executed in the Improve Phase:

1. Identify improvement opportunities and plans
2. Design future state
3. Establish performance targets and project scorecard
4. Pilot improvements
5. Train and execute.

Identify Improvement Opportunities and Plans and Design Future State

Gaps from Best Practice

Best practice principles are not being applied consistently in the pre-screening outpatient surgical process. The following best practice principles based on systems and management engineering and Lean concepts are:

Table 6.5 Eight wastes analysis

Waste type	Wastes in pre-screening process
Transportation: any material movement that is unnecessary	• Inefficient work layout causing additional walking • Searching for and moving charts
Over production: producing more than needed, or faster than needed	• Touching the charts multiple times to verify information • Having to investigate the status of charts due to lack of visual control and chart tracking to easily know the status of each chart • No standard surgeon order set
Motion: any movement of people that does not add value	• Reaching for forms • Reaching for documents from the fax machines that are placed too high • Walking across the office for forms • Walking around old, extra equipment (fax machines, printers)
Defects: not meeting customers' requirements; repairing mistakes in the process	• Surgeons not sending orders • Missing information • Results not available
Delays: people waiting in the process. Information not available when needed	• Lack of synchronization of resources, no patient appointments for pre-screening and assessment • Documentation not provided by surgeons • Orders not provided by surgeons causing delays in screening information • Insurance verification group not performing their work early enough to have the appropriate registration status and electronic assessment available when needed by the pre-screening nurses, causing the nurses to wait until the registration status is changed
Inventory: any excess material or information; disorganized materials or information	• Storing of chart information is not segregated by patient but grouped by surgery date • Storing excessive forms in the immediate area (more than is needed for a week's worth of work) • Excessive clutter due to poor layout, and extra equipment, and materials in the work area that are not used frequently
Processing: efforts which add no value in the process leading to inefficiencies	• No aggregation of duties by pre-screening nurses causing multiple people reviewing the same charts
People: waste of not using people's mental, creative, and physical abilities	• People performing activities at inappropriate skill levels, i.e., nurses and physicians applying chart/form labels

Table 6.6 Issues and root causes

Issues	Root causes
Lack of organization of chart information	• No pre-screening checklist
Getting information for same case separately across time and from surgeon and physicians	• No efficient process for gathering documentation from surgeon and primary care physician (PCP)
Missing key patient information	• Faxing technology; no standard cover sheet
Having to search for PCP name, phone numbers, etc.	• No set process for gathering and consolidating information
No standard surgeon order sheet: surgeon doesn't always provide orders timely and without requesting them	• No policy for a standard order sheet
Lack of visual control of status of charts in pre-screening area	• No mechanism for visual control or tracking chart status
Constant interruptions during processing of pre-screening data by phone calls (surgeons' offices, PCPs, patients), patient walk-ins, and request for status on charts, anesthesia questions	• No policy for scheduling patients for pre-screening visits or phone calls • No standardization or measurement for anesthesia seeing outpatients or reviewing charts, or seeing inpatients prior to surgery
Not working far enough in advance on charts and preparing them prior to the day before the surgery	• No standard process to ensure orders are performed far enough in advance, that documentation is received at least 3 days prior to the surgery
Can't enter assessment in system because insurance certification department hasn't entered the patient in the system or completed the insurance verification	• Insurance certification department not working far enough in advance
Rework and handling patient charts multiple times	• No electronic access to lab results • No standard process for the surgeons' office and PCPs to provide documentation timely, and at least 3 days prior to the surgery
Need to put labels on chart on the day of the surgery; don't have medication orders scanned to pharmacy prior to the day of the surgery	• No process for who should print and apply labels prior to the day of the surgery to enable the chart to be complete prior to the day of the surgery and to scan medication orders to pharmacy prior to the day of the surgery
Multiple phone calls to reach the patient for pre-screening; wrong phone numbers or no phone numbers	• No policy for scheduling patients for pre-screening visits or phone calls • Patient demographics not consistently provided or updated from surgeons' offices
Frequent OR schedule changes for time of cases causes patient confusion, need to re-call the patient to provide the surgery tie, convenience of surgeon	• Need to align OR staff, equipment, and supplies • No policy to limit schedule changes
Lack of privacy; noise in office; need to travel to retrieve forms and obtain patient chart status	• Poor office layout

(continued)

Table 6.6 (continued)

Issues	Root causes
Need to expedite add-ons late in the day prior to surgery	• No policy to limit add-ons after a set time the day prior to surgery
Lack of process control	• Few control mechanism or metrics to control the process

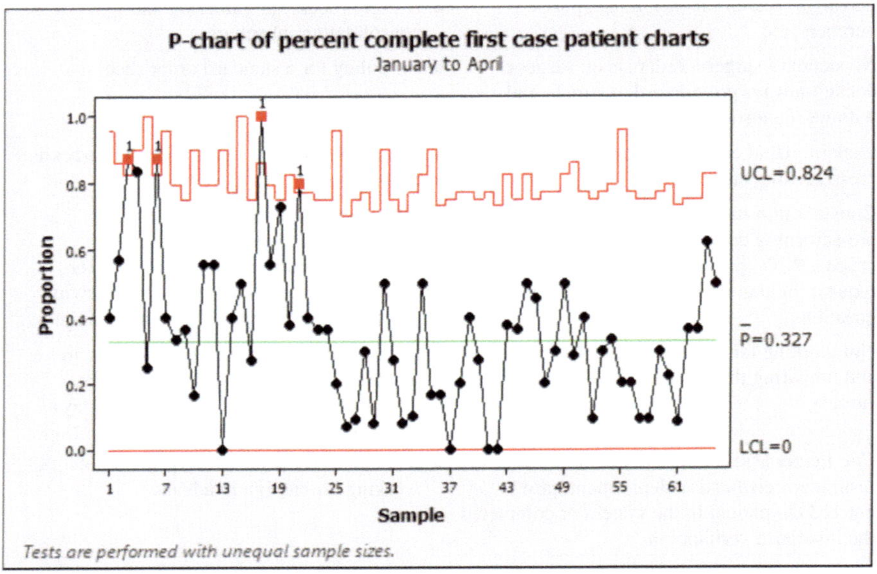

Fig. 6.7 Percent charts complete control chart

- *Reduce waste and rework* to touch the charts as few times as possible.
- *Optimize the entire value stream* from start to end, do not suboptimize functions within the value stream.
- Implement *visual controls* to be able to easily assess process status.
- Implement *mistake proofing*, such as check sheets to assess chart readiness, and reduce errors.
- Implement *metrics* to assess process control.
- *Standardize* and *document* the process to ensure process consistency and reduce process variability.
- *Pull work* based on *patient demand*, implementing *signals* for work.
- Organize *the work layout* to reduce unnecessary motion and travel, and eliminate unnecessary documents, equipment, and materials in the work area.
- Understand the patients' *requirements* and build your processes to meet their *needs*.

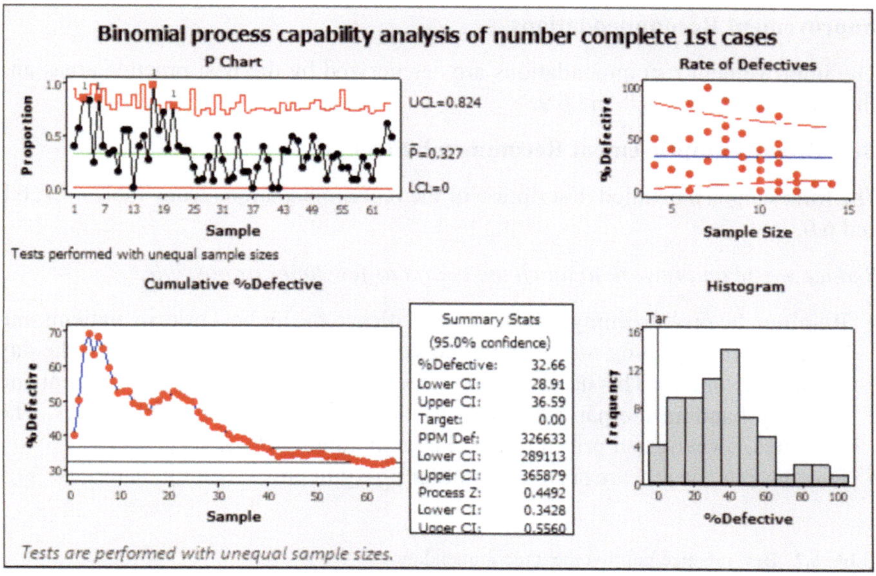

Fig. 6.8 Process capability

The team developed improvement recommendations based on eliminating the identified root causes and incorporating best practices. The best practice categories are shown in the affinity diagram in Fig. 6.9.

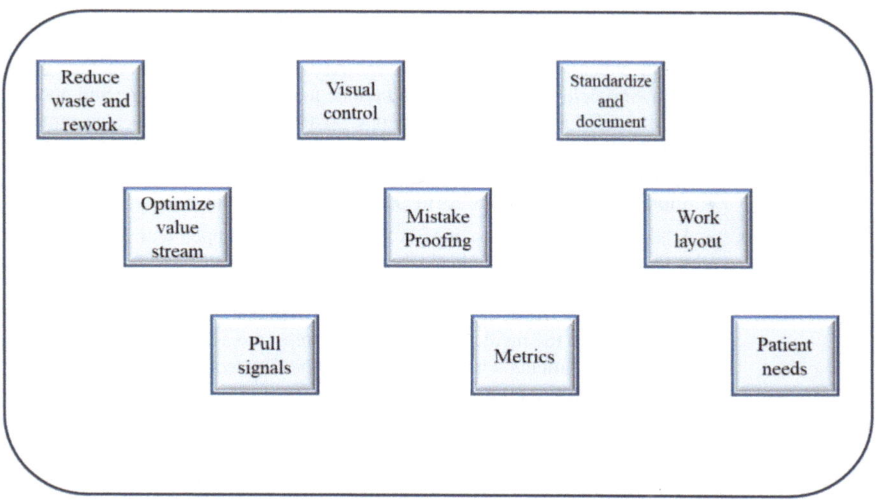

Fig. 6.9 Affinity diagram of best practice process improvements

Improvement Recommendations

The improvement recommendations are categorized by the best practice areas and shown in Tables 6.7, 6.8 and 6.9.

Best Practice Improvement Recommendations

The following is a detailed description of the recommendations from Tables 6.7, 6.8 and 6.9.

Reduce waste and rework to touch the charts as few times as possible.

- Re-align the pre-screening staff work activities to the highest priority patients and have the pre-screening nurses track the same patients from 1 week until the day before the surgery. This helps the pre-screening nurse get to know the patient and not keep handing the patients off between all of the pre-screening nurses. The activities, owners, and priorities are shown in Table 6.10.
- Do consents, healthcare surrogate, and/or proxy in pre-screening materials.

Table 6.7 Best practice improvement recommendations

Best practice area	Improvement recommendations
Reduce waste and rework	• Re-align the pre-screening staff work activities to the highest priority patients and have the pre-op nurses track the same patients from 1 week until the day before the surgery • Do consents, healthcare surrogate and/or proxy in pre-screening materials • Align the staffing to the pre-screening visits from 7 am to 5 pm • Print and put the labels on the charts prior to the day of the surgery. Ensure there is a convenient printer in the pre-screening office to print labels • Purchase an automatic hole punch for the charts • Reduce OR schedule changes • Create a customized OR schedule for the pre-screening area to include the following fields: patient name and phone number, surgeon name and phone number, PCP name and phone number, whether the patient has a pacemaker, whether the patient is on beta blockers, reference lab where patient is getting pre-surgical labs performed • Implement electronic access to Quest and LabCorp lab results
Visual controls	• All documentation that comes in should be paper clipped with the pre-screening readiness check sheet • Implement a simple pre-screening patient tracker visual control to easily see what is missing from the patients' charts. Investigate downloading the OR schedule from the medical information system into a spreadsheet to track the status of the patients' charts, and share with all appropriate staff: pre-screening, pre-op holding, anesthesia • Implement a simple surgical inpatient tracking mechanism to ensure that the inpatients are ready for surgery
Mistake proofing	Implement a pre-screening readiness check sheet of all chart requirements, created when the patient case is scheduled
Metrics	Implement metrics to control the pre-screening process

Table 6.8 Best practice improvement recommendations (continued)

Best practice area	Improvement recommendations
Optimize the value chain	• Synchronize admitting, insurance verification and pre-screenings process • Have an outpatient registrar register surgical patients for their pre-screening visit, and day of surgery (only those that did not visit) • Provide a list of surgical patients when they are scheduled by the OR scheduler to insurance verification so that they process insurance timely • Require pre-admitting patient visits. If the patient cannot come into the hospital for pre-screening visit, then schedule a telephone interview with the pre-screening nurse for the nursing assessment and medication reconciliation. The pre-op visit will include a nursing pre-screening assessment, anesthesia evaluation, social worker assessment for those patients that will need home health and/or rehabilitation services, nursing will provide education and documentation on the discharge instructions and information on the surgery, and completion of needed orders • Anesthesia should see all inpatients the day before surgery and perform an anesthesia evaluation • Anesthesia should review outpatient charts to evaluate high-risk patients (4 days out from surgery date) • The OR scheduler should schedule the pre-screening surgical visits and the telephone nursing assessments when they schedule the surgical case with the surgeon's office. The OR scheduler should enter the reference lab where the patient is getting their pre-op labs performed in the medical information system to print on the OR schedule. The patients should be scheduled to come in 1 week from their scheduled surgery date (no more than 3 patients per hour). In the future the hospital can investigate whether the pre-screening surgical visit prior to 4–7 days before surgery will be billed separately and not bundled with the surgical visit. However, admitting would need to determine how the registration would be done and how many account numbers are needed • The pre-screening secretary should handle the patient technology reminders of the pre-op visit and the telephone nursing assessments, and any necessary re-scheduling

- Align the staffing to the pre-screening visits from 7 am to 5 pm, based upon the average number of outpatient surgical cases and the standard deviation.
- Print and put the labels on charts prior to the day of the surgery. Outpatient registration can print the labels when the patient is registered during the pre-screening visit. For the patients who do not come in for a pre-screening visit, the pre-screening secretary can print the labels prior to the pre-screening telephone appointment. Ensure there is a convenient printer in the pre-screening office to print labels.
- Purchase an automatic hole punch for the charts.
- Reduce OR schedule changes. This would require a process improvement project to identify the reasons for the schedule changes.
- Create a customized OR schedule for the pre-screening area to include the following fields: patient name and phone number, surgeon name and phone number, PCP name and phone number, whether the patient has a pacemaker,

Table 6.9 Best practice improvement recommendations (continued)

Best practice area	Improvement recommendations
Standardize and document	• Implement desktop work instructions • Require the surgeons to provide their orders on the standard order set. Track surgeon office compliance to this requirement (see pre-screening patient tracker). Send letter to surgeons and PCPs with pre-screening requirements • Require the surgeon's office to provide the pre-op documentation: orders, results, medical clearance, etc. no less than 3 days prior to surgery. Track surgeon office compliance to this requirement (see pre-screening patient tracker) • Implement a standard fax cover sheet to provide necessary case information. Include data fields: patient name; patient phone number; date of surgery; procedure; surgeon name and phone number, PCP and phone number • Implement a time limit for add-ons the day before surgery (4 pm). If add-ons are necessary after 4 pm, then schedule the surgeries for late the next day • Register the patients for surgery during the pre-screening visit so that they do not have to register on the day of the surgery • Ensure charts are complete the day prior to the surgery • Hire a process engineer to continue the process improvements, and track process metrics across the hospital • For the pre-op holding area, put labels on inpatient forms in the afternoon prior to the day of the surgery
Pull work based on patient demand	• Apply the pre-screening patient tracker and the pre-screening chart readiness checklist as tools to identify the patients who still have outstanding items for their charts. Use this list to determine the work priorities for the pre-screening staff. Implement signals for work
Organize work layout	• 5S the office to remove equipment and supplies not needed • Re-design the pre-screening area for privacy, flow, and quietness
Build process to meet patients' needs	• Implement a surgical patient survey specific for the surgical experience, separate from the inpatient experience • Implementing the above defined recommendations will better meet the patients' needs to help them to be better prepared for surgery. Based on the evidence in the literature this will help to reduce patients and their families' anxiety and enhance patient satisfaction

whether the patient is on beta blockers, and reference lab where patients are getting presurgical labs performed. This would need to be an IT project.

- Implement electronic access to Quest and LabCorp lab results. This would need to be an IT project.

Optimize the entire value stream from start to end, do not sub-optimize functions within the value stream.

Synchronize admitting, insurance verification, and pre-screening's processes.

- Have an admitting registrar register surgical patients for their pre-screening visit, and on the day of the surgery (only those that did not visit). Labels should be printed

Table 6.10 Pre-screening activities, owners, and priorities

Priority	Pre-screening nurse 1	Pre-screening nurse 2	Pre-screening nurse 3	Pre-screening nurse 4	Secretary	Nurse manager	Outpatient registration
First	Pre-screening assessment appointments	Pre-screening assessment appointments	Day of surgery patient chart preparation	Patient chart review for patients 1 day from surgery; process add-ons	Manage pre-screening visit calendar and call patients for appointment reminder	Coordinate day of surgery readiness; coordinate pre-screening visit, nursing assessments	Register patient for pre-screening visits and day of surgery
Second	Patient chart review for patients, keeping the same patients form 1 week out until the day before the surgery, alternating weeks with pre-screening nurse 2	Patient chart reviews for patients, keeping the same patients form 1 week out until the day before the surgery, alternating weeks with the pre-screening nurse 1	Ensure H&Ps have been completed; review charts with anesthesia	Call patients with surgical case times	Complete pre-screening patient tracker. Follow-up with surgeons and PCPs for late documentation (3 days out)	Coordinate pre-screening nurse activities	Print and apply chart labels, prior to the day of the surgery, and for inpatients

by outpatient registration when the patient is registered for the pre-screening visit, or by the pre-screening department secretary at least one day prior to the pre-screening appointment for the pre-screening telephone call.

- Have the insurance verification department pull the OR schedule daily and ensure that the insurance verifications are performed no less than 10 days from surgery and within 1 day from being added if added less than 10 days from surgery.
- Require pre-admitting patient visits if possible for all cases. If the patient cannot come in prior to the surgery, then schedule a telephone interview with the pre-screening nurse for a nursing assessment and medication reconciliation. The pre-screening in person surgical visit will include a nursing pre-screening assessment, anesthesia evaluation, social worker assessment for those patients that will need home health and/or rehabilitation services, nursing to provide education and documentation on the discharge instructions and information on the surgery, and completion of needed orders. A study by Pearson et al. (2004) reviewed the evidence of best practice in preparing the patient for day surgery. The study found that both preoperative telephone screening or questionnaires, or a pre-screening clinic a few days prior to surgery were both effective in enhancing patient satisfaction, reducing cancellations, and reducing anxiety for patients and their families. The following suggestions are recommended.
  - Appointments should be set up with the OR Scheduler. Either a shared Outlook calendar can be used or the medical information system scheduling function to schedule the patients' pre-screening visit appointments is used. IT can set this up.
  - All patients should be encouraged to come to the hospital one week prior to their surgical date for a pre-screening surgical visit.

The pre-screening surgical visit will include the following activities:

Admitting

- Admitting registration for the pre-screening visit and surgery.

Pre-screening Nursing

- Nursing pre-screening assessment
- Any labs or outstanding orders would be completed
- Missing items would be identified on the pre-screening check sheet and the readiness summary sheet
- Provide education and documentation on the discharge instructions and information on the surgery and completion of needed orders.

Anesthesia

- Anesthesia evaluation would be performed.

Social Worker

- Social worker assessment for those patients that will need home health and/or rehabilitation services.

If a patient cannot come in for a visit, the pre-screening telephone nursing assessment interview will include the nursing assessment and medication reconciliation. It is important that the pre-screening visit includes an anesthesia evaluation. Evidence has shown that the presurgical anesthesia evaluation significantly reduced surgical cancellation rates (Lew et al., 2004). The paranesthesia clinic also reduces the length of the hospital stay by performing discharge planning during the clinic, and by providing appropriate patient information regarding the surgery.

- Anesthesia should see all inpatients the day before surgery and perform an anesthesia evaluation.
- Anesthesia should review outpatient charts to evaluate high risk patients (4 days out from the surgery date).
- The OR scheduler should use either the medical information system scheduler or a shared Outlook calendar to schedule the pre-screening visits and the telephone nursing assessments when they schedule the surgical case with the surgeon's office. The OR scheduler should enter the reference lab where the patient is getting their pre-screening labs performed in the medical information system to print on the OR schedule. The patients should be scheduled to come in 1 week prior to their scheduled surgery date. In the future, the hospital can investigate whether they will bill separately for the pre-screening surgical visit. However, admitting would need to determine how the registration would be done and how many account numbers are needed.
- The pre-screening secretary should handle the patient telephone reminders of the pre-screening visits and the telephone nursing assessments, and any necessary re-scheduling.

Implement visual controls to be able to easily assess process status.

- All documentation that comes in should be paper clipped with the pre-screening readiness check sheet that is created when the first piece of documentation comes into the pre-screening office.
- Implement a simple pre-screening Patient Tracker visual control to track whether the surgeons, anesthesia, and the primary care physicians are adhering to our pre-screening requirements. Investigate downloading the OR schedule from the medical information system into a spreadsheet to track the status of the patients' charts, and share with all appropriate staff: pre-screening, pre-op holding, and anesthesia. Track the percentage of charts ready by 5 pm prior to the day of the surgery. The chart readiness tracking metrics are shown in Fig. 6.10. The Outpatient pre-screening Patient Tracker is shown in Fig. 6.11.
- Implement a simple surgical inpatient tracking mechanism to ensure that the inpatients are ready for surgery. The Surgical Inpatient Tracker is shown in Fig. 6.12.
- Implement a tracker to assess the compliance of anesthesia in evaluating the outpatients and/or reviewing the charts, shown in Fig. 6.13.

Implement mistake proofing, such as check sheets to assess chart readiness, and reduce errors.

PRE-SCREENING CHART READINESS DASHBOARD

Date	Day	# of 1st Cases	# of Complete 1st Case Charts	% Complete	Week	Month
3-Jan	Monday	5	2	40.00%	2	1
4-Jan	Tuesday	7	4	57.14%	2	1
5-Jan	Wednesday	8	7	87.50%	2	1
6-Jan	Thursday	6	5	83.33%	2	1
7-Jan	Friday	4	1	25.00%	2	1
10-Jan	Monday	8	7	87.50%	3	1
11-Jan	Tuesday	5	2	40.00%	3	1
12-Jan	Wednesday	9	3	33.33%	3	1
13-Jan	Thursday	11	4	36.36%	3	1
14-Jan	Friday	6	1	16.67%	3	1
17-Jan	Monday	9	5	55.56%	4	1
18-Jan	Tuesday	9	5	55.56%	4	1
19-Jan	Wednesday	6	0	0.00%	4	1
20-Jan	Thursday	10	4	40.00%	4	1
21-Jan	Friday	4	2	50.00%	4	1
24-Jan	Monday	11	3	27.27%	5	1
25-Jan	Tuesday	7	7	100.00%	5	1
26-Jan	Wednesday	9	5	55.56%	5	1
27-Jan	Thursday	11	8	72.73%	5	1
28-Jan	Friday	8	3	37.50%	5	1
31-Jan	Monday	10	8	80.00%	6	1
1-Feb	Tuesday	10	4	40.00%	6	2

Average of % Complete	
Week	Total
2	58.60%
3	42.77%
4	40.22%
5	58.61%
6	42.55%
7	20.79%
8	22.33%
9	20.79%
10	24.77%
11	34.81%
12	26.48%
13	18.08%
14	38.71%
15	30.22%
16	26.26%
17	28.67%

Fig. 6.10 Percent charts complete by 5 pm prior to day of surgery data input form

Outpatient Pre-Screening Tracker

Patient Name	Date of Birth	Date of Surgery	Pre-Visit Appointment	Surgeon Name	Primary Care Physician	Date Receive Surgeon Preop Orders	Date Receive Surgeon Preop Docs	Date Receive Medical Clearance	Seen by Anesthesia/ Chart Reviewed (1 = Y, 0 = N)	Add on? (1 = Y, 0 = N)	Chart Ready 5 PM (1 = Y, 0 = No)	Pre-Visit Complete (1 = Y, 0 = N)	Date Pre-Visit Complete	Week
Surgical, Patient	9/6/1954	3/9/2012	3/6/2012	Dr. Surgeon	Dr. Primary	3/1/2012	3/6/2012	3/6/2012	1		1	1	3/6/2012	10
Surgical, Patient	9/6/1954	3/9/2012	3/7/2012	Dr. Surgeon	Dr. Primary	3/1/2012	3/6/2012	3/6/2012	0		0	0	3/6/2012	10
Surgical, Patient	9/6/1954	3/9/2012	3/6/2012	Dr. Surgeon	Dr. Primary	3/1/2012	3/6/2012	3/6/2012	1		1	1	3/6/2012	10
Surgical, Patient	9/6/1954	3/9/2012	3/17/2012	Dr. Surgeon	Dr. Primary	3/1/2012	3/6/2012	3/6/2012	1		1	0	3/6/2012	10
Surgical, Patient	9/6/1954	3/9/2012	3/10/2012	Dr. Surgeon	Dr. Primary	3/1/2012	3/6/2012	3/6/2012	0		0	0	3/6/2012	10
Surgical, Patient	9/6/1954	3/9/2012	3/6/2012	Dr. Surgeon	Dr. Primary	3/1/2012	3/6/2012	3/6/2012	1		1	1	3/6/2012	10
Surgical, Patient	9/6/1954	3/9/2012	3/9/2012	Dr. Surgeon	Dr. Primary	3/1/2012	3/6/2012	3/6/2012	1		1	1	3/6/2012	10
Surgical, Patient	9/6/1954	3/9/2012	3/6/2012	Dr. Surgeon	Dr. Primary	3/1/2012	3/6/2012	3/6/2012	1		1	1	3/6/2012	10

Fig. 6.11 Outpatient pre-screening tracker

Implement a *pre-screening Chart Readiness Checklist* of all chart requirements, created when the patient case is scheduled. This is shown in Fig. 6.14.

Implement metrics to assess process control.

- Implement metrics to control the pre-screening process. The metrics are included in the pre-screening inpatient and outpatient trackers (Figs. 6.11 and 6.12).

Surgical Inpatient Tracker

Patient Name	Date of Birth	Date of Surgery	Inpatient Floor/Unit	Inpatient Nurse	Surgeon Name	Primary Care Physician	Date Receive Surgeon Preop Orders	Date Receive Surgeon Preop Docs	Date Receive Medical Clearance	Seen by Anesthesia/Chart Reviewed (1 = Y, 0 = N)	Chart Ready 5 PM (1 = Y, 0 = No)	Inpatient Report Complete Prior to Report Out (1 = Y, 0 = N)	Week
Surgical, InPatient	9/6/1954	3/9/2012	2S	Imma Nurse	Dr. Surgeon	Dr. Primary	3/1/2012	3/6/2012	3/6/2012	1	1	1	10
Surgical, InPatient	9/6/1954	3/9/2012	3S	Imma Nurse	Dr. Surgeon	Dr. Primary	3/1/2012	3/6/2012	3/6/2012	0	0	1	10
Surgical, InPatient	9/6/1954	3/9/2012	3S	Imma Nurse	Dr. Surgeon	Dr. Primary	3/1/2012	3/6/2012	3/6/2012	1	1	0	10
Surgical, InPatient	9/6/1954	3/9/2012	2S	Imma Nurse	Dr. Surgeon	Dr. Primary	3/1/2012	3/6/2012	3/6/2012	1	1	0	10
Surgical, InPatient	9/6/1954	3/9/2012	2S	Imma Nurse	Dr. Surgeon	Dr. Primary	3/1/2012	3/6/2012	3/6/2012	0	0	1	10
Surgical, InPatient	9/6/1954	3/9/2012	3S	Imma Nurse	Dr. Surgeon	Dr. Primary	3/1/2012	3/6/2012	3/6/2012	1	1	0	10
Surgical, InPatient	9/6/1954	3/9/2012	3S	Imma Nurse	Dr. Surgeon	Dr. Primary	3/1/2012	3/6/2012	3/6/2012	1	1	1	10
Surgical, InPatient	9/6/1954	3/9/2012	3S	Imma Nurse	Dr. Surgeon	Dr. Primary	3/1/2012	3/6/2012	3/6/2012	1	1	1	10

Fig. 6.12 Inpatient pre-screening tracker

% Anesthesia Seeing Inpatients & Reviewing Outpatient Charts

Date	Day	# of Cases	# Seen by or Reviewed by Anesthesia	% Compliant	Week	Month
3/5/2012	2	11	11	100%	10	3
3/6/2012	3	12	10	83%	10	3
3/7/2012	4	16	12	75%	10	3
3/8/2012	5	12	8	67%	10	3
3/9/2012	6	9	8	89%	10	3
3/12/2012	2	12	11	92%	11	3
3/13/2012	3	14	13	93%	11	3
3/14/2012	4	16	14	88%	11	3
3/15/2012	5	18	17	94%	11	3
3/16/2012	6	19	17	89%	11	3
3/19/2012	2	20	16	80%	12	3
3/20/2012	3	22	22	100%	12	3
3/21/2012	4	22	22	100%	12	3
3/22/2012	5	18	17	94%	12	3
3/23/2012	6	19	17	89%	12	3
3/26/2012	2	21	19	90%	13	3
3/27/2012	3	25	20	80%	13	3
3/28/2012	4	24	23	96%	13	3
3/29/2012	5	22	21	95%	13	3
3/30/2012	6	24	23	96%	13	3
4/2/2012	2	24	18	75%	14	4
4/3/2012	3	23	21	91%	14	4
4/4/2012	4	21	20	95%	14	4
4/5/2012	5	19	18	95%	14	4
4/6/2012	6	20	20	100%	14	4

NOTE: Enter data in yellow cells only; Do not include add ons

Average of % Compliant	
Week	Total
10	83%
11	91%
12	96%
13	92%
14	91%
15	#DIV/0!
16	#DIV/0!
17	#DIV/0!
18	#DIV/0!
19	#DIV/0!
20	#DIV/0!
21	#DIV/0!
22	#DIV/0!
23	#DIV/0!
24	#DIV/0!
25	#DIV/0!
26	#DIV/0!
27	#DIV/0!
28	#DIV/0!
29	#DIV/0!
30	#DIV/0!
31	#DIV/0!
32	#DIV/0!
33	#DIV/0!

Fig. 6.13 Percent anesthesia seeing inpatient and reviewing outpatient charts data input form

- Implement a daily huddle with key OR staff to identify and resolve issues with on-time starts (Fig. 6.15). Where there are identified issues with pre-screening, these issues can be provided to the pre-screening department.

Standardize and document the process to ensure process consistency and reduce process variability.

- Implement desktop work instructions that can be used to ensure that the procedures are followed.

Pre-Screening Chart Readiness Checklist

1st Case: Y / N Start Time: ____________

SURGERY DATE: Surgery Time: IP 23-Hr Obs OP AM

PATIENT: SURGEON:

PATIENT PHONE:

PROCEDURE:

ALLERGIES:

ANESTHESIA: General Local MAC IV Block Other

REFERENCE LAB:

	RECEIVED?: YES	N/A
Insurance Authorized		
ORDERS		
CONSENT		
H & P		
LABELS		
NURSING ASSESSMENT		

CLEARANCE	RECEIVED?: YES	N/A
Medical		
Cardiac		
Pulmo		
Other: ______		

Height / Weight ______ / ______

PACEMAKER: Y / N

BETA BLOCKERS: Y / N

ENTERED IN EMR	RECEIVED?: YES	N/A
Height / Weight		
Allergies		
VTE		
SCIP		

PREOPS NEEDED: Y / N

	RECEIVED?: YES	N/A
CBC		
UA		
PT		
PTT		
PLAT		
T & C		
T & S		
BUN/CREATINE		
GLUCOSE		
HCG QUAL		
BLEEDING TIME		
XRAY:		
CHEST		
IVP		
BE		
EKG		
OTHER		
MRI		
CT		
US		

MET W/ ANESTHESIA: Y / N

ABNORMAL RESULTS: Y / N

NEW ORDERS RECEIVED: Y / N

ADMISSION STATUS: Y / N

NOTES: ****SEE NURSE'S NOTES*****

Chart Complete @ 5 PM: Y / N

Did you collect results from Quest LabCorp

Fig. 6.14 Pre-screening chart readiness checklist

- Require that the surgeons provide their orders on the standard order set. Track the surgeon office compliance to this requirement (see the pre-screening Patient Tracker). Send a letter to the surgeon's office and PCPs with the pre-screening requirements.
- Require the surgeon's office to provide the pre-op documentation: orders, results, medical clearance, etc. no less than 3 days prior to the surgery. Track the surgeon office compliance to this requirement (see pre-screening Patient Tracker).
- Implement a standard fax cover sheet to provide the necessary case information.

Daily OR Improvement Huddle					
Date	Delay Reasons	Immediate Action	Root Cause Category	Systemic Improvement	Comments
3/6/2012	MD Late	Address with Surgeon	People	Track Surgeon On Time	
3/6/2012	Anesthesia Late	Address with Anesthesia	People	Track Anesthesia On Time	
3/6/2012	Type & Screen	Perform T&S	Method	Improve T&S process	
3/7/2012	Patient Late	Discuss with patient	People	Preadmission procedure	
3/7/2012	Patient Late	Discuss with patient	People	Preadmission procedure	
3/7/2012	Equipment	Find equipment	Equipment	Kaizen	
3/8/2012	Room not ready	Discuss with OR Staff	Method	Kaizen	
3/8/2012	Patient Late	Discuss with patient	People	Preamission procedure	
3/8/2012	MD Late	Address with Surgeon	People	Track Surgeon On Time	

Fig. 6.15 OR daily huddle issues tracking

- Implement a time limit for add-ons the day before the surgery (4 pm). If an add-on is necessary after 4 pm, then schedule the surgery at the end of the next day.
- Register the patients for surgery during the pre-screening visit so that they do not have to register on the day of the surgery.
- Ensure charts are complete the day prior to the surgery.
- Hire a process engineer to continue the process improvements, and track process metrics across the hospital.
- For the pre-op holding area, put labels on the inpatient charts in the afternoon prior to the day of the surgery.

Pull work based on patient demand, implementing signals for work.

Apply the pre-screening Patient Tracker as a tool to identify the patients who still have outstanding items for their charts. Use this list to determine the work priorities for the pre-screening staff.

Organize the work layout to reduce unnecessary motion and travel and eliminate unnecessary documents, equipment, and materials in the work area.

5S is a Lean technique that helps to organize the work area. This recommendation includes implementing a 5S improvement project in the pre-screening office to remove equipment and supplies that are not needed. If possible, re-design the pre-screening area for privacy, flow, and quietness.

The 5S steps would include:

- *Sort*. Sort out the unnecessary equipment, forms, and supplies that are not used every day from the immediate area. There are broken or old fax machines that can be removed. There are a large number of forms in the office area. Only keep the forms that are used on a weekly basis within the office area. Store the other forms in a cabinet that can easily replenish the working area on a weekly basis.
- *Straighten*. Organize and straighten the pre-screening area after removing the unneeded items from the sort activity. Identify storage areas for the items that aren't used daily. Keep only a week's worth of forms and supplies and replenish the forms from the storage areas on a weekly basis. Identify a place for all of

the equipment, supplies and forms. Have a place for everything and ensure that everything is in its place. Label and mark the equipment, forms, and supplies clearly. Determine the following for each item (equipment, forms, and supplies):

  - Where it will be used
  - Where it will be placed
  - How much of each item will be placed
  - How it will be placed.

- *Shine*. Identify how often the area will be cleaned and organized, as follows:
  - Create an organizing schedule
  - Have a person responsible for each area
  - Remove root causes for disorganization
  - See the workplace through the patients' eyes!!
- *Standardize*. Standardize the organization.
  - Color code areas according to categories
  - Create "shadow boards"; Outline areas where things should be
  - Visual Management: Do you know what's going on in your area in 30 s or less?
- *Sustain*. Discipline. The following ideas will help sustain the organization of the pre-screening area.
  - Keep the processes going through training, communication, and organization structures
  - Allocate time for maintaining the organization
  - Create awareness of improvements
  - Management support for maintaining
  - Training and rewards
  - Daily check sheets and a monthly 5S Scorecard can be used to assess the organization (see chapter on 5S case study)
  - Make a person responsible for each area.

Another project could be performed to re-design the pre-screening area for privacy, flow, and quietness.

Understand the patients' requirements and build your processes to meet their needs.

- Implement a surgical patient survey specific for the surgical experience, separate from the inpatient experience.
- Implementing the recommendations in this report will better meet the patients' needs to help them to be better prepared for surgery. Based on the evidence in the literature this will help to reduce patients and their family's anxiety and enhance patient satisfaction.

Table 6.11 Performance targets and scorecard

Performance scorecard		
Metric	Target(%)	Weekly value(%)
% charts ready by 5 pm	100	95 (amber)
% pre-visits complete	90	90 (green)
% anaesthesia seen / received	95	99 (green)
% surgeon's orders received on time	95	80 (red)
% surgeon's docs received on time	95	75 (red)
% medical clearances received on time	95	80 (red)

Design Future State

The future state process maps were developed, and vetted with the process owners, the Director of Surgery and the Chief Nursing Officer.

Establish Performance Targets and Project Scorecard

The team established perform targets for the key metrics, which are shown in the Scorecard Table 6.11. The data are collected on the pre-screening Patient Tracker and the pre-screening Chart Readiness Checklist. Administration can easily see whether the areas are meeting their performance targets.

Pilot Improvements and Train and Execute

We began piloting the activities as soon as the Chief Nursing Officer and The Director of Surgery approved the recommendations. We began the implementation with the execution of the 5S project. This enabled the pre-screening area to clean up, get rid of unnecessary equipment, supplies, and forms, and put their area in order. This was extremely successful, as the Black Belt worked daily with the pre-screening area staff for 2 weeks to clean up the area. She ensured that owners were assigned to re-stock forms and incorporate the "shine" discipline so that daily the pre-screening staff ensured that everything was back in its place. This enhanced the patient's experience when they came to the office for their pre-screening appointments. Longer term, there will be a re-design of the office space to enhance patient comfort and privacy.

The Black Belt then worked with the pre-screening staff to implement the pre-screening Chart Readiness Checklist and the pre-screening Patient Tracker. The pre-screening manager was trained on the checklist and the pre-screening patient tracker and began immediately tracking the desired metrics. The Black Belt trained the pre-screening nurses on how to use the Chart Readiness Checklist. She helped them to re-arrange the work aligning with the priorities identified in the improvement recommendations. She helped them to re-organize the patient assignments to the nurses, so that the nurses could follow the patient throughout their pre-screening process. This ensured better consistency and a better patient experience.

The Lean Six Sigma Master Black Belt developed a desktop instruction manual that enabled the processes to be standardized, and become repeatable, ensuring that everyone performed the work activities in a standard way. The Chart Readiness Checklist was an immense time saver, by enabling anyone to pick up the chart to readily see what pre-screening activities remained for the patient to be ready for surgery, and what orders, results, and documents were needed to complete the chart. This helped in the hand-offs between departments, especially to the holding area on the day of the surgery.

IT helped the pre-screening area get on-line access to patients' lab results.

Requiring almost all of the patients to come to the hospital for their pre-screening visit reduced the number of missing orders, results, and documentation. Anesthesia more consistently saw the inpatients the day prior to the surgery and reviewed high-risk charts and saw those patients as necessary. Tracking these critical metrics helped to change the behaviors and enhance the patient experience.

Control Phase

The purpose of the Control Phase is to implement control mechanisms and procedures to ensure that the process maintains the levels of improvements attained during the pilot activities.

The following activities were executed in the Control Phase:

1. Measure results and manage change
2. Report scorecard data and create process control plan
3. Identify replication opportunities
4. Develop future plans.

Measure Results and Manage Change

The results of this project were highly successful and had great value to the patient experience and to the organization. The graphs for the percentage of patient charts ready by 5 pm and the percent of patients seen and/or charts reviewed by anesthesia are shown in Figs. 6.16 and 6.17. The percent of patient charts complete the day before for the first cases was hovering around 95%, which is an excellent result. The anesthesia percentage was hovering around 92% by the end of the project. Another excellent result and validation that Lean Six Sigma works!

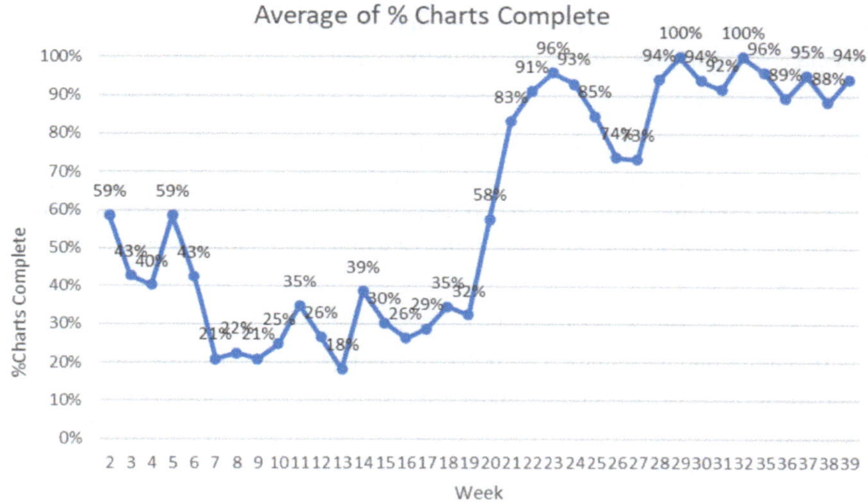

Fig. 6.16 Results of the percent patient charts ready by 5 pm

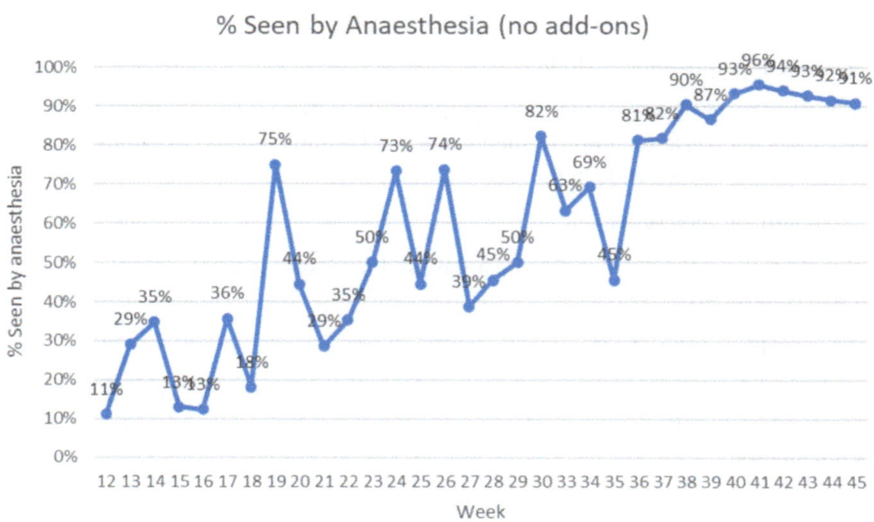

Fig. 6.17 Results of the percent of patients/charts seen or reviewed by anesthesia

Report Scorecard Data and Create Process Control Plan

A control plan provides methods to control the process and ensure that the standard process is adhered to and incorporates metrics to track improvement. The following items are part of the control plan that have been discussed throughout the report.

- Outpatient Preadmission Tracker

This spreadsheet can be used to track compliance by the surgeons, the primary care physicians and anesthesia to the preadmission requirements.

- Surgical Inpatient Tracker

The Surgical Inpatient Tracker can be used to ensure that the inpatient nurses are providing the appropriate surgical readiness information during their reports, and that they are available for the report and transfer of the patient to surgery. It also can be used to track anesthesia compliance in evaluating the inpatients prior to the day of the surgery.

- Preadmission Readiness Checklist

The Preadmission Readiness Checklist is used to track the patient readiness requirements and any outstanding items prior to the day of the surgery.

- Metrics:
  - The Percent of Charts Complete Prior to the day of the surgery spreadsheet and chart can be used to track the improvement in readying charts for the surgery.
  - The Percent of Cases that Anesthesia Evaluates and Reviews can be tracked on the "% Anesthesia Seeing Inpatients and Reviewing Outpatient Charts" spreadsheet and chart.

- Daily OR Improvement Huddle

The Daily OR Improvement Huddle spreadsheet can be used to identify issues and root causes and systemic improvements that can be implemented to improve the on time starts and turnarounds in the OR.

- Standard surgeon order set

A standard surgeon order set form can be used to ensure that the surgeon's orders are received timely.

- Standard fax cover sheet

The standard fax cover sheet can be used by the surgeon's and the primary care physician's offices to provide consistent information with the patient's chart documentation.

- Desktop work instructions with future state process maps

The desktop work instructions manual with the future state process maps can be used to train the preadmission and admitting staff on the new processes.

Identify Replication Opportunities

The Lean Six Sigma problem solving approach can and has been replicated in other areas in the hospital. Many of the improvements can be replicated in other hospitals.

Develop Future Plans

The pre-screening department will continually improve their pre-screening processes.

Project Critical Success Factors

The following key factors contributed to the success of this project:

- Chief Nursing Officer's and Director of Surgery's visible commitment; engagement; and *gentle encouragement* towards goals
- Key team members selected to participate
- Experienced facilitator with technical and change management skills
- Ability to implement new data collection and measurement techniques
- Team monitoring performance on a weekly basis
- Staff willing to pilot improvements
- Surgeons and anesthesia as key process innovators.

Even though the process changes will not be directly replicated on another project for the same hospital, many of the best practices can be shared with other hospitals in the hospital system as well as other hospitals through this book and journal articles.

Summary

This case study provides a thorough analysis of the preoperative screening processes and identifies best practice recommendations to streamline the processes for a more satisfactory patient experience. Approximately 80% of the recommendations were implemented during this project, which is an extremely effective project implementation. The results were stellar and consistently maintained after the project.

References

http://healthtools.aarp.org/galecontent/elective-surgery/2 (2012).

Lew, E., Pavlin, D. J., & Amundsen, J. (2004). Outpatient preanaesthesia evaluation clinics. *Singapore Medical Journal, 45*(11), 509.

Pearson, A., Richardson, M., & Cairns, M. (2004). Best practice in day surgery units: A review of the evidence. *Journal of Ambulatory Surgery, 11*(1–2), 49–54.

Chapter 7
Appropriate Use of CT Scan in the Emergency Department

Abstract This chapter describes a case study in a community-based acute care hospital that was able to successfully reduce the Computed Tomography (CT) utilization in their Emergency Department by 8.7% for 1 year, and 13.8% since the prior 2 year period. The reduction was especially important because 10% of the U.S. population gets a CT scan each year, and use of CT imaging is growing more than 10% per year, and with it, increased long-term exposure can increase potential cancer risks. The cross-functional team applied the Six Sigma methodology and tools to define the problem, measure the current state, identify root causes, and develop medical research-based and evidence-based CT guidelines to reduce CT utilization. The information presented in this case study is not designed or intended to provide medical advice or recommendations.

Keywords Computed tomography · Emergency department · Compute tomography scan · Emergency department physician · Compute tomography exam

Introduction

"Computed Tomography" (CT) scanning, also called computerized axial tomography (CAT) scanning, is a medical imaging procedure that uses X-rays to show cross-sectional images of the body. A CT imaging system produces cross-sectional images or "slices" of areas of the body, like the slices in a loaf of bread. These cross-sectional images are used for a variety of "diagnostic and therapeutic purposes" (www.fda.gov).

"With the exception of patients who were allergic to dye or had renal insufficiency, CT was considered completely safe" (www.archinternmed.com). It is estimated that more than 62 million CT scans per year are currently obtained in the United States, including at least 4 million for children (www.archinternmed.com). Another study estimates the use of CT scans in the United States has increased more than threefold since 1993 to approximately 72 million scans annually (www.crcpd.org). "Each year,

S. Furterer, *Lean Six Sigma for the Healthcare Enterprise*,
https://doi.org/10.1007/978-1-4471-7617-6_7

10% of the U.S. population gets a CT scan, and use of this imaging is growing more than 10% per year" (Mettler et al., 2008a).

There are many diagnostic benefits to CT scan, however, there are risks as well, mainly due to the radiation exposure from the CT scans. The risk to individuals is likely to be small, but because of the larger number of CT scans and number of people exposed, even small risks could translate into a considerable number of future cancers (Mettler et al., 2008a). CT delivers much higher radiation doses than do conventional diagnostic X-rays. For example, a chest CT scan typically delivers more than 100 times the radiation dose of a routine frontal and lateral chest radiograph (de González et al., 2009; Smith-Bindman, 2012).

"Radiation doses from commonly performed diagnostic CT examinations are higher and more variable than generally quoted, highlighting the need for greater standardization across institutions." (Mettler et al., 2008a). The effective radiation dose varied significantly within and across institutions, with "… a mean 13-fold variation between the highest and lowest dose for each study type" (Mettler et al., 2008a).

One article estimated that approximately 29,000 future cancers could be related to the number of CT scans performed in the US in 2007. The risks were higher as the CT scans performed increased with age at exposure up to the age of 45 years, and nearly one-third of the scans (30%) were estimated to be performed in adults aged 35–54 years, 13% in those aged 18–34 years, and 7% in persons younger than 18 years. About 60% of the scans were estimated to be performed in females. The risk per 10,000 scans varied with scan type, with higher risks for chest or abdomen CT angiography and whole-body CT. The risks were higher in females because of additional risk of breast and lung cancer. The scan types of highest risks were the ones also used most frequently, those of the abdomen and pelvis, head and chest angiography. The highest cancer types were lung, colon, and leukemia (Mettler et al., 2008a). Another study found the increased cancer risk in younger females. The authors also found the cancer risks declined with age for males (Mettler et al., 2008b). They estimated one in every 270 40 year-old women undergoing a CT coronary angiogram will develop cancer from the procedure.

In a study by Kirsch et al. the 2006 Emergency Department CT utilization rate was 16.7% for adults. The CT utilization rate ranged from 11.3% for adults aged 20 to 29–24.6% for adults over 65 years. The CT utilization rate varied by patient age, ED volume, training background of the provider, and disposition status of the patient. The number of CT scans ordered or performed in the ED has grown by 200 to 400% over a 5 year period (Kirsch et al., 2010). As the number of CT scans increase the cancer risks are also increasing. There is growing consensus and concern about the risk of CTs and overexposure of radiation due to this growing use of CT scans.

The growing concern related to patients' long-term radiation exposure due to the increased number of CT scans was the impetus for the formation of a Six Sigma project to reduce the number of CT scans ordered and performed in the Emergency Department (ED) at the hospital. The hospital is a community-based hospital serving a diverse demographic of patients. Six Sigma was used to develop guidelines and recommendations to reduce the number of CT scans, and to more appropriately use CT scans for diagnostic testing. There was a statistically significant reduction (8.7%) in the number of CT scans from the baseline time period compared to after improvements were implemented. There was a 13.8% reduction in CT scans from one year prior to the baseline to one year after the baseline. This chapter will describe this very successful project.

Define Phase

In the Define phase the team was formed, and appropriate stakeholders were identified. The following steps were executed in the Define Phase:

1. Develop project charter.
2. Perform stakeholder analysis.
3. Identify Critical to Satisfaction (CTS).
4. Select team and launch the project.
5. Create project plan.

The project charter is shown in Table 7.1.

Develop Project Charter

Problem Statement

The radiologists, ED physicians and the radiology department, following the current literature that identified the increasing number of CT scans in the patient population, especially impacted by visits to the Emergency Department, decided to focus on an effort to reduce the use of CT scans in the Emergency Department at the hospital. About 60% of the admissions to the hospital come through the ED. About 31% of the patients received a CT scan during their visit during the time period of January through December. The highest frequency of CT scans for diagnostic purposes in the ED are (1) spine; (2) abdomen/pelvis; (3) chest; (4) renal; (5) orbital/facial. There could also be a problem with ordering first CTs without contrast and then having to order another CT with contrast. Additionally, patients are not currently consistently screened for their histories of prior CT scans.

Customer/Stakeholders. The customer is the ultimate ED patient. The other project stakeholders are ED physicians, Radiologists, Radiology technicians, ED Nurses (including the triage nurses), the Nurse Educator, Lab, and Radiology Department.

What is Important to These Customers—Critical to Satisfaction (CTS).

Number of CT tests; frequency of CT tests; appropriate use of CT; Decrease over-utilization of tests; appropriate diagnosis, decrease in the radiation that patients receive from CT scans.

Table 7.1 Project charter

Project Charter
Project Overview, Problem Statement, & Goals
Provides a description of the problem to be solved. Be as descriptive as possible. Include improvement goals that are SMART (Specific, Measurable, Attainable, Realistic, and Time-Based).
Project Name: Appropriate use of CT scan in the Emergency Department. Problem Statement: The problem involves the appropriate use of CT for diagnostic purposes in the ED including spine, abdomen / pelvis, chest, renal and facial / orbits. This can be a safety issue involving overexposure or repeated exposure to radiation. The problem also involves the use of non-contrast CT versus contrast CT. Using non-contrast might result in the need for another CT with contrast. Patients are not screened for histories of previous CT. Goal(s): To reduce radiation exposure, to reduce overall number of tests, to utilize the tests appropriately; decrease in the radiation that patients receive from CT scans. Scope Statement: To focus on the ED patients who receive CT scans of the spine, abdomen/pelvis, chest, facial/obits. We are assuming a correlation between the CT scans and level of radiation (CT scans give highest level of radiation of scans given in ED). Not focusing during this project on long term radiation exposure, other than that the number and frequency of tests contributes to long term exposure. Projected Financial Benefits: Reduction of unnecessary radiation due to overuse of CT scans.

Goals of the Project. The goal of the project is to reduce the radiation exposure from CT scans, to reduce the overall number of CT scans, and to utilize the CT scans appropriately based on the symptoms presented.

Scope Statement. To focus on the ED patients who receive CT scans of the spine, abdomen/pelvis, chest, renal and facial/orbits. We are assuming a correlation between the CT scans and the level of radiation (CT scans give the highest level of radiation of scans given in the ED). We are not focusing during this project on long-term radiation exposure, other than that the number and frequency of tests contributes to long-term exposure.

Project Milestones. The project started in September and was completed in December of the following year.

Perform Stakeholder Analysis

The team performed the stakeholder analysis by defining the stakeholders impacted by the project, and what their concerns are, and how they are impacted by the project. The ED physicians assess the patient and order labs and diagnostic exams. They are concerned about the timely turnaround on diagnostic tests; the number of tests ordered for a patient, especially those exposing the patient to radiation; the frequency of the CT tests that a patient has encountered, increasing the cumulative radiation effects; that the CT exams are ordered appropriately for their symptoms; and that the patient is provided the appropriate diagnosis based on the results of the tests.

The radiologists read the radiology exams, including the CT scans, and provide the results of the exams. They are concerned about timely turnaround on the diagnostic tests; appropriate use of CT scans; appropriate diagnosis; radiation safety of patients; and number of CT scans ordered and performed on patients.

The next stakeholder group consists of the radiology technicians who perform the exams and the lab and radiology department staff who provide lab and radiology exams. They are concerned with timely turnaround on diagnostic tests; appropriate diagnosis and appropriate test delivery providing the appropriate level of radiation to the patient.

The patients are the critical customer stakeholders who receive diagnostic tests, results, and ED physician diagnosis and disposition. They are concerned with radiation exposure and appropriate diagnosis, as well as timely turnaround.

The stakeholders are shown in Table 7.2.

Suppliers-Inputs-Processes-Outputs-Customers

The SIPOC is used to ensure the scope of the project is well understood. The SIPOC is shown in Table 7.3. The processes in the scope of this project are the ED physicians ordering the diagnostic tests, including CT scans, performing the diagnostic tests, and the radiologists providing the test results to the ED physicians. The patients present themselves to the Emergency Department and describe their symptoms. The ED physician assesses the patient, and orders lab and diagnostic exams. The orders are entered into the ED Bed board system and into the medical record and are then performed. Some lab test results are required before being able to perform certain radiology exams, including the CTs, for example pregnancy tests in women of childbearing age. The radiologists will read the exams and provide the results via the PACS radiology system, which the ED physicians then review to determine the patient disposition, to discharge or admit the patient to the hospital.

Table 7.2 Stakeholder analysis

Stakeholder analysis					
Stakeholder	Type	Primary role	Impacts/concerns	Initial receptivity	Future receptivity
ED physicians	Primary	Physicians who order tests and provide diagnoses in the ED	Timely turnaround on diagnostic tests Number of tests Frequency of tests Appropriate use of CT Appropriate diagnosis	Strongly support	Strongly support
Radiologist	Primary	Radiologists who provide radiology services/ diagnosis	Timely turnaround on diagnostic tests Appropriate use of CT Appropriate diagnosis Radiation safety Number of CT scans	Moderate support	Strongly support
Radiology technicians Lab and radiology staff	Primary	Staff who provide radiology and lab testing services	Timely turnaround on diagnostic tests Appropriate diagnosis Appropriate test delivery	Moderate support	Strongly support
Patients	Primary	Patients who receive diagnostic tests	Radiation exposure Appropriate diagnosis Timely turnaround	Neutral	Neutral

Table 7.3 Suppliers-inputs-process-outputs-customers (SIPOC)

SIPOC				
Suppliers	Inputs	Process	Outputs	Customers
Patient Medical staff	Symptoms Patient info	Order diagnostic tests	Order tests	Medical staff Patients
Patient Medical staff	Patient information Orders for tests	Perform diagnostic tests	Test info	Patient Medical staff
Medical staff	Test info	Provide test results	Test results	Patient ED physicians

Identify Critical to Satisfaction

There are several characteristics that are critical to the satisfaction of the patients and the project stakeholders, as follows.

- Number of CT tests.
- Frequency of CT tests.
- Appropriate use of CT.
- Decrease over-utilization of tests.
- Appropriate diagnosis.
- Decrease the radiation to patients.

The goal of the project is to reduce the number of CT tests and how frequently the patients have CT tests over time. It is important that the physicians order the tests appropriate to what is supported in the medical research literature and is evidence-based. There is also a focus to decrease over-utilization of tests, and providing the appropriate diagnosis, even as the CT tests are reduced. Reducing the number and frequency of the CT tests should also decrease the amount of radiation that the patients are exposed to.

Select Team and Launch the Project

The team members consisted of the internal stakeholders, including ED physicians, a radiologist, the radiology department director, the ED nurse educator, the Continuing Medical Education (CME) coordinators, and a Six Sigma Master Black Belt (the case's author). We launched the project on October 1st.

Create Project Plan

We developed the project milestones for each phase of the DMAIC problem solving methodology, shown in Fig. 7.1.

Once the project charter was complete, and the stakeholders were defined the team moved into the Measure Phase.

Measure Phase

The Measure Phase included the following activities

1. Define the current process.
2. Define detailed Voice of Customer (VOC).
3. Define the Voice of Process (VOP) and current performance.

Project Plan - Milestones

	Estimated Dates	Activities
D	9/14 – 10/1	Define project charter and scope, stakeholders and initial Critical to Satisfaction (CTS).
M	10/1 - 11/1	Measure and document current As-Is process; baseline throughput time; collect data on current process
A	11/1 - 1/7	Analyze problems and their root causes; identify waste and non-value-added activities.
I	1/7 - 6/17	Identify improvement ideas; pilot improvements; assess improvement
C	6/1 – 12/31	Identify and implement control mechanisms; create procedures and training for new improvements

Fig. 7.1 Project milestones

Define the Current Process

The first step was to map the current process by creating detailed process maps of the current diagnostic processes in the Emergency Department. The process maps are shown in Figs. 7.2 and 7.3. The patient arrives to the Emergency Department and signs in at the kiosk. If an ED room is available, the patient is placed in the ED room and is triaged by the nurse. Prior to improving the ED throughput, since the time to see a doctor was quite lengthy, averaging 93 min, the triage nurse assessed and triaged the patient. The ED physician will see and assess the patient, and then order diagnostic tests, including labs and radiology exams. The orders are requested by the ED physician by entering them into the ED Bed Board system. The unit associate then enters the orders into the medical information system. If labs are needed, the lab specimens are drawn by the nurse and/or phlebotomist. The lab tests are run and results obtained. If oral contrast is needed for the CT scan, the nurse or radiology technician gives the oral contrast to the patient. If oral contrast was given to the patient, the patient must wait for 60 min prior to the CT exam. The radiology technician retrieves the patient and moves them to the CT scan and performs the exam. The radiologist reads the exam and provides the results to the ED physician. The ED physician then diagnoses the patient and provides the disposition to either discharge or admit the patient to the hospital.

The team developed the data collection plan to determine the voice of the process and the current baseline. The data collection plan is shown in Table 7.4.

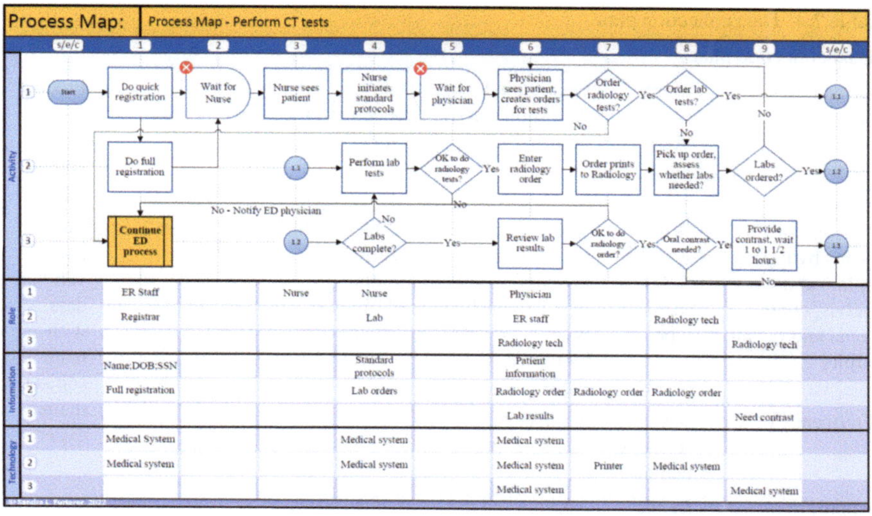

Fig. 7.2 Process map, perform CT tests page 1

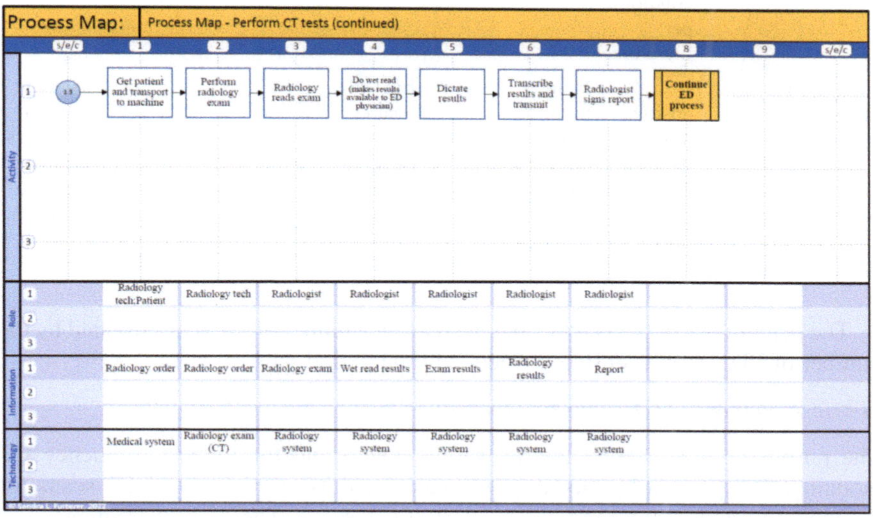

Fig. 7.3 Process map, perform CT tests page 2

Operational Definitions of the Metrics:

The operational definitions were developed for each metric.

Metric 1: Percent of CT Scans

The purpose of the first metric was to measure the number of CT scans.

Table 7.4 Data collection plan

Data collection plan						
Critical to satisfaction (CTS)	Metric (short title)	Operational definition (metric description)	Data collection source	Analysis mechanism	Sampling plan	Process to collect and report
Number of CT scans by type: spine, abdomen/pelvis, chest, renal and facial/orbits	% CT scans % CT scans by type	% CT scans: # CT scans/# ED patients % CT scans by type: # CT scans by type/# CT scans	ED bed board system; radiology system	Statistics	Jan–Dec	Verify and validate data with team and subject matter experts
Frequency of tests	% CT scans per patient	Number of CT scans per patient				
Appropriate use of CT (over-utilizations)	Better exam available for condition per literature	Appropriate use per medical research	Medical research			
Appropriate diagnosis	# positive tests per patients across time frame	Number positive tests for pulmonary embolism per patients across timeframe	ED bed board system; radiology system			Number negative d-dimers Number positive d-dimers

The percent of CT scans was calculated as the number of CT scans divided by the number of ED patients, during the defined time period.

Metric 2: Percent of CT scans by Modality

The purpose of the second metric was to understand the highest frequency CT modality. The percent by modality was calculated as the number of CT scans for the modality divided by the total number of CT scans. The highest frequency CT exams included those of the spine, abdomen and pelvis, chest, renal and facial/orbits.

Metric 3: Appropriate use of CT scans

The third metric was to assess the appropriate use of CT scans for the types that we were assessing. This was assessed through a review of the medical research.

Metric 4: Appropriate Diagnosis

The last metric was to determine if the appropriate diagnosis was attained for those patients who used the CT scan and were at risk of having a pulmonary embolism. This metric was defined as the number of positive tests per patient diagnosed with pulmonary embolism.

We used a baseline period of January through December of the first year, and also January through April of the same year for the analyses, per the data available.

Define Detailed Voice of Customer

For the voice of the customer, we reviewed the ED Bed Board data that identified the number and types of CT scans across the patient population for the January through December time frame. The Pareto Chart by Modality is shown in Fig. 7.4.

The highest CT modality is CT of the brain without contrast, consisting of 32% of the CTs performed. CT of the abdomen and CT pelvis without contrast were each 16% of the CT exams. CT pelvis and abdomen with contrast consisted of 7.1% each of the exams. The CT Spine without contrast was 5.9%.

Thirty-three percent (33%) of the patients visiting the ED during the baseline year had at least one CT exam, shown in the bar chart in Fig. 7.5. For the patients who

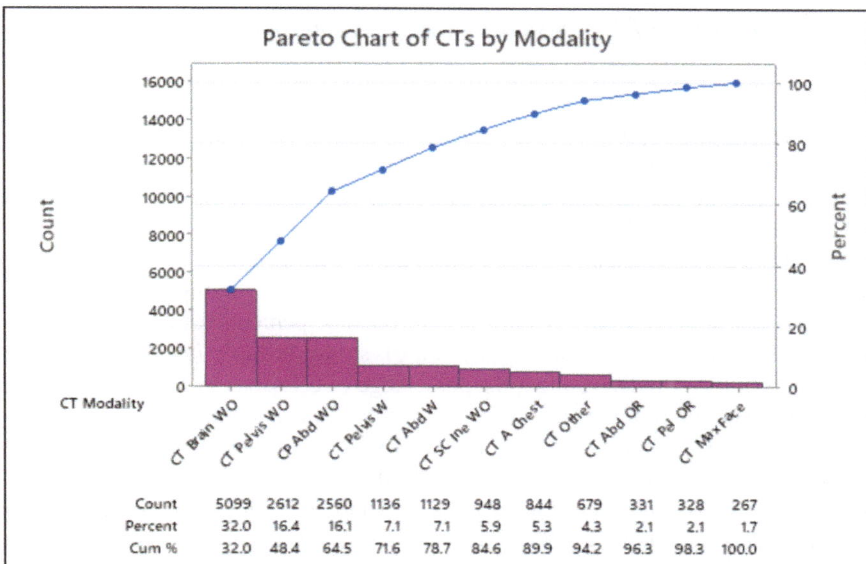

Fig. 7.4 Frequency of CT scans by modality

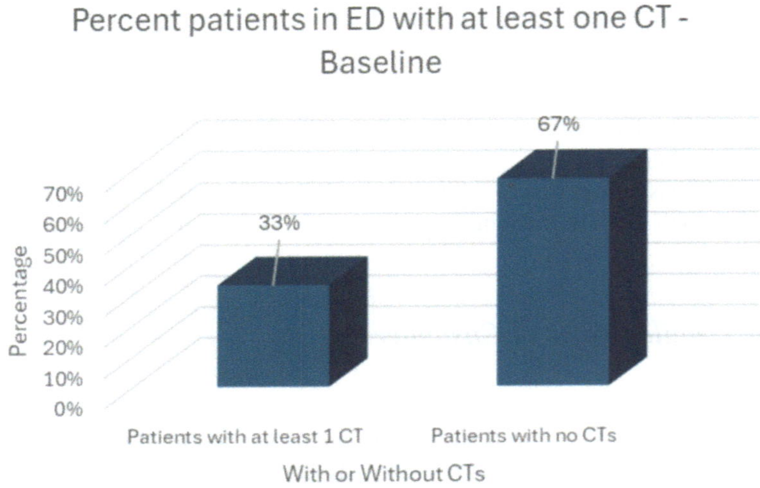

Fig. 7.5 Bar chart of CT scans

were admitted to the hospital after visiting the ED, 47% of the admitted patients had at least one CT exam.

Define the Voice of Process and Current Performance

The physicians searched the literature for each CT scan type (spine, abdomen/pelvis, chest, face/orbits) for guidelines for appropriate use of CT scans.

The average number of CT scans ordered by the ED physicians varied from 10 to 92% across the ED physician group, showing quite a bit of variability in how the ED physicians order CT scans, as shown in Table 7.5.

We also assessed whether the CT utilization was different by age and found that CT scans were a higher percentage compared to the overall population for the older patients in age ranges 50–65 and over 65. The percentages of CT scans and percentages of the ED population by age range is shown in Table 7.6. We performed a Chi square analysis to determine if the percentage of CT scans by age range differed from the ED population percentage by age range. We found that the percentages of CT scans by the age ranges for <18, 18–29, 30–39, and 40–49 were lower than the ED population percentages by age range. We also found that the percentages of CT scans for the 50–65 and over 65 were higher than the ED population percentages by age range. The p value was 0.000. The observed versus expected values are shown in Fig. 7.6.

The Chi square output follows:

Table 7.5 CT ordering patterns across physicians

ED physician	Percent of patients with at least 1 CT (%)
EDP Y	92
EDP E	81
EDP Q	72
EDP D	63
EDP B	58
EDP H	58
EDP V	56
EDP R	55
EDP A	54
EDP C	54
EDP G	52
EDP J	50
EDP M	50
EDP T	48
EDP I	46
EDP F	34
EDP W	31
EDP K	30
EDP O	22
EDP N	21
EDP U	21
EDP L	19
EDP X	18
EDP S	10

Table 7.6 % CT scans by age range compared to ED population

Age range	Count of patients with at least 1 CT scan	% patients with CT scans by age range (%)	% in age range of ED population (%)
<18	131	1	4
19–29	856	8	14
30–39	912	9	13
40–49	1,225	12	13
50–65	1,751	17	16
>65	5,202	52	40
Grand total	10,077		

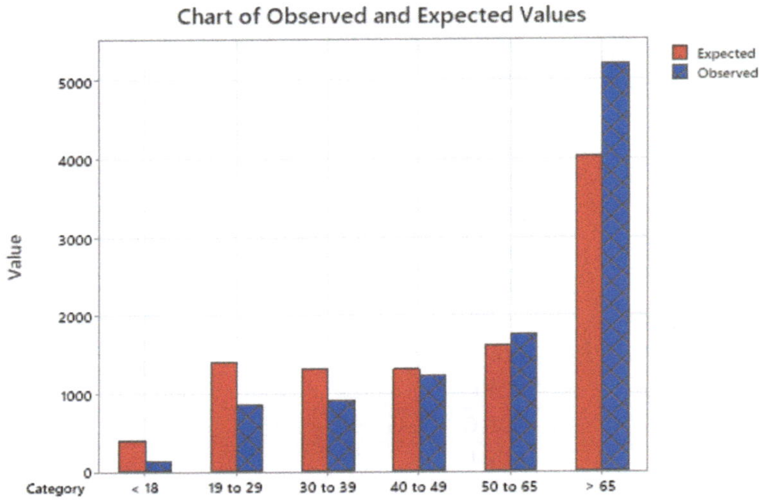

Fig. 7.6 Chi square analysis for % CT scans and ED population by age range

Observed and Expected Counts

Category	Observed	Historical counts	Test proportion	Expected	Contribution to chi-square
<18	131	403	0.039992	403	183.583
19 to 29	856	1411	0.140022	1411	218.303
30 to 39	912	1310	0.129999	1310	120.919
40 to 49	1225	1310	0.129999	1310	5.515
50 to 65	1751	1612	0.159968	1612	11.986
>65	5202	4031	0.400020	4031	340.174

Chi-Square Test

N	DF	Chi-Sq	P-Value
10,077	5	880.480	0.000

A D-Dimer lab test screens for a potential pulmonary embolism. If a positive result for a possible pulmonary embolism is received, the ED physician assesses whether a CT scan should be performed. For those patients where a D-Dimer test was given, along with a CT scan, 27% of the diagnoses were a Pulmonary Embolism.

We then moved into the Analyze Phase.

Analyze Phase

The following activities were performed for the Analyze Phase to identify and validate the root causes of appropriate use of CT scan.

1. Develop cause and effect relationships.
2. Determine and validate root causes.

Develop Cause and Effect Relationships and Determine and Validate Root Causes

The root causes were quite straightforward for the potential overuse of CT scans, and several were supported in the literature as well. The following root causes were identified as contributing to the increasing and sometimes high usage of CT scans by particular ED physicians, as follows:

- There were no standard guidelines for when to order CT scans across the ED physicians. The ED physicians used their experience, knowledge, and judgment to determine when to order CT scans. All three of these elements could widely vary. Table 7.5 shows the wide variability in the ordering patterns across the ED physicians. The chart shows the percent of patients who had CT scans for each ED physician. This ranged from 10 to 92% across the ED Physicians.
- There was no measurement of the amount of radiation exposure due to CT scans. Prior to this project, the radiology department did not measure the amount of radiation exposure to the patient from the CT scans.
- There was little focus on the appropriate use of CT scans in the Emergency Department either with the nurses, or the ED physicians.
- Patients may complain if there is no diagnostic test. "A popular current paradigm for health care presumes that more information, more testing, and more technology inevitably leads to better care" (www.archinternmed.com). "Patients who pressure doctors for tests they may not need are part of the problem, writes Dr. Bruce J. Hillman of the University of Virginia, and Jeff Goldsmith, president of Health Futures Inc., a Charlottesville, Va.-based health information policy and analysis company" (www.archinternmed.com).
- The type of patient insurance could have an impact on prescribing CT scans. We looked at the type of insurance that the patients had, or if they were a self-pay. For the most frequent insurance types, within the ED population, we analyzed the percent of patients compared to the percent of patients with at least one CT scan performed. For the Medicare patients, they made up 37% of the patients, yet they had a higher percent (43%) of patients with at least one CT scan. This was also true for patients with private insurance. They made up 28% of the patients, yet they had 31% of the patients with CT scans. The self-pay patients had a lower percent of patients with CT scans (17%) compared to them making up 24% of the patient population. The HMO patients also had a lower frequency of CT scans

Table 7.7 Insurance breakdown by patients and number of CT scans

Insurance type	Percent of patients (%)	Percent patients with CTs (%)	Statistically significant difference?
HMO	11	9	Yes (p-value = 0.001)
Medicare	37	43	Yes (p-value = 0.000)
Private	28	31	Yes (p-value = 0.001)
Self-pay	24	17	Yes (p-value = 0.000)

(9%) compared to the percent of patients with HMO insurance (11%). The table summarizing this data is shown in Table 7.7.

- ED physicians have a high potential for getting sued for malpractice. Miami-Dade County in south Florida has one of the highest malpractice rates in the country. The latest available data from the Medical Liability Monitor (2008, survey of malpractice rates), reports Dade County, Fla., as having the highest rates in the country with OB/Gyns and general surgeons paying as much as $238,728 for $1–$3 million coverage. Internists in Dade County also pay the most, as much as $54,710 (www.amednews.com, 2009).

We then moved into the Improve phase to develop the guidelines, based upon the current medical literature.

Improve Phase

The following steps were executed in the Improve Phase:

1. Identify improvement opportunities and plans.
2. Design future state.
3. Pilot improvements.
4. Train and execute.

Identify Improvement Opportunities and Plans and Design Future State

The medical literature suggested three ways to reduce the overall radiation dose from CT (Marchione, 2010).

(1) Reduce the CT-related dose in individual patients. Using automatic exposure-control options on the latest generation of scanners can help.
(2) Replace CT use with other types of technology, such as ultrasound (US) and magnetic resonance imaging (MRI), which do not expose the patient to radiation. Depending upon the patient's symptoms, US and MRI can replace some of the

CT scans, but not necessarily all of them. Additionally, depending upon the patient's Body Mass Index (BMI) and percent of body fat, CT can be a better diagnostic choice.

(3) Decrease the number of CT exams ordered.

CT scanning has become an integral diagnostic modality in the management of ED patients.

Table 7.8 shows the radiation dose for different CT procedures, along with the equivalent number of X-rays for the equivalent effective dose, and the time period for the equivalent effective dose from the natural background radiation. A CT examination with an effective dose of 10 millisieverts (abbreviated mSv; 1 mSv = 1 mGy in the case of X-rays) may be associated with an increase in the possibility of fatal cancer of approximately one chance in 2000. This increase in the possibility of a fatal cancer from radiation can be compared to the natural incidence of fatal cancer in the U.S. population, about one chance in five. Cancer risks and radiation exposure from computed tomographic scans (2009).

The team developed guidelines for prescribing diagnostic tests for the highest frequency types of CT scans ordered for the patient symptoms presented.

Guidelines to Order a CHEST CT Angiogram (Institute for Clinical Systems Improvement)

The chest CT angiograms are in the majority of cases ordered when a D Dimer is elevated, and there is a potential pulmonary embolism. Therefore, it seems reasonable to follow Well's Criteria to order a D Dimer when predicting a pulmonary embolism (PE).

The team suggested that the following guidelines be observed when considering ordering a D-Dimer for suspected PE. Table 7.9 shows the Wells Clinical Model for predicting the pretest probability of pulmonary embolism.

Table 7.8 Radiation dose by CT modality

Diagnostic procedure	Typical effective dose (mSv)	Number chest X-rays (PA film) for equivalent effective dose	Time period for equivalent effective dose from natural background radiation
Chest X-ray (PA film)	0.02	1	2.4 days
Skull X-ray	0.1	5	12 days
Lumbar spine	1.5	75	182 days
I.V. urogram	3	150	1.0 year
Upper G.I. exam	6	300	2.0 years
Barium enema	8	400	2.7 years
CT head	2	100	243 days
CT abdomen	8	400	2.7 years

Table 7.9 Wells clinical model for predicting the pretest probability of pulmonary embolism

Variable	Points assigned
Clinical signs and symptoms of deep venous thrombosis	3.0
An alternative diagnosis is less likely than pulmonary embolism	3.0
Heart rate > 100 beats/min	1.5
Immobilization or surgery in the previous 4 weeks	1.5
Previous deep venous thrombosis or pulmonary embolism	1.5
Hemoptysis	1.0
Malignancy (on treatment, treated in the last 6 months, or palliative)	1.0
Score	*Clinical assessment probability*
<2 points	Low probability
2–6 points	Intermediate probability
>6 points	High probability

Keeping in mind that a VQ scan delivers 10 times less radiation than a CT scan of the chest, consider obtaining VQ Scans on the patients with a low pretest probability, a negative Chest X-ray, and a positive D Dimer (i.e., a typical chest pain in a young female with SOB).

Guidelines for Abdominal CT Scans and Abdominal Imaging (BMJ, 2009; **Stoker et al.,** 2009)

The literature supports plain CT scans of the abdomen for patients presented to the ED with abdominal pain. However, in the following are the situations when the contrast will help the diagnostic process and yield better information to the radiologist and treating physician.

1. BMI less than 25 (low fat patients) use PO contrast.
2. Suspected Pancreatitis: If doing CT, do it with IV contrast to help stage.
3. For All PO contrast, wait 60 min. For distal colon 90 min may be better.
4. Aortic dissection, need IV contrast.
5. Consideration of Diverticulitis: PO contrast will yield more diagnostic information and potentially help in long-term management (abscesses, perforation, etc.).
6. Abdominal Trauma—give IV contrast.
7. Known malignancies: PO and IV contrast.
8. Pelvic US is useful in young females with lower abdominal pain as the first line study.
9. Pregnant Patients with suspected Appendicitis: Consider MRI.

Guidelines for Cervical Spine Imaging

NEXUS Low-Risk Criteria

- No tenderness at the posterior midline of the cervical spine
- No focal neurologic deficit
- Normal level of alertness
- No evidence of intoxication
- No distracting injury.

Canadian C-Spine Rules

- Is there any high-risk factor that mandates radiography (age > 65, dangerous mechanism, paresthesias in extremities)?
- Is there any low-risk factor present that allows safe assessment of range of motion (simple rear-end MVC, delayed onset of pain, absence of midline tenderness)?
- Is the patient able to actively rotate neck 45° to the left and right?

When Should We CT the C-Spine Instead of Using Plain Radiograph?

High Energy Mechanism

- Hi-speed MVC.
- MVC with death on scene.
- Fall from height >10 ft.

High Risk Clinical Parameter

- Neurologic signs suggestive of c-spine injury.
- Significant head injury.
- Age > 50 with moderate-energy mechanism.

Guidelines for Imaging Patients Presenting to the ED with Suspected Renal Colic (Baumgarten et al., 2007; Katz et al., 2006)

Purpose: to limit the amount of radiation the patient is receiving

Consider US and KUB as the first line of testing

- Patients with a history of uncomplicated kidney stones, less than 50 years of age.

CT scan

- First time presentation with suspected renal colic.
- History of renal stone presenting with suspected renal colic over 50 years of age (concern is AAA).
- Patients with a history of complicated kidney stones (lithotripsy, stents, large stones).
- If an operative procedure is contemplated.
- For pregnant patients start with renal and bladder US.

An important part of this equation is patient education concerning potential cumulative effects of radiation long term.

In the younger population check PACS for old prior studies.

Orbital CT/Facial Bones Sinus Imaging Recommendations

CT should be performed for a patient with orbital trauma with any of the following findings:

- Evidence of a fracture on physical examination.
- Limitation of extra ocular motility.
- Decreased visual acuity.
- Severe pain.
- Inadequate examination (usually because of soft tissue swelling), particularly in patients with altered mental status.
- Sinusitis: *Always* do plain sinus films first.
  - Clouding of the maxillary sinus indicates the possibility of blood or orbital fat in the sinus as the result of an orbital floor fracture.
  - Plain sinus films and a Waters view constitute a good screening tool with a minor injury.

The sensitivity of plain radiographs for detecting orbital fractures is poor (about 50%). In addition, plain radiographs are often difficult to interpret and are inadequate to determine whether or not a patient requires operative intervention.

They may be useful as a screening tool when CT is unavailable or for children with a minor mechanism of injury who would require sedation for CT.

Guidelines for LS Spine Imaging

The initial Study for Low back pain and trauma should be Plain Films.

LS Spine CT should be limited to fractures. Any other imaging of the LS spine (Pain Tumors Discs) should be done via MRI.

The guidelines were reviewed with other physicians in the hospital to gain their buy-in and input.

Pilot Improvements

The ED physicians began piloting the new guidelines in January after the baseline year. They held a Grand Rounds for the hospital in March to share the guidelines. There was some reduction in the percent of CT scans in April to 41% from the average percent of CT scan in the baseline of 48.3%, shown in Fig. 7.7. The percentage of CT scans went up through September of the following year. The team held a more extensive review of the CT scan guidelines in another Grand Rounds session in mid-October. The Director and Co-Directors again worked with the ED physicians to encourage the use of the CT scan guidelines. The percentage of CT scans was again reduced to an average of 39.6% for October through December of the following year, shown in Table 7.10. This was an 8.7% reduction in the percentage of CT scans ordered in the ED. We performed a two-proportion test, which was statistically significant with a p-value of 0.000.

There was a 13.8% reduction in the percent of CT scans since the year prior to the baseline year to October through December of the improvement year, with a statistical significance and a p-value of 0.000.

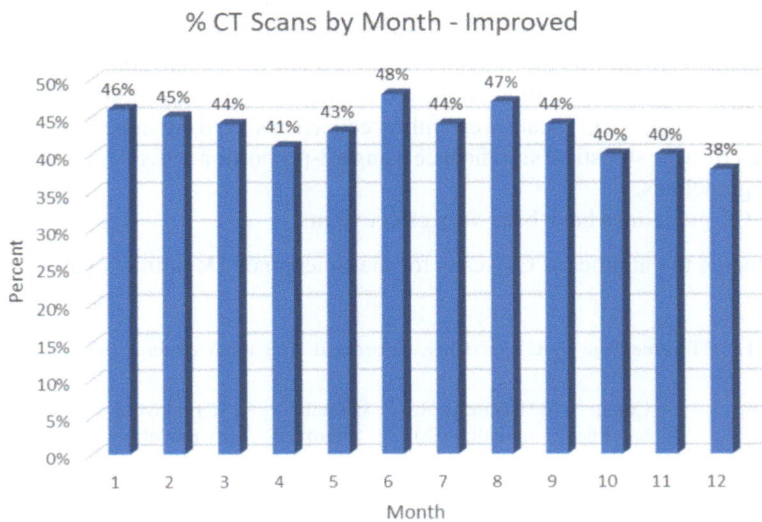

Fig. 7.7 Percent of CT by month with improvements

Table 7.10 Percent of CT scans in ED patient population, for January through June of the baseline year, compared to January through June of the improvement year

Variable	Baseline—1 year	Baseline year	Improved year (Jan–Sep)	Improved year (Oct–Dec)
Number CT scans	17,427	15,933	12,085	3,701
Total number patients	32,637	32,963	26,817	9,346
% CT scans of ED patients (%)	53.4	48.3	45.1	39.6
Count of patients with CT scans	10,794	10,079	7,674	2,352
% patients with CT scans (%)	33	31	29	25
Reduction in % CT scans (%)	5.1	3.3	8.7	13.8

The percentages by patient for the number of CT scans also was reduced from the baseline, as shown in Table 7.11. Six percent (6%) of patients had three to a maximum of seven different CT modalities in the time period of the baseline year. For the improvement year, only 5% of patients had 3–7 different CT modalities in this time period.

For the highest frequency of CT modalities, all percentages of CT exams per count of patients decreased or remained the same, except for the CT abdomen with contrast and CT pelvis with contrast increased from 7 to 8%; the CT abdomen and pelvis with and without contrast increased as well. This data is shown in Table 7.11.

The Pareto charts before and after improvements by CT modality type are shown in Figs. 7.8 and 7.9. The highest modality by occurrence before and after improvements is the CT Brain without contrast, followed by the CT abdomen without contrast.

The percentages of CT scans were either reduced or stayed the same for each physician, based on the statistical significance using a 2-proportion test, after implementing the CT guidelines.

The CT guidelines have been very effective in:

- Reducing the number of CT scans for the Emergency Department patients.

Table 7.11 CT percentage by CT modality, before and after improvements (highest frequency CTs)

Modality	Count baseline	% Baseline by modality (%)	Count improved	% Improved by modality (%)	P-Value
CT brain WO	5,099	32	4,792	30.4	Reduced, 0.002
CT pelvis WO	2,612	16.4	2,558	16.2	Same, 0.68
CT Abd WO	2,560	16.1	2,508	15.9	Same, 0.694
CT Abd W	1,129	7.1	1,325	8.4	Increased, 0.000
CT pelvis W	1,136	7.1	1,318	8.3	Increased, 0.000
CT spine WO	948	5.9	821	5.2	Reduced, 0.004
CT A Chest	844	5.3	815	5.2	Same, 0.607
CT Abd OR	331	2.1	392	2.5	Increased, 0.015
CT Pel OR	328	2.1	389	2.5	Increased, 0.015
CT max face	267	1.7	260	1.6	Same, 0.494
CT chest two	158	1.0	179	1.1	Same, 0.228
Other	540	3.4	429	3	Reduced, 0.001
Total	15,951		15,786		

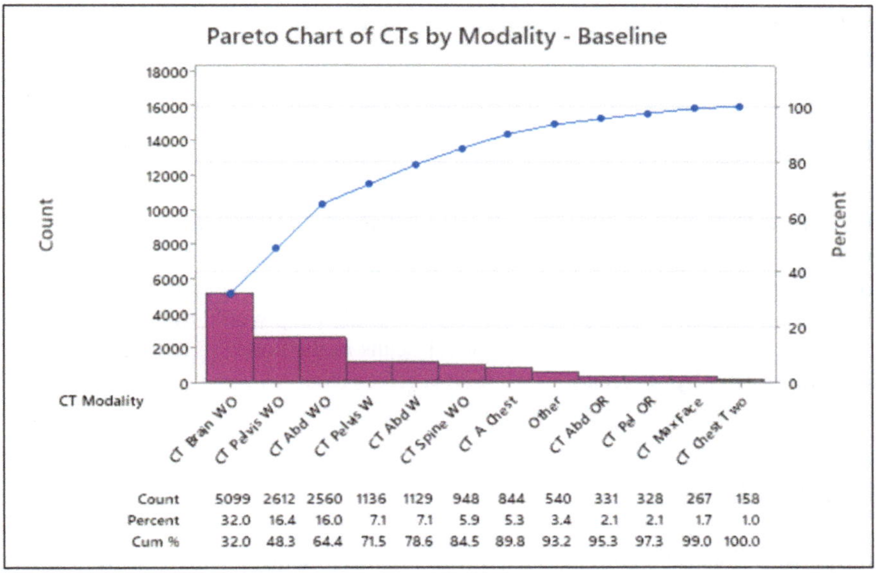

Fig. 7.8 Pareto chart by modality before improvements

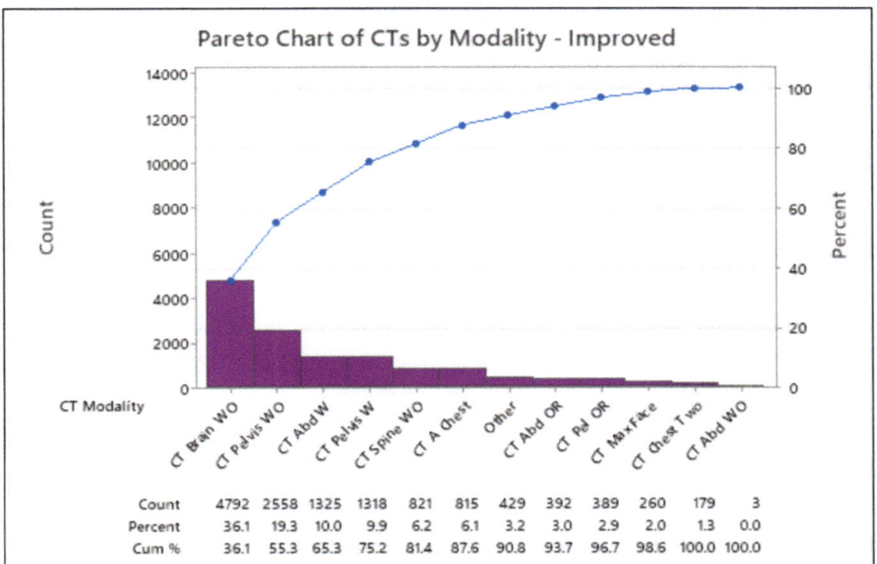

Fig. 7.9 Pareto chart by modality after improvements

- Reducing their radiation exposure due to the CT scans.
- Using CT scans more appropriately.

Train and Execute

The Nurse Educator developed training to share the CT guidelines with the ED nurses. The Radiologist created a memo to ensure that the radiology technologists view the patient history prior to performing the ordered CT scan to assess whether the patient has had recent CT scans in the hospital. They would then notify the ED physician if there was a history of prior CT scans in the hospital, to determine if the CT should still be performed.

Control Phase

The purpose of the Control Phase is to implement control mechanisms and procedures to ensure that the process maintains the levels of improvements attained during the pilot activities.

The following activities were executed in the Control Phase:

1. Measure results and manage change
2. Report scorecard data and create process control plan
3. Identify replication opportunities
4. Develop future plans.

Measure Results and Manage Change

We measured the results and continue to manage the implementation of the CT guidelines.

Report Scorecard Data and Create Process Control Plan

The control plan is to continue to monitor the following metrics:

(1) Percent of CTs across the Emergency Department patients.
(2) Frequency of CTs by modality.
(3) Average CTs per patient.
(4) Percent of CTs by physician.
(5) Radiation exposure to patients. This metric had not yet been collected, but there are plans to capture the radiation received during the CT exams.

Variable	Baseline minus 1 year	Baseline Year	Improved Year (Oct - Dec)
Number CT scans	17,427	15,933	3,701
Total number patients	32,637	32,963	9,346
% CT scans	53%	48%	40%
Count of patients with CT	10,794	10,079	2,352
% Patients with CT scans	33%	31%	25%
	Baseline -1 to Baseline Year	Baseline to Improved	Basline minus 1 year to Improved
Reduction in % CT scans	5%	9%	14%

Modality	Count Baseline	% Baseline by Modality	Count improved	% Improved by modality
CT Brain WO	5,099	32%	4,792	30.4%
CT Pelvis WO	2,612	16.4%	2,558	16.2%
CT Abd WO	2,560	16.1%	2,508	15.9%
CT Abd W	1,129	7.1%	1,325	8.4%
CT Pelvis W	1,136	7.1%	1,318	8.3%
CT Spine WO	948	5.9%	821	5.2%
CT A Chest	844	5.3%	815	5.2%
CT Abd OR	331	2.1%	392	2.5%
CT Pel OR	328	2.1%	389	2.5%
CT Max Face	267	1.7%	260	1.6%
CT Chest Two	158	1.0%	179	1.1%
Other	540	3.4%	429	3%
Total	15,951		15,786	

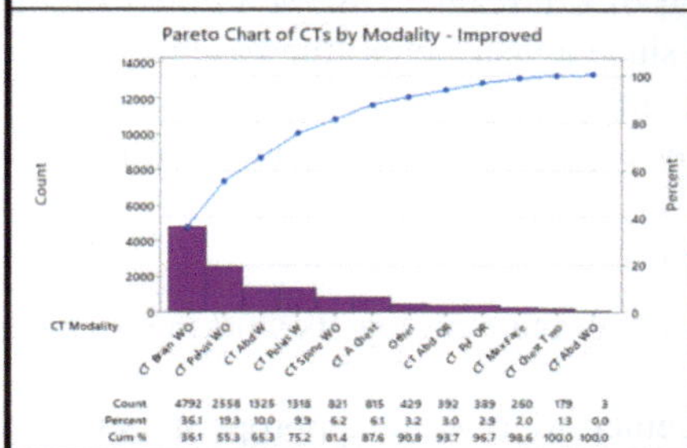

ED Physician	Percent of Patients with at least 1 CT
EDP Y	92%
EDP E	81%
EDP Q	72%
EDP D	63%
EDP B	58%
EDP H	58%
EDP V	56%
EDP R	55%
EDP A	54%
EDP C	54%
EDP G	52%
EDP J	50%
EDP M	50%
EDP T	48%
EDP I	46%
EDP F	34%
EDP W	31%
EDP K	30%
EDP O	22%
EDP N	21%
EDP U	21%
EDP L	19%
EDP X	18%
EDP S	10%

Fig. 7.10 Dashboard

An example dashboard is shown in Fig. 7.10. It includes the percentage of CTs comparing the baseline year, one year prior to the baseline year, and the improvement time period. It also shows a Pareto Chart by CT modality, a table of the percentage of CTs by modality and by ED physician.

Identify Replication Opportunities

There are potentially other opportunities to implement the CT guidelines for the inpatients, as well as in other hospitals.

Develop Future Plans

Physicians can now receive up to 20 Continuing Medical Education requirements for performance improvement education and projects, as the physicians did for this project. They plan to continue other efforts to enhance practice performance using the Lean Six Sigma methodology.

Conclusions and Critical Success Factors

This project was very successful in applying the Six Sigma methodology to reduce the number of CT scans ordered and performed for the Emergency Services patients at the hospital. While nationally, Emergency Departments are increasing the usage of CT scans by about 10% per year, this project demonstrates that applying CT technology appropriately provides satisfactory medical diagnoses, without jeopardizing patient safety. We successfully reduced the percent of CT scans by **8.7%** in the October through December timeframe, and **13.8%** since a year before the baseline period. While many departments are increasing the number of CT scans, we have successfully reduced the utilization of CT scans to better align physician CT orders to medical guidelines.

The critical success factors of this project were:

- Physician specialists including ED physicians and radiologists working together to develop appropriate CT guidelines.
- Physicians and librarians working together to research appropriate medical guidelines for CT utilization.
- Application of the Six Sigma methodology in understanding the current state, the root causes, analysis of data to show and maintain improvement and change management to encourage buy-into the improvement ideas.

References

Baumgarten, D. A., Francis, I. R., Bluth, E. I., Bush Jr., W. H., Casalino, D. D., Curry, N. S., Israel, G. M., Jafri, S. Z. H., Kawashima, A., Papanicolaou, N., Remer, E. M., Sandler, C. M., Spring, D. B., & Fulgham, P. (2007). *Expert panel on urologic imaging: 14 American college of radiology ACR appropriateness criteria acute onset flank pain—suspicion of stone disease.* Date of origin: 1995 Last review date: 2007.

Cancer risks and radiation exposure from computed tomographic scans how can we be sure that the benefits outweigh the risks? (2009). (Reprinted) *Archives of Internal Medicine, 169*(22), 2049. December 14/28 www.archinternmed.com American Medical Association. All rights reserved. Retrieved December 17, 2009, from www.archinternmed.com

de González, A. B., Mahesh, M., Kim, K.-P., Bhargavan, M., Lewis, R., Mettler, F., & Land, C. (2009). Projected cancer risks from computed tomographic scans performed in the United States in 2007. *Internal Medicine, 169*(22), 2071–2079 www.archinternmed.com

Imaging strategies for detection of urgent conditions in patients with acute abdominal pain: Diagnostic accuracy study. *BMJ, 339*, b2431 (2009). https://doi.org/10.1136/bmj.b2431

Institute for Clinical Systems Improvement. (2009). *Pulmonary Embolism (PE) diagnosis algorithm* (9th ed.).

Katz, S. I., Saluja, S., Brink, J. A., & Forman, H. P. (2006). Genitourinary imaging. Original research radiation dose associated with unenhanced CT for suspected renal colic: Impact of repetitive studies. *AJR, 186*, 1120–1124. 0361-803X/06/1864-1120 (American Roentgen Ray Society).

Kirsch, T. D., Hsieh, Y.-H., Horana, L., Holtzclaw, S. G., Silverman, M., & Chanmugam, A. (2010). Computed tomography scan utilization in emergency departments: A multi-state analysis. *The Journal of Emergency Medicine, 7*, 771–779.

Marchione, M. (2010). CT scans pose a growing danger to Americans. In *Experts call for increased regulation to improve safety*. Associated Press.

Medical Liability Monitor. (2008). *Rate survey*. Published in the January 5, 2009 issue of American medical news www.amednews.com

Mettler, F. A., Huda, W., Yoshizumi, T. T., & Mahesh, M. (2008a). Effective doses in radiology and diagnostic nuclear medicine: A catalog. *Radiology., 248*(1), 254–263.

Mettler, F. A., Thomadsen, B. R., Bhargavan, M., Gilley, D. B., Gray, J. E., Lipoti, J. A., ... & Mahesh, M. (2008b). Medical radiation exposure in the US in 2006: preliminary results. *Health physics*, *95*(5), 502–507.

Smith-Bindman, R. (2012). Environmental causes of breast cancer and radiation from medical imaging: Findings from the institute of medicine report. *Archives of Internal Medicine, 172*(13), 1023–1027.

Stoker, J., van Randen, A., Laméris, W., & Boermeester, M. A. (2009). Imaging patients with acute abdominal pain. *Radiology*, *253*(1), 31–46 (Radiology.rsna.org).

http://www.fda.gov/RadiationEmittingProducts/RadiationEmittingProductsandProcedures/MedicalImaging/MedicalX-Rays/ucm115317.htm

Chapter 8
Applying Lean Six Sigma to Optimize Light Levels in the NICU

Abstract The neonatal intensive care unit (NICU) is an intensive care nursery that specializes in the care of ill and/or premature infants. These infants are fragile and sensitive to their environments, including light, sound, heat and cold. The goal of this study was to reduce light exposure for the newborns within the NICU setting. The LSS DMAIC methodology was applied to identify improvements to reduce and minimize newborns' light exposure, which can help improve their healing.

Keywords NICU · Newborns · Environmental factors · Light reduction · Lean Six Sigma DMAIC methodology

Introduction

The neonatal intensive care unit (NICU) is an intensive care nursery that specializes in the care of ill and/or premature infants. These newborn infants are not only fragile, but also incredibly sensitive to their environments. Everything these newborns smell, see, hear, taste, and touch during their time in intensive care can have a huge impact on their development. The healthcare industry is working to create and foster a setting within NICUs that will heal and nurture infants quickly and effectively to send them home with their families happy and healthy. This project particularly focuses on light reduction within the NICU setting.

The project was conducted at a pediatric healthcare institute. The institute provides the most comprehensive care for newborns in the United States, Midwest region. They have a level III NICU, which means they are capable of caring for newborn infants with extreme prematurity or who are critically ill or require surgical intervention.

Lighting reduction is an important goal for NICUs around the country because light plays a huge factor in the development of newborn visual systems, as well as circadian rhythm, sleep quality, and even comfort. On the other hand, the medical professionals require adequate lighting to perform and execute the required care for the infants. The goal of this project was to observe, analyze, and determine a solution

S. Furterer, *Lean Six Sigma for the Healthcare Enterprise*,
https://doi.org/10.1007/978-1-4471-7617-6_8

for the NICU lighting that would promote the development of newborn infants, as well as provide a sufficient work environment for the medical professionals.

Methodology

The team applied the Six Sigma DMAIC (Define-Measure-Analyse-Improve-Control) methodology and tools to identify improvements to the NICU related to lighting reduction as shown in Table 8.1. This was a one-semester capstone project within the Industrial and Systems Engineering undergraduate program. The Six Sigma methodology was used to improve an existing process within the NICU of the hospital to optimize lighting conditions for the infants and the medical staff. The Measure phase was critical in this project to assess the current environment and lighting conditions from both a Voice of Process (VOP) and a Voice of Customer (VOC) perspective. The Six Sigma methodology incorporated collecting both quantitative data on the VOP, and qualitative data such as the VOC Survey. Design of Experiments was applied to identify critical factors that impact lighting levels in the NICU. A pilot study was run to assess the impact of some of the proposed changes, showing a reduction in the lighting levels both during the day and during the night, when lights were cycled, and turned lower during the night versus the day lighting levels.

Table 8.1 Six Sigma tools applied

DMAIC phase	Tools applied
Define	• Problem definition • Benchmarking • Critical to quality (CTQ)
Measure	• Voice of the customer (VOC) • Statistical process control (SPC) • Time series plots • Process capability • Bar charts
Analyse	• Cause and effect analysis • Design of experiments and analysis of variance • Pareto charts
Improve	• Improvement recommendation • Pilot study • Standard operating procedures
Control	• Inferential statistics • Control plan

DMAIC Methodology Applied

Define Phase

Variables

The Critical to Quality (CTQs) characteristics were chosen to assess and measure both the current state of lighting in the NICU and the changes in lighting after implementation. These CTQs included the average and peak brightness at the babies' eyes. Brightness was measured in lux, using a light meter, at different times throughout the day and night. In addition, a survey of healthcare professionals provided information on the Voice of the Customer (VOC) to assess perceived ease of work at the current and future lighting conditions.

The proposed input variables that were used to control the CTQs included the wattage of light bulbs, a cover over the baby, duration of light during the day, the brightness of the windows, and the reflectiveness of surfaces. There was also a need to care for the development of the infants' circadian cycles. The type of light bulb, and in turn the spectrum of light emitted, plays a role in this development.

Preliminary Methods

The team began benchmarking research to discover what other hospitals' NICUs were doing to reduce lighting. The research pointed out that there were many factors that go into the lighting levels of NICUs, most notably the color appearance of the light and glare. Additionally, the circadian rhythm of the patients, the hospital staff, and families needed to be considered.

The color appearance of lighting is an important factor in a hospital NICU. The color of a baby's skin can tell you important information about their health. A lighting source is considered poor for a NICU if it changes the appearance of the objects that it is illuminating. There are three general hues of lighting: cool, warm, and neutral. The correlated color temperature (CCT) is a metric that puts the hues of lighting into these categories. Other metrics for determining the hue of a lighting source are the Colour Rendering Index (CRI) and the Gamut Area Index (GMI), measuring the difference between the light source and the saturating capabilities, respectfully.

There are two types of glare that have an impact in a hospital's NICU: disability and discomfort. Disability glare refers to light that is reflected off of surfaces such as walls, floors, screens, and windows. Discomfort glare refers to the brightness of light. There are two major design features that can impact the effects of glare. The first is the contrast between the light source and its background, and the second is how well the light source is shielded.

The Circadian Rhythms are changes in physiology and behavior which are regulated by a biological clock that uses external time cues to keep on track. Each person's rhythm is dynamic and can be changed, for example, when the person changes time zones. The phototransduction mechanism is what regulates the circadian rhythm by converting light signals to neural signals. This plays into the production of melatonin, a hormone for sleep. Melatonin is produced at night and in the presence of darkness. Production of melatonin can be interrupted when light reaches a certain level.

Generally, the lighting in a NICU is operated without regard for the circadian rhythm of the patients and employees. However, several studies have shown that cycled light has many benefits to the health of infants, including better weight gain, less time on the ventilator, improved motor coordination, and less fussing. In addition, disrupting the circadian rhythm of night shift employees can be devastating to their health. While remaining awake and alert throughout the night is essential for their work, there are steps that can be taken to have a minimal impact on their day-night cycle.

Overall, the benchmarking research has shown that the lighting solution that is chosen should have individualized lighting to support the needs of the individual patients, that the lighting should be cycled as much as possible, that the lighting should meet set standards in terms of brightness (lux) and hue (CRI and GMI), and that the lighting source should be shielded.

Measure Phase

Control Charting

To determine the current state of the NICU at the institute, the team used light meters to record the lux (lx) at 4 different locations for 24 h periods. During those 24 h periods, nurses recorded every time that they adjusted the lighting. Two of the locations were on the B Side of the NICU and two were on the C Side. At each location one meter was placed on the wall as close to the baby's head as possible and one was placed at the nurse's workstation.

The data was downloaded from the meters. Initially the team was going to use an I-MR chart to analyze the data and compare to the upper limit for lux during various times of the day. However, as the team began charting, they realized that because data points were taken every second, the readings were autocorrelated. The I-MR control charts were developed for each meter and the *maximum reading during each hour*, shown in Figs. 8.1 and 8.2. A time series plot was used to better understand the data over time. The time series plots of the maximum readings each hour are shown in Figs. 8.3 and 8.4.

The team analyzed the charts and came to the conclusion about the current lighting state in the NICU. The first thing that the team noticed is that the average lighting is well within the range that is safe for infants. The acceptable ranges of lux for infants are approximately 100 lx for ambient lighting, approximately 300 lx for observation, and approximately 500 lx for treatments. The maximum readings that are harmful to the infants happen infrequently but are very high lux levels. An observation of the team was that the hospital staff indicated that they used cycled lighting, however, the

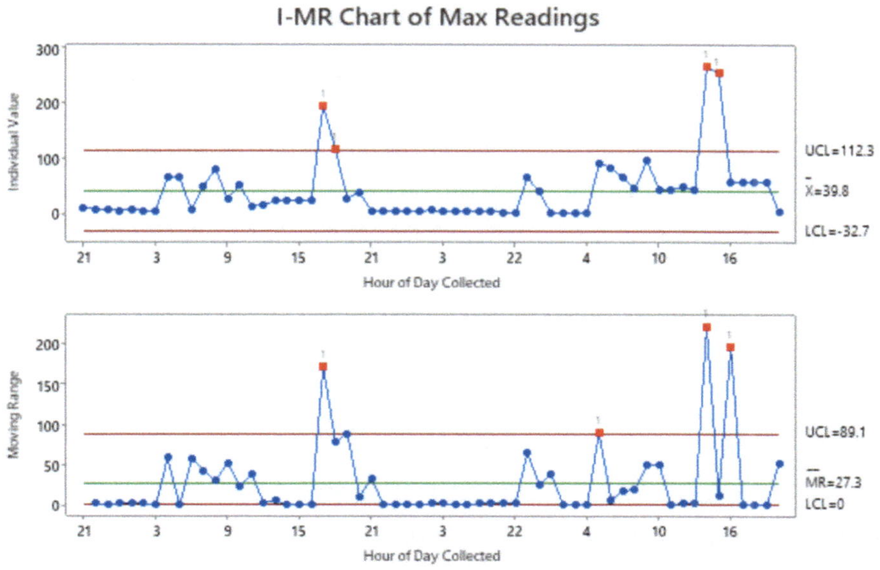

Fig. 8.1 I-MR control charts of maximum readings per hour—meter 1

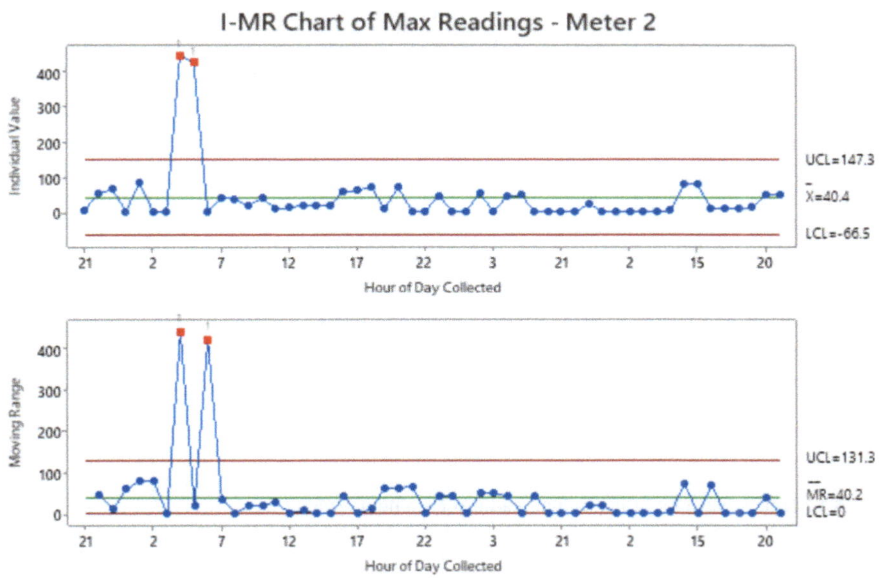

Fig. 8.2 I-MR control charts of maximum readings per hour—meter 2

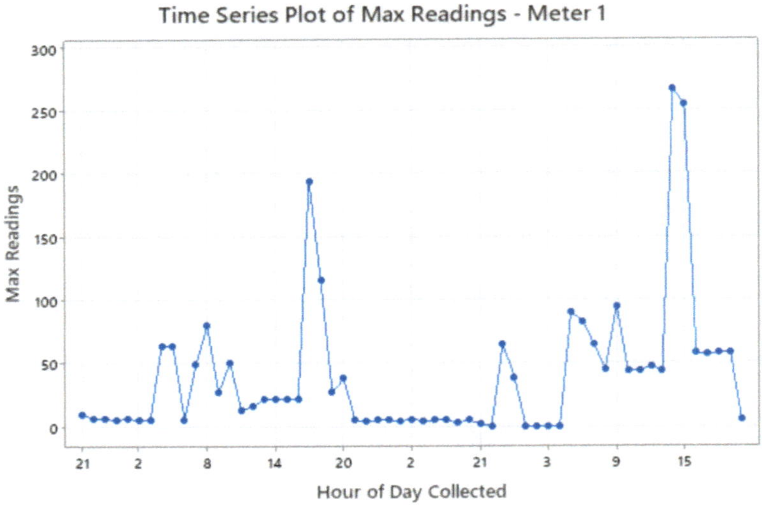

Fig. 8.3 Time series plot of maximum readings each hour—meter 1

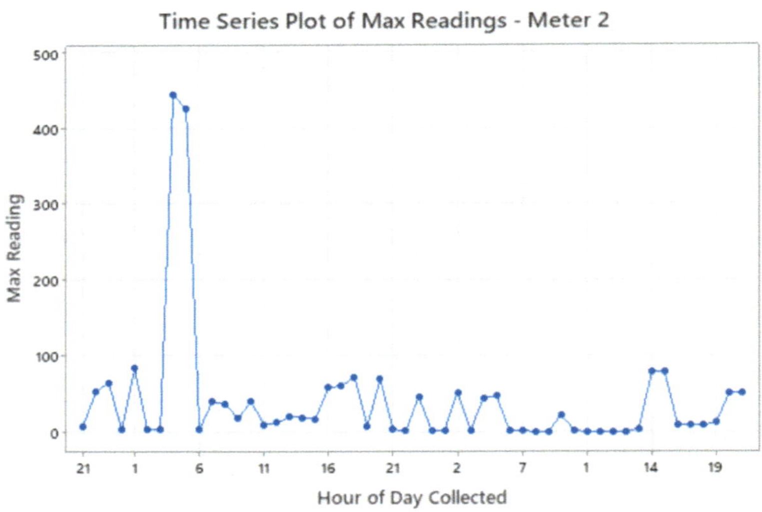

Fig. 8.4 Time series plot of maximum readings each hour—meter 2

meters did not pick up on changes from daytime to night-time. The overall reading shows fluctuations from 0 to 50 lx with peaks reaching to the 450 lx range. It should be noted that while the meters were placed as close to the infant's eyes as possible, the infants are receiving a much more direct lighting, and some of the infants are covered with quilts.

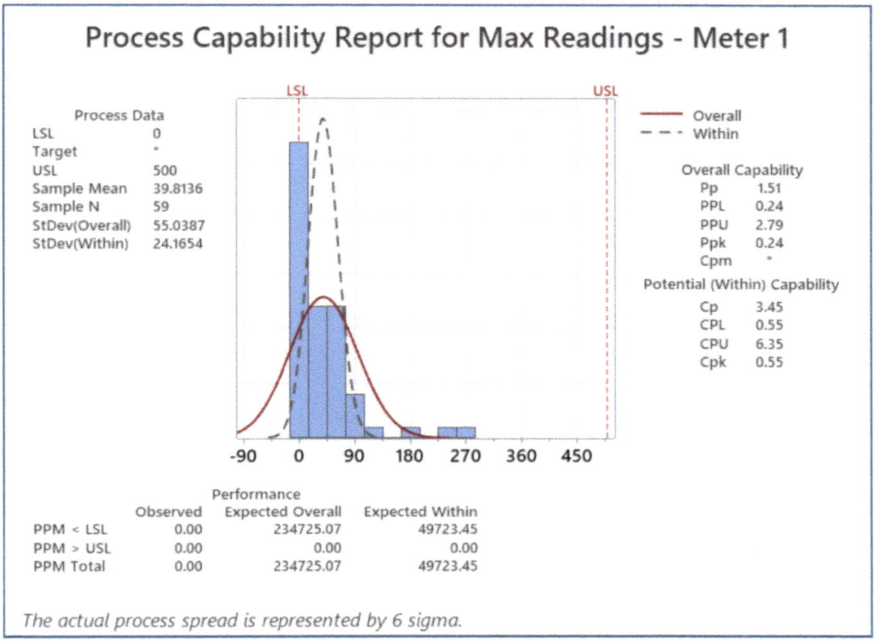

Fig. 8.5 Process capability analysis—meter 1

A process capability analysis was performed for the maximum readings for each meter, using a lower specification limit of 0 and an upper specification limit of 500, shown in Figs. 8.5 and 8.6. The Cps were 3.45 and 2.34 for meters 1 and 2, respectively, which are very good capabilities to ensure that the infants are not exposed to dangerous light levels. Additionally, the Cpk values were 0.55 and 0.38, showing that most of the max readings were skewed toward the lower specification limit.

VOC Survey

In addition to collecting the data from the light meters, a survey was given to the NICU staff, specifically the patients' nurses. These questions were asked to gain a better understanding of the conditions the NICU staff is working under. The key deliverable for the project was to not only provide better lighting for the infants, but the NICU staff as well. The survey asked questions in which the staff would give a rate for their response:

- Is there enough lighting during medical procedures?
- Is there enough lighting for non-medical work?
- Is it easy to use the lighting for both medical and non-medical work?

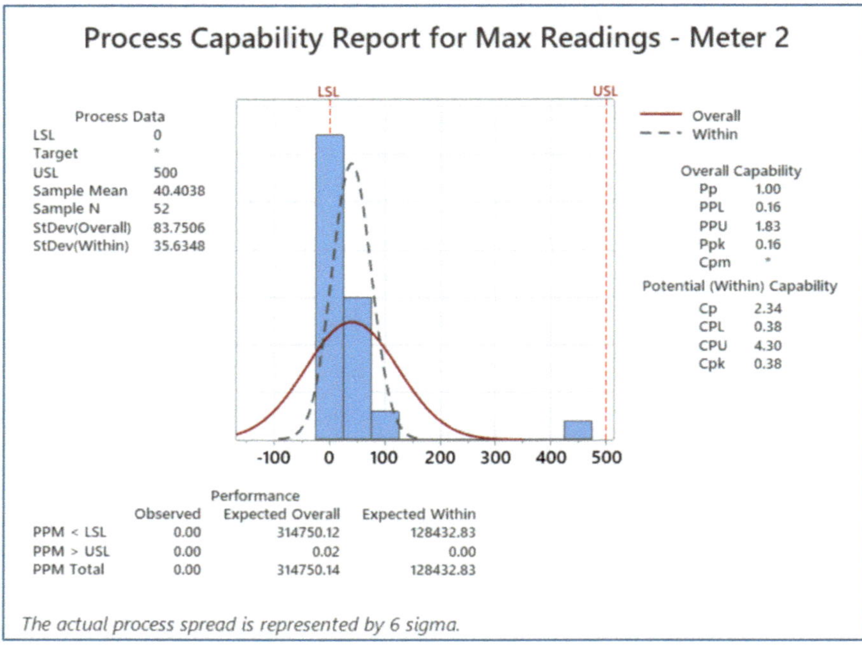

Fig. 8.6 Process capability analysis—meter 2

- The survey also asked the staff to give any optional feedback or to share their opinion of the current lighting environment in the NICU.

The full survey and results can be seen in the Appendix.

The results of the survey (Fig. 8.7) showed that for some of the questions there was a wide range of responses, while other questions were overwhelmingly answered similarly. Unfortunately, the majority of survey responses came from the C side of the NICU and ideally there would have been more of an equal representation from both the B and C side. Moving forward the data recorded from the surveys would continue to be used for developing potential solutions for the NICU lighting.

The lowest rated questions (2.31 out of 5), "it is easy to see during medical procedures," "it is easy to see during non-medical work" (2.38), highlights that there is indeed an environmental lighting problem.

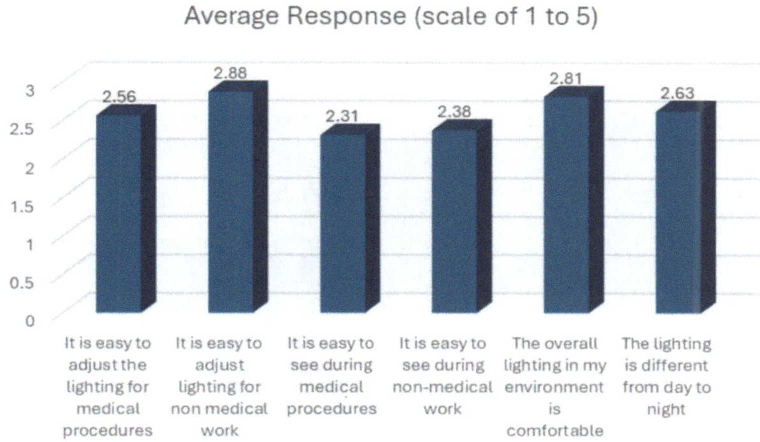

Fig. 8.7 VOC survey results

Analyze Phase

Design of Experiment with a Full Factorial Analysis and Regression Analysis

A full factorial analysis was used to determine which factors had a significant impact of light brightness (lux). Two out of the six initial input factors (X's), light duration and type of light bulb, do not impact brightness, and therefore were not considered in this analysis. This left the other four factors, light brightness, window brightness, reflectiveness, and a cover, to be used. With four factors a full factorial can be done with only 16 runs and allows for no aliasing in our interactions.

The experiment was performed in a room with adjustable lighting and windows with adjustable shades. The light brightness levels were low or all on. Window levels were with the shades down or up. The levels for a cover were without the cover or with a piece of white paper approximately 6 inches above the meter to simulate the cover. Reflectiveness was set with either a low-reflective, dark fabric surrounding the meter or without it.

The analysis found three of the four factors to be significant, lights, reflectiveness, and the cover, along with a significant second-order interaction between the lights and the cover (Fig. 8.8). Window brightness was not found to be significant.

The regression model, shown below in Eq. (8.1), for this data is not strictly applicable for generalization to a different room than the one tested in. However, it does provide insight into how each factor affects the lux of an area. In general, the model shows that as the lights in a room get brighter, the lux increases. A highly reflective surface also increases the lux, and the addition of a cover decreases the lux

Fig. 8.8 Pareto chart of significant factors

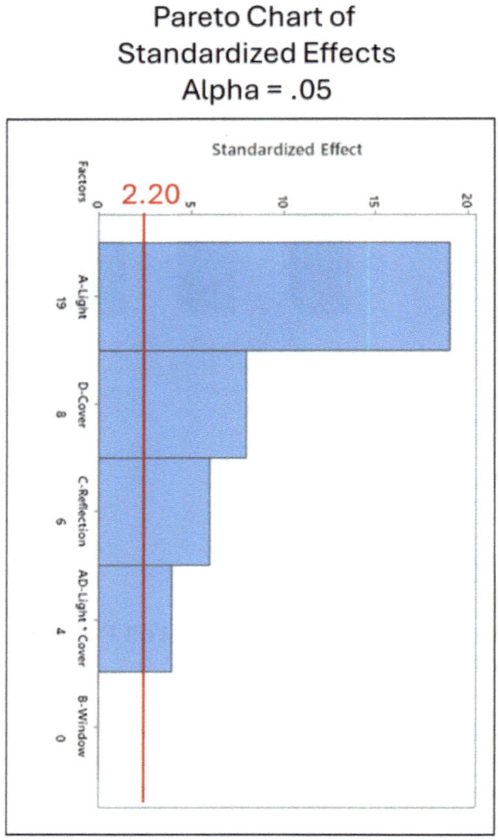

$$\text{Lux} = 452.6 + 240.1 * \text{Lights} + 66 * \text{Reflectiveness} - 104.6 * \text{Cover} - 50.1 * \text{Lights} * \text{Cover} \quad (8.1)$$

Fishbone Diagram

To find the root cause of the poor lighting conditions in the NICU in regard to the infants, the group used a fishbone diagram (Fig. 8.9) to generate potential root causes and further analyze them. The potential causes were broken down into five main categories: man, machine, environment, methods, and materials. Within those categories, the group was able to list and identify the top four potential root causes based on what was controllable and what was observed in our other analyses.

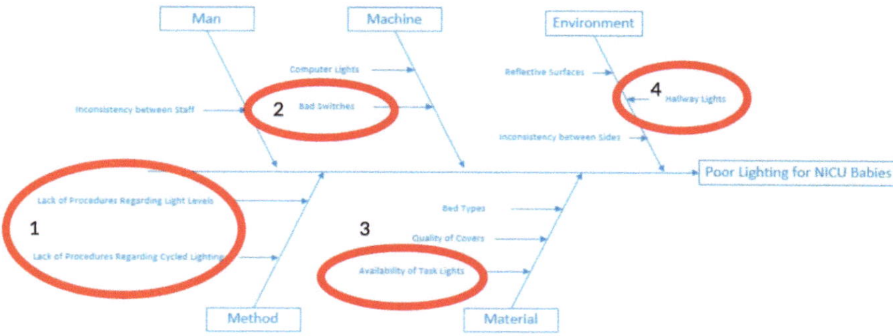

Fig. 8.9 Fishbone diagram of potential root causes

The four main causes identified were (1) Lack of procedures regarding cycled lighting/light levels, (2) Bad/defective switches, (3) Availability of task lights, and 4. Hallway lights. To address these causes, the group both brainstormed and discussed potential solutions with the NICU medical staff. The solutions included a standardized write-up of procedures, replacement and implementation of step light switches, cycled lighting implementation, and poka-yoke labels for the light switches to maintain the cycled schedule. The solutions are further discussed and explained in the improve phase.

Pareto Chart

A Pareto chart (Fig. 8.10) was used to pinpoint the assignable causes. The majority of the causes of peak lighting were due to the medical staff performing procedures/caring for the patients. It should be noted that the data showed that much of the causes falling under "Care," were within a low-level light reading, it should also be noted that there were brief spikes in the readings during the 24 h periods, however, they were negligible and were determined to not be a main cause. Some of the patient care activities and the lux level were recorded and shown in Table 8.2.

After analyzing the data, further research, and additional discussion with the NICU staff, it was determined that the current lighting levels were within the recommended ranges, however, the data did lead to the discussion of cycled lighting. So, while the main cause of light was patient care, the present lighting did not show to be harmful to the infants. The report will later explain the significance of cycled lighting rather than light intensity, and how it plays an important part in the infants' circadian rhythm.

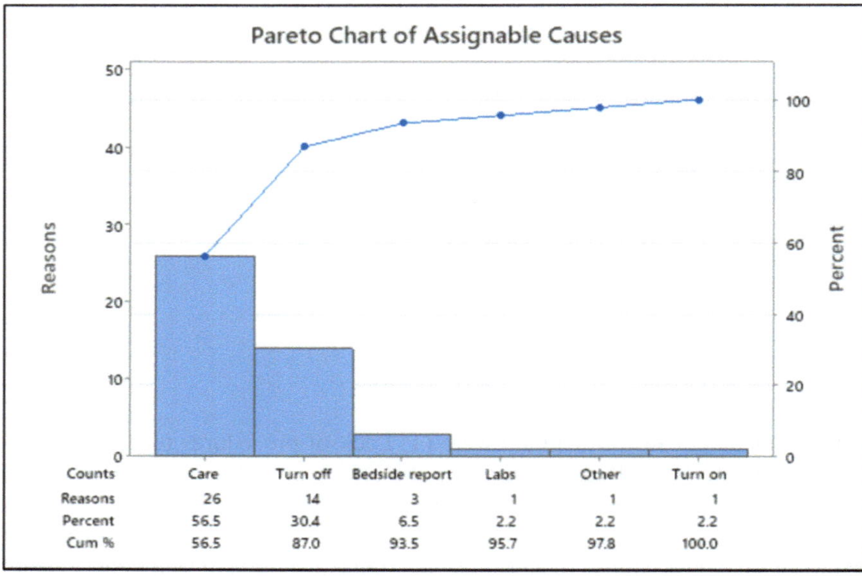

Fig. 8.10 Pareto chart of assignable causes

Table 8.2 Activities and lighting lux

Activity	Lux level (lx)
Care	50–80
Bedside report	25
Lights on	15
Labs	300
Other (computer sensor)	450

Improve Phase

Solutions

Cycled lighting was the main focus for improvements. These improvements were achieved with new procedures and signs to make lighting use easier and more deliberate. Research found that infants need their longest sleep period at night to be between 2.5 and 4 h (Adair & Bauchner, 1993, p. 155). This led to the desire to design a nighttime sleep period that could sufficiently encompass that time period. The nighttime cycled lighting period requires all non-essential and light intensive care procedures to be performed before and after the darkness period. The new NICU Cycled Lighting Procedures and Reminders are listed below.

NICU Cycled Lighting Procedures

1. Procedures requiring increased light should be completed before 10:00 pm.
2. All light switches marked cycled lighting should be turned off at 10:00 pm.
3. Procedures requiring increased light should be avoided when possible during the nighttime cycled lighting period.
4. Light switches marked cycled lighting can be turned back on at 5:00 am with nurses' discretion.
5. Indirect overhead unit lights can be turned on.
6. Normal working procedures can be resumed.

Other Reminders

- Curtains around the beds should stay closed when possible
- Follow the switch level for bedside lighting.

The addition of light safety gages can inform nurses and parents how bright to turn the lights on safely, as shown in Fig. 8.11. Cycled lighting signs, Fig. 8.12, would be posted on the non-essential lights that need to be turned off at night. Additional educational signage would also be provided to help remind healthcare professionals and patients' parents to observe the new lighting practices.

Fig. 8.11 Light safety gages

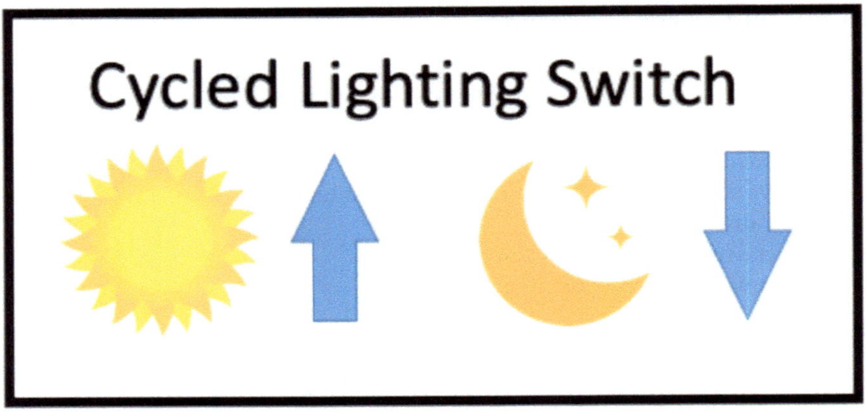

Fig. 8.12 Cycled lighting sign

Additional solutions pertained to improving the NICU's cycled lighting equipment. Light wavelength and color were addressed by using more red light in the NICU. The team noticed that all of the bedside computer screen backgrounds were blue, a simple switch to a red background can be less impactful to the babies' eyes. The inclusion of red LED lights in the ceiling that can be turned on at night would make it seem significantly darker to the babies, while still allowing for the professionals to be able to see and complete their work. These red lights would only need to be placed over the beds and not the central nurses' station so that they are not made to feel uncomfortable. Moreover, differing light limit guidelines could be put into place that help the nurses decide how much light to use in the units during the day, as well as a lower set of limits during the night.

Control Phase

Data Collection

The switch plate covers were implemented on four of the NICU C-side bed spots to test their usability and effectiveness. The nurses were trained on the new procedure. Once they were comfortable, the lux levels were measured. The measurements were taken at each of the bed-spots, one meter near the infant's head location and one at the nurse's workstation, for a 24 h period. Once completed, the data was downloaded and analyzed. The maximum and average lux levels were calculated over a 24 h period, and charted (Fig. 8.13).

The figure shows that the lux levels are lower during the nighttime hours between 10:00 pm and 5:00 am. The maximum levels are less than 10 lx, and the average was 1.6 lx. This is a reduction from before the improvement where the nighttime

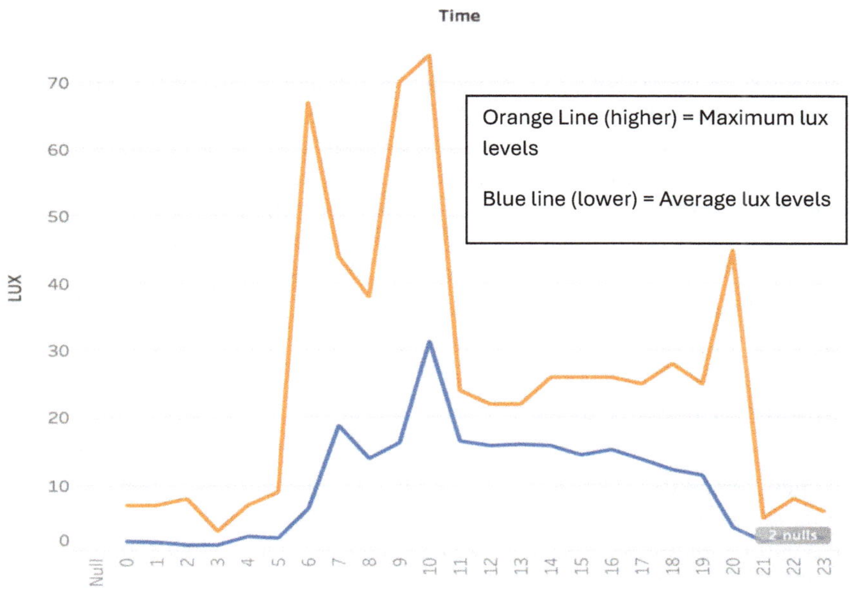

Fig. 8.13 Pilot—average and maximum lux over a 24 h period

lux levels were averaged at 9 lx, resulting in a reduction in average light lux levels of **82%**. Additionally, the daytime light levels standard deviation was reduced from 30 to 12 lx stabilizing the levels at an appropriate range, resulting in a reduction in standard deviation of **60%**. As a control plan, the I-MR statistical process control charts can be used to ensure that the light levels remain in control as additional improvements are implemented in the future.

Continued Education

With the success of the four-bed trial, the switch plate covers were rolled-out to the rest of the NICU, and all of the nurses were to be trained. In addition, the team created signage that was placed in high traffic areas that remind the staff to follow the new lighting procedures.

Future Recommendations

Throughout the project, the group took note of potential solutions and recommendations that could be implemented in future NICUs. Based on this project and previous capstone projects, the NICU environment is crucial to the development and health of premature infants. Ideally, future NICUs would be able to create unique and tailored environments according to each infant's needs while not affecting neighboring infants. Every day, steps are being made to move closer and closer to that kind of state-of-the-art NICU. In the meantime, it is important to create and foster an environment that will best suit the NICU as a whole. Future NICUs should implement step dimmer switches to create a standardized usage of different levels of light as well as create a cycled lighting schedule to promote the circadian rhythms of the infants and even the staff. Other recommendations include the use of hard covers (that can be easily sanitized) for the incubators to shield the infants from light, the implementation of computer blue light filters in the bays, and the use of flexible tasked lighting for the incubators.

Managerial Implications

This study achieved the important goal for NICUs to reduce lighting, that plays a large role in newborn development, their circadian rhythms, sleep quality and comfort. The proposed improvements included procedures to cycle the lights to lower levels at night. These improvements balanced the needs of the infants, and the medical professionals to be able to perform their tasks.

Lessons Learned and Key Challenges

The Six Sigma capstone project was a valuable learning experience for the Industrial and Systems Engineering students to apply the Six Sigma DMAIC methodology and tools to a real-world and important healthcare setting. It was important throughout the study for the students to work closely with the nurses and physicians in the NICU to understand their voice, expectations and needs, and to gain their buy-in for the proposed improvements. The pilot study was critical in demonstrating the value of the improvement recommendations in significantly reducing the lighting levels. The statistical data analysis, application of statistical process control charts, process capability analysis, Design of Experiments, VOC survey, and root cause analysis were all important tools that helped the team understand the current environment and the current lighting levels, as well as to build the improvement case for implementing the proposed improvement recommendations. Setting up data collection methods to capture the CTQ metrics was also critical to collect data and perform the Design of Experiments to identify factors that impact lighting. A key challenge on the project was that the students needed to have an escort to gain access to the facility for observation and data collection, which made it difficult to plan for the site visits. Another challenge was that the team could not perform invasive data collection nor propose invasive recommendations, due to the nature of the fragile infants. Overall, the Six Sigma DMAIC method and tools was highly effective in solving the problem to optimize lighting levels for newborns in the NICU.

Conclusions

This chapter described the use the Lean Six Sigma DMAIC methodology within a NICU to optimize the lighting levels. Techniques applied included statistical process control charts, process capability analysis, time series charting, Pareto charting, Design of Experiments and Analysis of Variance (ANOVA) for a full factorial analysis, benchmarking, and poka-yoke. The implementation of a procedure and switch plate covers as a pilot study reduced cycled lighting and lighting levels. This project also gave insights into further large-scale improvements that could be made down the line or in future hospitals.

Appendix of Results

Survey Distributed

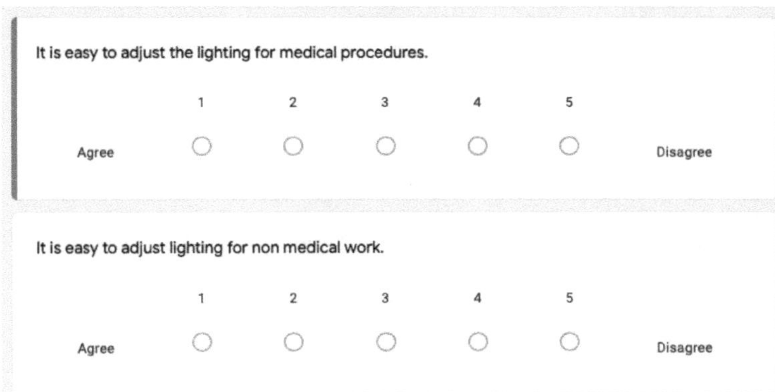
It is easy to adjust the lighting for medical procedures.

	1	2	3	4	5	
Agree	○	○	○	○	○	Disagree

It is easy to adjust lighting for non medical work.

	1	2	3	4	5	
Agree	○	○	○	○	○	Disagree

It is easy to see during medical procedures.

	1	2	3	4	5	
Agree	○	○	○	○	○	Disagree

It is easy to see during non medical work.

	1	2	3	4	5	
Agree	○	○	○	○	○	Disagree

The overall lighting in my work environment is comfortable.

	1	2	3	4	5	
Agree	○	○	○	○	○	Disagree

The lighting is different from day to night.

	1	2	3	4	5	
Agree	○	○	○	○	○	Disagree

Which NICU area do you work in?

- ○ B Side
- ○ C Side

What improvements can be made to lighting in the NICU to ease your comfort and/or working ability?

Long answer text

Survey Results

Timestamp	Adjust medical	Adjust non-medical	See medical	See non-medical	Comfortable	Different day/night	NICU area
10/4 14:57	1	4	1	2	3	1	C Side
10/4 15:31	3	3	3	3	3	1	C Side
10/4 16:30	4	5	4	5	5	5	C Side
10/4 16:38	4	4	2	2	2	4	C Side
10/4 16:53	5	5	5	5	5	5	B Side
10/4 18:41	4	2	4	2	4	2	C Side
10/4 19:54	1	3	2	1	3	1	C Side
10/5 10:24	3	3	1	2	4	5	C Side
10/5 10:29	1	1	1	1	1	1	C Side
10/5 23:48	1	2	1	3	2	2	C Side
10/7 10:56	3	3	3	2	2	2	C Side
10/7 16:31	1	1	1	1	1	2	C Side
10/8 7:46	1	1	1	1	1	5	C Side
10/9 16:01	4	4	3	3	5	3	C Side
10/14 11:06	1	1	1	1	1	1	C Side
10/26 6:30	4	4	4	4	3	2	C Side
	2.56	2.88	2.31	2.38	2.81	2.63	

Survey Comments

Response number	What improvements can be made to lighting in the NICU to ease your comfort and/or working ability?
1	
2	Dimmable lights at every bedside

(continued)

(continued)

Response number	What improvements can be made to lighting in the NICU to ease your comfort and/or working ability?
3	
4	Lights that are truly able to dim, spotlights that can be repositioned, lights that more closely resemble natural light
5	Switches near the bedside, a true dimmer not dark and surface of the sun, a better "work" light over computer areas that would not interfere with cycled lighting for neonate
6	More natural light
7	
8	Most of the dimmer switches at our bedsides don't work, so it is hard to adjust the lighting to somewhere in the middle (now, many of the lights can either be off or VERY bright). Also, some of the hallway lighting, we have been told, cannot be turned off. This can be disturbing for some babies who are older and need to have dark/be sleeping at night, because this is appropriate for them developmentally. Nurses have resorted to hanging additional blankets on the netted areas of the curtains to block out the light from the babies' bedsides. It makes it very difficult to simulate a nighttime environment for any of the babies. Also, it would be great to have more natural lighting in our setting. Even some of the private rooms on the wall don't have windows. This is limited by our current setting, though (we can't create windows easily)
9	
10	
11	It would be great if all the switches worked all the time
12	We should be able to adjust the lights that run along the center of each POD!
13	Be better at making sure all lighting works properly
14	Less fluorescent light
15	
16	Sometimes our bedside lights have a strobe light affect, not developmentally okay for our babies

Reference

Adair, R. H., & Bauchner, H. (1993). Sleep problems in childhood. *Current Problems in Pediatrics, 23*(4), 147–170. https://doi.org/10.1016/0045-9380(93)90011-z

Bibliography

Berson, D. M., Dunn, F. A., & Takao, M. (2002). Phototransduction by retinal ganglion cells that set the circadian clock. *Science, 295*(5557), 1070–1073.

Boo, N. Y., Chee, S. C., & Rohana, J. (2002). Randomized controlled study of the effects of different durations of light exposure on weight gain by preterm infants in a neonatal intensive care unit. *Acta Paediatrica, 91*(6), 674–679.
Brandon, D. H., Holditch-Davis, D., & Belyea, M. (2002). Preterm infants born at less than 31 weeks' gestation have improved growth in cycled light compared with continuous near darkness. *Journal of Pediatrics, 140*(2), 192–199.
Leproult, R., Holmback, U., & Van Cauter, E. (2014). Circadian misalignment augments markers of insulin resistance and inflammation, independently of sleep loss. *Diabetes, 63*(6), 1860–1869.
Mann, N. P., Haddow, R., Stokes, L., Goodley, S., & Rutter, N. (1986). Effect of night and day on preterm infants in a newborn nursery: Randomised trial. *British Medical Journal (Clinical Research Edition), 293*(6557), 1265–1267.
Miller, C. L., White, R., Whitman, T. L., O'Callaghan, M. F., & Maxwell, S. E. (1995). The effects of cycled versus non-cycled lighting on growth and development in preterm infants. *Infant Behavior and Development, 18*(1), 87–95.
Nurminen, T. (1998). Shift work and reproductive health. *Scandinavian Journal of Work, Environment & Health*, 28–34.
Rea, M. S. (2013). *Value metrics for better lighting*. SPIE.
Rea, M. S., & Freyssinier, J. P. (2010). Color rendering: Beyond pride and prejudice. *Color Research & Application, 35*(6), 401–409.
Rea, M. S., & Freyssinier, J. P. (2011). White lighting. *Color Research & Application, 38*(2), 82–92.
Rea, M. S., & Freyssinier-Nova, J. P. (2008). Color rendering: A tale of two metrics. *Color Research & Application, 33*(3), 192–202.
Smith, M. R., Fogg, L. F., & Eastman, C. I. (2009). A compromise circadian phase position for permanent night work improves mood, fatigue, and performance. *Sleep, 32*(11), 1481–1489.
Van Cauter, E., Spiegel, K., Tasali, E., & Leproult, R. (2008). Metabolic consequences of sleep and sleep loss. *Sleep Medicine, 9*(1), S23–S28.

Chapter 9
Sepsis Protocol Design Project

Abstract Sepsis is a systemic infection that is the 10th leading cause of death in the United States. The sooner that the antibiotic protocols are started, the better is the patient's outcome. This case study applied the Lean Six Sigma methodology and tools to develop and implement Sepsis protocols in the emergency department of a community-based hospital to improve patients' sepsis mortality by 9.6%, while improving the time that the sepsis protocols were administered from the time that the patient arrived in the emergency department. There is still improvement to be made in the consistency of administering the sepsis protocols for suspected sepsis patients in the ED. This chapter is not intended to provide medical advice or recommendations.

Keywords Emergency department · Septic shock · Severe sepsis · Mean arterial pressure · Process capability

Introduction

"Sepsis is defined by a systemic inflammatory response syndrome, such as fever, tachypnea, tachycardia, or leukocytosis in response to a culture-proven or clinically suspected infection" (Morrell et al., 2009). The grading of sepsis includes severe sepsis, septic shock, and refractory septic shock (Annane et al., 2003; Levy et al., 2003). Severe sepsis includes the previously mentioned clinical criteria for sepsis as well as at least one sign of organ hypoperfusion or dysfunction. Septic shock is defined by severe sepsis with a systemic mean blood pressure <60 mm Hg (or 80 mm Hg if previous hypertension) after an attempt at adequate fluid resuscitation or a need for vasopressors to maintain a systemic mean blood pressure >60 mm Hg (or >80 mm Hg if previous hypertension). Refractory septic shock exists if dopamine >15 μg/kg/min or norepinephrine or epinephrine >0.25 μg/kg/min is required to maintain a mean blood pressure >60 mm Hg (80 mm Hg if previous hypertension) (Morrey et al., 2009).

Severe sepsis and septic shock continue to be a major healthcare problem. Overall mortality from severe sepsis or septic shock ranges from 30 to 60% and accounts

S. Furterer, *Lean Six Sigma for the Healthcare Enterprise*,
https://doi.org/10.1007/978-1-4471-7617-6_9

for 9.3% of all deaths in the United States (Annane et al., 2003; Angus et al., 2001; Sasse et al., 1995). Severe sepsis and septic shock are the 10th leading cause of death in the United States (Kung et al., 2008). More than 750,000 cases of sepsis occur each year in the United States. The average cost per case of sepsis or septic shock is $22,000. Annual total costs related to patients with sepsis are estimated to be near $17 billion (Angus et al., 2001).

Due to the continuing persistence of sepsis and septic shock as a healthcare problem, the hospital's emergency department decided to use the Six Sigma approach to implement standardized protocols with the goals to reduce mortality, identify septic patients promptly, and to initiate treatment immediately.

This chapter will describe the application of the Six Sigma problem-solving methodology regarding the better management of severe sepsis and septic shock, when presented in the Emergency Department (ED).

Define Phase

The following steps were executed in the Define phase:

- Develop project charter.
- Perform stakeholder analysis.
- Perform initial Voice of Customer (VOC) and identify Critical to Satisfaction (CTS).
- Select team and launch the project.
- Create project plan.

Develop Project Charter

The team developed the project charter that consisted of defining the problem, identifying the stakeholders, what is critical to satisfaction, the project goals, and the potential benefits.

The project charter is shown in Table 9.1, project charter.

Perform Stakeholder Analysis

The emergency department physicians are key stakeholders in identifying and treating septic patients in the ED. They are concerned with timely identification and treatment of septic patients and reducing the sepsis mortality rate of these patients. Intensivists are physicians who care for patients in the Intensive Care Unit (ICU). They are concerned with the patients' sepsis mortality rate and with patient care. The ED nurses who provide patient care in the ED and the Nurse Educator who provides nurse education for ED nurses are concerned with patient care, patient mortality, and timely identification and treatment of sepsis/septic patients. The patients who come to the ED with sepsis are concerned with patient care, their mortality from sepsis and

Table 9.1 Project charter

Project Overview, Problem Statement, and Goals
Provides a description of the problem to be solved. Be as descriptive as possible. Include improvement goals that are SMART (Specific, Measurable, Attainable, Realistic, and Time-Based)
Project Name: Sepsis Protocol Design Project Problem Statement: There is a national initiative to improve sepsis mortality. Sepsis mortality is one of the core CMS measures. We want to provide excellent care to our sepsis/septic patients Goal(s): To reduce mortality, identify patients promptly, and initiate treatment immediately Scope Statement: All the patients presented with a potential septic problem to the emergency department. We are focusing on the five most common anatomical sources of infection, which are lungs, skin, abdomen, urinary tract, and nervous system. This includes the process of identifying the septic patients and providing prompt treatment. Also includes inpatients related to sepsis Projected Financial Benefits: Reduction of the length of stay in the hospital. Reduction in severity of illness, reduction of cost of care

their health. Hospital administration is a project stakeholder and is concerned with sepsis mortality rate across the hospital, not just in the ED, but when patients are admitted into the hospital's intensive care unit. The stakeholder analysis is shown in Table 9.2

Suppliers-Inputs-Processes-Outputs-Customers

A description of the high-level process follows. The patient arrives to the emergency department with symptoms of sepsis or septic shock. The patient is registered, triaged, and assessed by the ED physician. The ED physician orders diagnostic tests which can include labs and imaging exams to assess the patient's condition. The diagnostic tests are performed. If sepsis is suspected or confirmed, the patient is admitted to the intensive care unit. In the current state, there are no sepsis protocols that are initiated in the emergency department. The Suppliers-Inputs-Processes-Outputs-Customers (SIPOC) is shown in Table 9.3.

Table 9.2 Stakeholder analysis

Stakeholder analysis					
Stakeholder	Type	Primary role	Impacts/ concerns	Initial receptivity	Future receptivity
ED physicians	Primary	Physicians who treat ED patients for sepsis	Timely identification and treatment Sepsis mortality rate	**Strongly support**	**Strongly support**
Intensivists	Primary	Intensive care physicians who care for patients in ICU	Sepsis mortality rate Patient care	**Neutral**	**Moderate support**
ED nurses (includes triage nurses); nurse educator	Primary	ED nurses who provide ED patient care and who triage the patients; Nurse educator who provides nursing education	Patient care Mortality Timely identification and treatment	**Moderate support**	**Strongly support**
Patients	Primary	Patients who have or may have sepsis	Patient care Mortality Health	**Neutral**	**Neutral**
Administration	Secondary	Hospital administration	Sepsis mortality rate	**Strongly support**	**Strongly support**

Table 9.3 SIPOC

SIPOC				
Suppliers	Inputs	Process	Outputs	Customers
Patient ED staff ED physician	Patient condition Patient information	Identify patient with sepsis/septic shock	Patient info	ED staff Patient
ED staff Patient	Patient info	Order diagnostic	Diagnostic orders	ED staff Radiology Lab
ED staff Radiology Lab	Diagnostic orders	Perform diagnostics	Results	ED staff ED physician
ED staff ED physician Admitting physician	Sepsis protocols Admission orders	Admit patient to ICU	Patient condition Admission orders	ICU staff Intensivists Patient Admitting physician

Perform Initial Voice of Customer and Identify Critical to Satisfaction

The critical to satisfaction criteria for this project are follows:

1. Reduced mortality rate of sepsis patients.
2. Time to identify patients with sepsis.
3. Time to identify diagnostic results.
4. Time to administer treatment to identified patients with sepsis, within 1 h of arrival (identification).
5. Time to source control (identify source of the infection).
6. Percent of patients receiving sepsis protocols.

These CTS' will be operationally defined in the Measure phase.

Select Team and Launch the Project

The team consisted of ED physicians and nurses, as well as the ED nurse educator, intensivists, Sepsis specialists, and a Lean Six Sigma Master Black Belt. It was important to have the extensive knowledge and buy-in of the intensivists and sepsis specialists involved on the project.

Create Project Plan

The project was started October 3, 2009. The Define phase was completed by October 22, 2009. The Measure phase was completed on November 7, 2009. The Analyze phase ended November 27, 2009. The Improve phase ended December 31, 2010. The Control phase was completed with assessing improvement and implementing the control plan in January 2011. The project milestones are shown in Fig. 9.1.

After the team completed the Define phase, we moved into the Measure phase.

Measure Phase

The Measure phase included the following activities:

- Define the current process.
- Define detailed Voice of Customer (VOC).
- Define the Voice of Process (VOP) and current performance.
- Validate measurement system.

Project Plan - Milestones

	Estimated Finish Date	Activities
D	10/22	Define project charter and scope, stakeholders and initial Critical to Satisfaction (CTS).
M	11/7	Measure current process and baseline. Mortality data for benchmarking. Identify improvement ideas, protocols.
A	11/27	Identify and analyze problems or issues that impede the process.
I	12/31	Prepare an initial protocol Move protocol through Committees Create order set Mortality data for benchmarking Create a Sepsis Alert Sepsis response Pilot project
C	1/31	Implement monitoring and control mechanisms

Fig. 9.1 Project milestones

Define the Current Process

The first step was to understand the current process by creating detailed process maps shown in Fig. 9.2. A description of the process follows.

The patient arrives at the emergency department and signs into the kiosk in the lobby. If an ED bed is available, the patient will be moved immediately to an ED room. If an ED bed is not available, the patient will be triaged in the triage room. When the bed is available the patient will be moved to the ED room. The ED nurse will assess the patient and start any treatment, such as an IV as appropriate. The ED physician will assess the patient and order appropriate diagnostic tests, including imaging exams and lab tests. The imaging and lab departments will perform the exams and provide the results. The ED physician will determine the patient's disposition. If sepsis or septic shock is diagnosed, the ED physician will contact the admitting physician for admission orders. The admitting physician will provide the admission orders to the intensive care unit. The patient will be admitted to ICU and moved to the unit. If sepsis is not diagnosed, the patient will be admitted or discharged as appropriate per the ED physician's diagnosis and disposition. In the current state process, there was no application of sepsis protocols for sepsis or septic shock patients while they are in the emergency department.

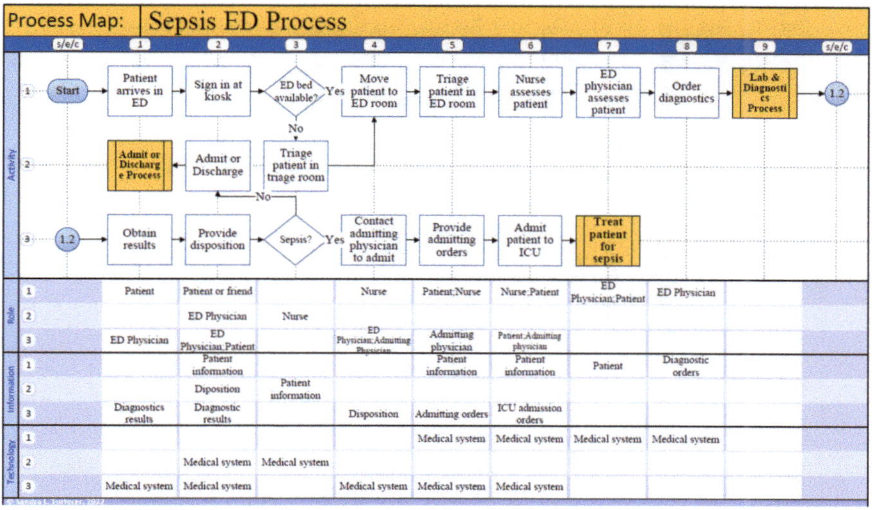

Fig. 9.2 Process map

Data Collection Plan

We developed the data collection plan to determine the metrics and operational definitions to measure the critical to satisfaction criteria. The data collection plan with the baseline metrics is shown in Table 9.4. The following metrics were defined.

Mortality Rate of Sepsis Patients

Defining the Measure. Mortality rate of hospital sepsis patients.

The percent of patients that expire after being diagnosed with sepsis, across the entire hospital (not just the emergency department).

Purpose. To understand the sepsis patient mortality rate across the hospital.

Measure the Process. Quality management has sepsis mortality rate information for the hospital, for Year-to-Date (YTD) from January 2009 through February 2010.

Time to Identify Patients with Sepsis

Defining the Measure. Time to identify patients with sepsis.

The time from when the patient arrives in the ED to when they are identified with sepsis.

Purpose. To understand the time to identify that the patient has sepsis. This will be measured by calculating the time from arrival to EDP disposition.

Measure the Process. This is not currently measured in the current state, but will be measured in the future, as the time from when the patient arrives in the emergency

Table 9.4 Data collection plan

Data collection plan						
Critical to satisfaction (CTS)	Metric (short title)	Operational definition (metric description)	Data collection source	Analysis mechanism	Sampling plan (size, frequency)	Process to collect and report
Mortality rate of hospital sepsis patients	Mortality rate for septic patient	Number patients expired from sepsis/ Number of patients diagnosed with sepsis	Director of quality, mortality rate: 13-month baseline	Statistics	13 months	Mortality rate (13 months) 21.8%
Time to identify patients with sepsis Time to source control	Time to identify patients with sepsis: Time from arrival to EDP disposition	Time from when patient arrives in EDP to sepsis identified	ED bed board			Average = 245 min Standard Deviation = 134 N = 111
Time to treatment	Time to begin treatment	Appropriate use per medical research	Do not currently measure			Do not currently measure
Patients receive sepsis protocols	Percent of patients receiving sepsis protocols	Percent of patients receiving sepsis protocols	Review of electronic medical record and patient charts for patients' initial diagnosis of sepsis for January 1 through June			Percent receiving sepsis protocol: 0%

department and signs into the kiosk to when the ED physician provides the diagnosis or disposition.

Time to Treatment

Defining the Measure. Time to treatment.

The time to begin sepsis treatment.

Purpose. To understand the time to start sepsis treatment.

Measure the Process. This is not currently measured in the current state, but will be measured in the future, as the time from when the patient arrives in the emergency department to when the sepsis protocol is started to be administered, but will be measured by the initial start time of the orders, making the assumption that the sepsis protocol is within the order set.

Percent of Patients Receiving Sepsis Protocols

Defining the Measure. Percent of patients receiving sepsis protocols.

The percent of patients receiving sepsis protocols of those that are diagnosed with sepsis.

Purpose. To understand the percent of the patients that have obtained the sepsis protocols.

Measure the Process. The baseline time period is January 1 through June 30, 2009, for those patients diagnosed with sepsis. We calculate the percent of those patients that received sepsis protocols in the emergency department. There were 28 patients diagnosed with sepsis that came through the ED, and none of them received a sepsis protocol in the ED, since none had been implemented at this time.

Define Detailed Voice of Customer

We met with the stakeholders to understand what is important to them regarding the patient care for septic patients. We derived the following critical to satisfaction criteria:

- Reduced mortality rate of sepsis patients.
- Time to identify patients with sepsis.
- Time to identify diagnostic results.
- Time to administer treatment to identified patients with sepsis, within 1 h of arrival (identification).
- Time to source control (identify source of the infection).
- Percent of patients receiving sepsis protocols.

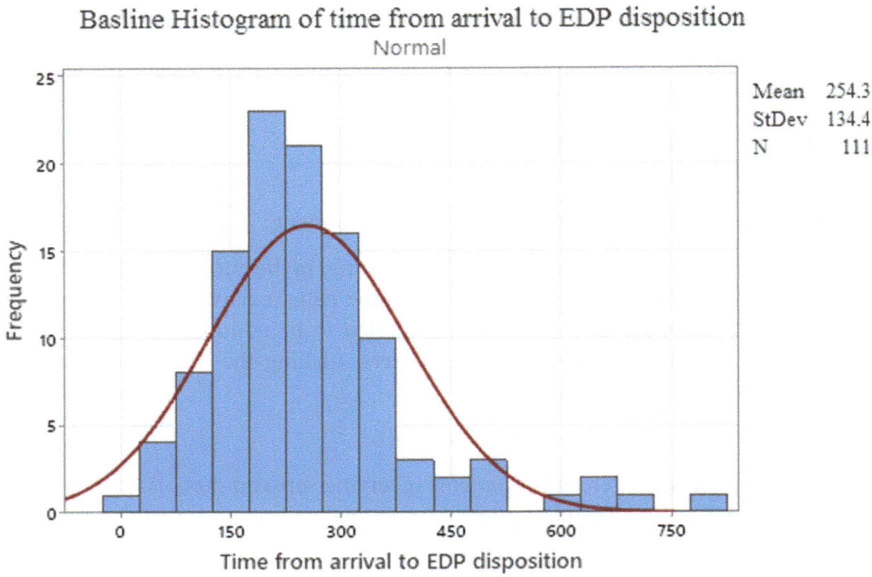

Fig. 9.3 Baseline—time to disposition (time to diagnose sepsis/time to identify source of infection) time from arrival to disposition

Define the Voice of Process and Current Performance

The current baseline performance for each metric is shown in Fig. 9.3.

We currently only measure the following metrics:

1. Sepsis mortality rate, which is 13.5% for all patients that were diagnosed with sepsis in the hospital (not just the emergency department). This is for the time period of January 2009 through February 2010.
2. The time to identify patients with sepsis is measured by the metric of time to ED physician disposition. The average time was 254 min, with a standard deviation of 134 min. The histogram of the distribution is shown in Fig. 9.4.
3. The percent of patients that received sepsis protocols in the emergency department for the time period of January 1 through June 2009, which was 0% of the 28 patients diagnosed with sepsis that presented in the emergency department. We did not currently have implemented sepsis protocols in the emergency department.

Validate Measurement System

We validated the ED Bed Board data for the sepsis patients by comparing the data to the patients' charts or electronic medical records.

We then moved into the Analyze phase of the project.

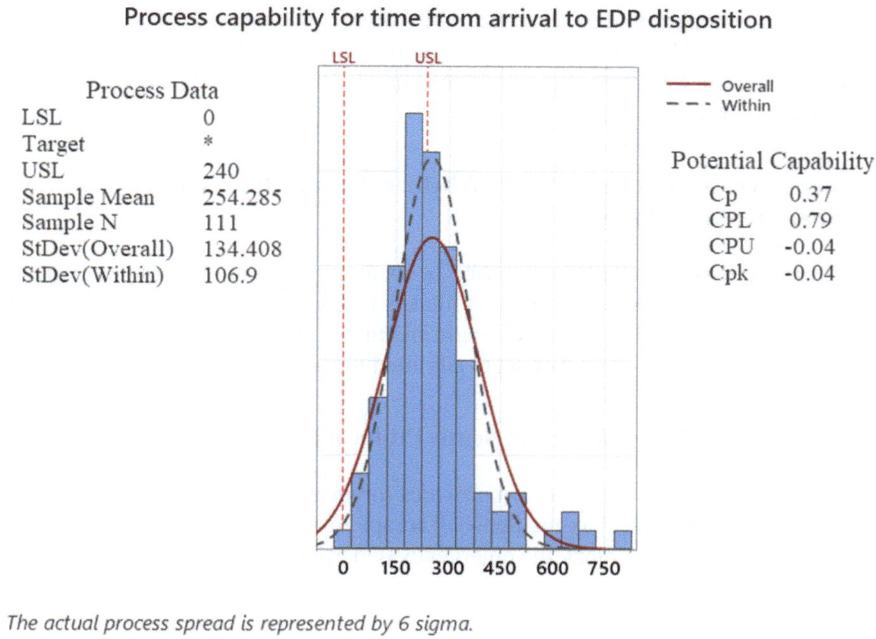

Fig. 9.4 Baseline process capability of time to disposition (time to diagnose sepsis/time to identify source of infection)

Analyze Phase

The following activities were performed for the Analyze phase to identify and validate the root causes and to assess whether the process is currently capable of meeting the desired specifications for the length of stay:

- Develop cause-and-effect relationships.
- Determine and validate root causes.
- Develop process capability.

Develop Cause-and-Effect Relationships and Validate Causes

The root causes were very obvious and were the reason for the project to be initiated.

Summary of Root Causes

- **There Were No Standard Protocols Defined for Sepsis in the Emergency Department**

1. No standard protocols had been developed or implemented in the emergency department for treating sepsis. The patients would be admitted to the Intensive Care Unit (ICU), but no consistent, standard protocols were used across the

emergency department to begin treatment prior to the patient being admitted to ICU. This would potentially lose precious treatment time for sepsis patients. The ED physicians used their experience and knowledge, but did not apply consistent, standard sepsis protocols.

- **There Were No Metrics to Assess Whether Patients Received Antibiotics and Other Treatment for Sepsis**

1. There were no standardized metrics or measures instituted in the emergency department to measure the time that it takes to assess, diagnose, and disposition a patient with sepsis, as well as begin treatment. The patients typically began treatment for sepsis once they were admitted to the ICU.

- **There Was No Use of a Formal Sepsis Protocol Database**

1. There is a computer program that provides for an assessment of symptoms using the MUST (Multiple Urgent Sepsis Therapies) protocol (Howell & Shapiro, 2004). Prior to this project, the emergency department did not use this program.

Develop Process Capability

We developed the process capabilities for three metrics:

- Time to identify patient with sepsis.
- Percent of patient receiving sepsis protocols.
- Sepsis mortality rate.

Time to Identify Patient with Sepsis (Time to ED Disposition)

We performed a process capability for the metric of the time to identify the patients with sepsis. The metric used was the time to ED physician disposition. The lower specification is 0 min, and the upper specification limit is 90 min. The process capability Cp index for the time to ED disposition was 0.37, which shows that the process is not capable. The Cpk index, which shows centering of the process between the upper and lower specification limits, is −0.04, showing that the process is also not centered. The process capability for the time to disposition is shown in Fig. 9.4.

Percent of Patients Receiving Sepsis Protocols

We then performed a process capability for the metric of percent of patients receiving sepsis protocols. For a proportion metric, the process capability is the percentage or proportion. The baseline percent of patient receiving sepsis protocols is 0%. The desired capability is 100%. The process is currently not capable of meeting this goal.

Sepsis Mortality Rate

The process capability of the sepsis mortality is 13.5%. The desired capability is 10%, with the lower the mortality the better. The process therefore is not currently capable of meeting the desired target.

Summary

We have established that the current process is not capable of meeting our goals of providing a diagnosis of sepsis in a timely manner, nor performing and obtaining the diagnostic tests timely as well, or in achieving the desired sepsis mortality rates. We also do not have current sepsis protocols that are administered in the emergency department and therefore will not be able to reduce sepsis mortality or administer sepsis treatment in a timelier manner, while the patients are still in the emergency department. We have much opportunity for improvement.

Improve Phase

The following steps were executed in the Improve phase:

- Identify improvement opportunities and plans.
- Design future state.
- Establish performance targets and project scorecard.
- Pilot improvements.
- Train and execute.

Identify Improvement Opportunities and Plans and Design Future State

The new high-level process steps for improving the process are shown in Table 9.5.

A new process step was incorporated after the sepsis diagnosis is identified by the ED Physician, "Implement Sepsis Protocol." In this step, the newly generated sepsis protocols will be implemented when the patient's disposition is sepsis, urosepsis, septic shock, or other related sepsis diagnoses.

The team also implemented the Sepsis Survival Alliance Sepsis Database that enables physicians for concurrent auditing and tracking of sepsis cases. It provides a decision tool to assess whether the appropriate sepsis survival protocols were applied.

The team implemented a sepsis protocol based on the Multiple Urgent Sepsis Therapies (MUST) Protocol. They also implemented a standard order set aligned to the Sepsis Protocol. They implemented a sepsis alert to implement the sepsis protocol quickly.

The following sepsis protocols were piloted first in the Emergency Department.

The Emergency Department physician identifies sepsis as a primary diagnosis and enters this into their disposition in the electronic medical record. They select the

Table 9.5 SIPOC for improved process

SIPOC				
Suppliers	Inputs	Process	Outputs	Customers
Patient ED staff ED physician	Patient condition Patient information	Identify patient with sepsis/ septic shock	Patient info	ED staff Patient
ED staff Patient	Patient info	Order diagnostic	Diagnostic orders	ED staff Radiology Lab
ED staff Radiology Lab	Diagnostic orders	Perform diagnostics	Results	ED staff ED physician
ED physician	Results Sepsis protocols	Implement sepsis protocols	Administered sepsis protocols	ED staff ED physician Patient
ED staff ED physician Admitting physician	Sepsis protocols Admission orders	Admit patient to ICU	Patient condition Admission orders	ICU staff Intensivists Patient Admitting physician

sepsis protocol orders which are then begun in the Emergency Department. Typically, the sepsis patients will be admitted to the hospital, many of them to the critical care unit.

Following is a description of the sepsis protocol and order sets:

<u>Patient Selection</u>

- SUSPECTED INFECTION <u>AND</u>
- Patients with:

Systolic blood pressure <90 <u>and/or</u> lactate over 4 or MAP <65

<u>And</u>

- Patients with 2 out of 4 of the following:

1. Fever (over 101) or temperature (under 96.8).
2. Heart rate >90 bpm.
3. Respiratory rate >20 or $PaCO_2$ <32.
4. WBC >12 or <4 <u>OR</u> bands >10,000/mm^3.

<u>STAT Laboratory Studies (Within 1 h of Identification)</u>

1. Serum lactate.
2. CBC and BMP.
3. ABG.
4. Obtain cultures (blood, urine and UA, sputum, etc.).
5. Type and Hold 2u PRBC.
6. PT, INR, PTT.

7. LFTs.
8. Pro-Calcitonin.

Diagnostic Procedures:

- Chest X-ray.

Interventions

- Fluid bolus 20 mL/kg.
- First dose antibiotics within 1 h (see below Empiric Antibiotic Choice per Source).
- Initiate ICU admission (*if no response after initial fluid bolus*).
- Airway and respiratory assessment.
- CVP placement, subclavian, or intrajugular (*if no response after initial fluid bolus*),
- Foley catheter,
- Arterial line, if on pressors.

Consults:

1. Intensivist.
2. Infectious disease, Dr. ______________________________

The Following Should Be in a CRITICAL CARE Setting
(no Response to Fluid Bolus)
Dysfunction Criteria for Severe Sepsis

1. Hypotension.
2. Bilateral infiltrates and requires supplemental O_2 to maintain saturation above 90.
3. PaO_2/FiO_2 <300.
4. Creatinine >2; urine output <0.5 mL/kg after fluid.
5. Platelets <100,000.
6. INR >5.
7. PT >20.
8. Bilirubin >2.
9. Altered mental status.
10. Glucose >120 mg/dL with no history of diabetes.

Early Goal Targets

- CVP 8–12 mm Hg.
- Mean arterial pressure (MAP) 65 mm Hg.
- Systolic blood pressure 90 mm Hg.
- Central venous oxygenation ($ScvO_2$) over 70% or mixed venous oxygen saturation (SvO_2) over 65%.
- Urine output > 0.5 ml/kg/hr.

Additional Orders

- Insulin per protocol (critical care hypoglycemia protocol).
- DVT prophylaxis: Heparin 5,000 units SC every 8 h

or ______________________________

- Stress ulcer prophylaxis: Famotidine 20 mg IV every 12 h

or ______________________________
Goal Directed Therapy (*antibiotics already given*).
Optimization of CVP—Step 1

- If CVP <8 mm Hg give sodium chloride 500 mL bolus over 15 min.
- Recheck CVP every 15 min and repeat boluses as above until target goal CVP of 8–12 mm Hg is achieved; then continue to Step 2.

Optimization of MAP—Step 2

- If MAP <65, start norepinephrine 2 mcg/min. May titrate by 2 mcg/min every 15 min to max 12 mcg/min to maintain MAP ≥65. Once MAP is achieved, continue to Step 3.

Optimization of $ScvO_2$—Step 3

- If $ScvO_2$ under 70% OR mixed $ScvO_2$ under 65% AND Hgb under 10 then transfuse 1 unit PBRC or as needed.
- If $ScvO_2$ under 70% OR mixed $ScvO_2$ under 65% AND Hgb over 10 then start dobutamine 2.5 mcg/kg/min. May titrate by 2.5 mcg/kg/min every 15 min to max of 20 mcg/kg/min until $ScvO_2$ over 70% or mixed $ScvO_2$ over 65%. Once $ScvO_2$ is achieved, goal directed therapy complete.

Empiric Antibiotic Selection by Source
First dose given with 1 h of admission

- *Suspected Community Acquired Pneumonia*

- Ceftriaxone 1 g IV every 24 h, first dose stat;
- azithromycin 500 mg IV every 24 h, first dose stat.

If allergic to penicillin or cephalosporins:
Levofloxacin 750 mg IV every 24 h, first dose stat.

1. *Suspected Healthcare Associated Pneumonia (HCAP) or Pseudomonas*
2. Cefepime 2 g IV every 8 h, first dose stat.
3. Levofloxacin 750 mg IV every 24 h, first dose stat.
4. Vancomycin 1 g IV every 12 h × 2, first dose stat.

If allergic to penicillin or cephalosporins:
Levofloxacin 750 mg IV every 24 h, first dose stat;
aztreonam 2 g IV every 8 h, first dose stat;
vancomycin 1 g IV every 12 h × 2, first dose stat.

- *Suspected Urinary Tract Infection*

1. Ceftriaxone 1 g IV every 24 h, first dose stat;

2. gentamicin 5 mg/kg IV stat.

   If allergic to penicillin or cephalosporins:
   Levofloxacin 750 mg IV every 24 h, first dose stat;
   gentamicin 5 mg/kg IV stat.

- *Suspected Intra-Abdominal Infection*

1. Levofloxacin 750 mg IV every 24 h, first dose stat;
2. metronidazole 500 mg IV every 8 h, first dose stat.

   If allergic to quinolones:
   ertapenem 1 g IV every 24 h, first dose stat.

- *Suspected Skin/Soft Tissue Infection*

1. Piperacillin/Tazobactam 4.5 g IV every 6 h, first dose stat;
2. vancomycin 1 g IV every 12 h × 2, first dose stat.

   If allergic to penicillin or cephalosporins:
   Levofloxacin 750 mg IV every 24 h, first dose stat;
   vancomycin 1 g IV every 12 h × 2, first dose stat;
   clindamycin 600 mg IV every 8 h, if diabetic foot ulcer or wound.

- *Suspected Severe Sepsis with Unknown Source of Infection*

- Piperacillin/tazobactam 4.5 g IV every 6 h, first dose stat;
- vancomycin 1 g IV every 12 h × 2, first dose stat.

  If allergic to penicillin or cephalosporins:
  Levofloxacin 750 mg IV every 24 h, first dose stat;
  metronidazole 500 mg IV every 8 h, first dose stat;
  vancomycin 1 g IV every 12 h × 2, first dose stat.
  Note The above dosages are calculated based on normal creatinine clearance.

Establish Performance Targets and Project Scorecard

The team established the performance targets for each metric as shown in Table 9.6.

Pilot Improvements, and Train and Execute

The sepsis protocol pilot began in March 2010 in the Emergency Department after going through the appropriate approval committees to gain buy-into the protocols. A

Table 9.6 Performance metric targets

Metric	Baseline	Target
Mortality rate for septic patient	Mortality rate: 21.8%	10%
Time to identify patients with sepsis: time from arrival to EDP disposition	Average = 245 min Standard deviation = 136 N = 111	90 min
Time to begin treatment	Do not currently measure	60 min
Percent of septic patients receiving sepsis protocols	Percent receiving sepsis protocols: 0%	100%

Grand Rounds information session was held with the physicians and other clinical staff to explain the sepsis protocols.

We initially assessed improvement of the pilots from March through May 2010, but did not have significant improvement in reducing the sepsis mortality rate, and realized that the sepsis protocol was not being used in the ED. We then restarted the pilot effort and then collected data to assess improvement for the May through September 2010 data.

We then moved into the Control phase where we assessed whether there was improvement in the metrics related to implementing the piloted changes.

Control Phase

The purpose of the Control phase is to implement control mechanisms and procedures to ensure that the process maintains the levels of improvements attained during the pilot activities.

The following activities were executed in the Control phase:

- Measure results and manage change.
- Report scorecard data and create process control plan.
- Identify replication opportunities.
- Develop future plans.

Measure Results and Manage Change

We collected the improved data for the metrics and performed the appropriate hypothesis tests to assess improvement.

We first assessed the sepsis mortality improvement for the time period of June through September. We performed the two-proportion test in Minitab®.

Table 9.7 Percent of patients receiving sepsis protocols—before and after

Metric	Baseline	Improved (June to September)
Sepsis mortality	66	9
Total cases	303	74
Percent mortality	21.8%	12.2%

Test and CI for Two Proportions: Mortality Rate

Sample	X	N	Sample p
1	66	303	0.217822
2	9	74	0.121622

Difference = p (1) − p (2)
Estimate for difference: 0.0962002
95% CI for difference: (0.00841781, 0.183983)
Test for difference = 0 (vs. not = 0): Z = 2.15 p-value = 0.032
Fisher's exact test: p-value = 0.074

The before and after mortality is shown in Table 9.7. The mortality was reduced by 9.6% from 21.8 to 12.2%, a statistically significant improvement, with a p-value of 0.032.

We then assessed whether there was a significant improvement in the time from arrival to EDP disposition. We performed a two-sample t-test as follows:

Two-Sample T-Test and CI: Time from Arrival to Disposition

Sample	N	Mean	St Dev	SE mean
1	111	254	136	13
2	64	197.0	87.0	11

Difference = mu (1) − mu (2)
Estimate for difference: 57.0
95% CI for difference: (23.7, 90.3)
T-Test of difference = 0 (vs. not =): T-Value = 3.38 p-value = 0.001 DF = 171

There was a significant improvement in the time from arrival to disposition; however, it is not meeting the target of 90 min. The average time improved from 254 to 197, with a *p*-value of 0.001. The histogram is shown in Fig. 9.5 for the time to disposition.

Since the sepsis protocols had not been used in the past, we could not assess improvement; however, we did collect data on the time from arrival to when the orders were initially begun. The average was 33.6, with a standard deviation of 28.1. This is better than the target of 60 min. The histogram of the time to the orders is shown in Fig. 9.6.

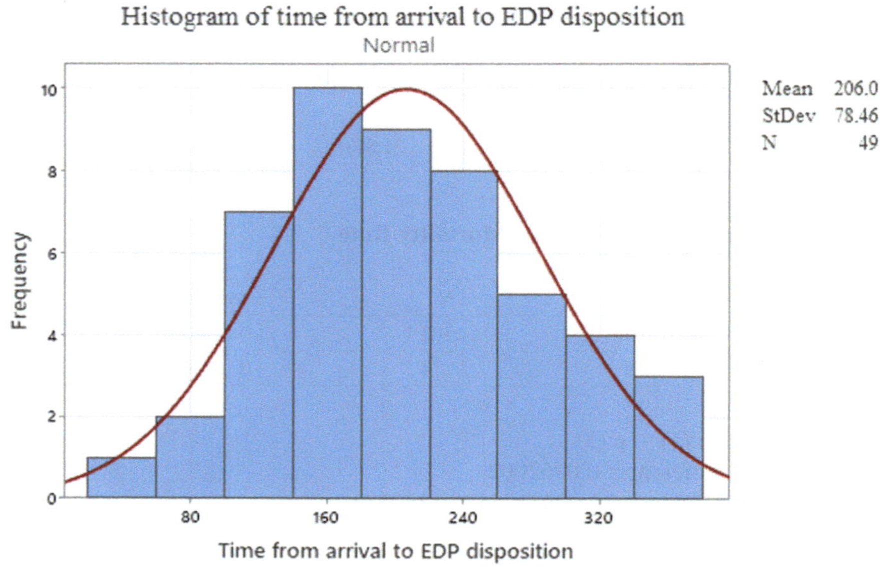

Fig. 9.5 Improvement—time to disposition

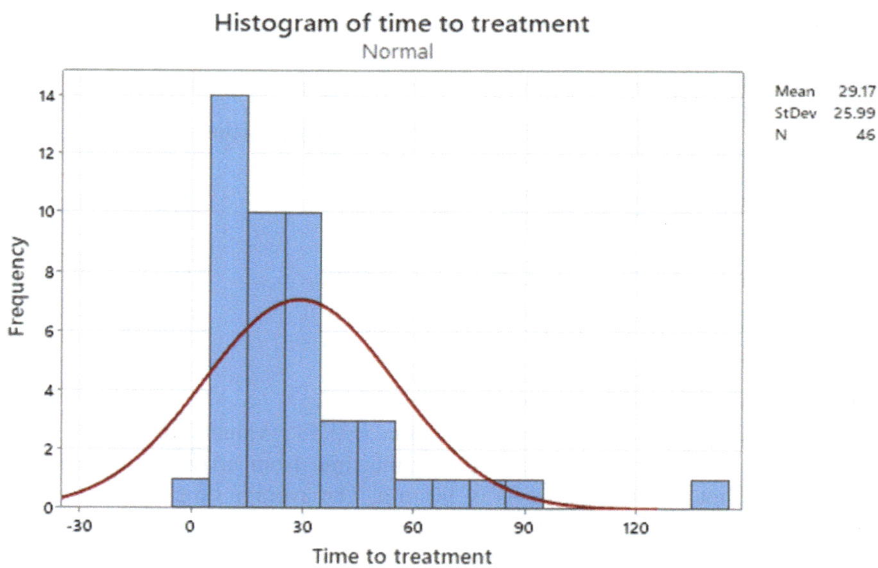

Fig. 9.6 Improvement—time to treatment

We then assessed whether there was a significant improvement in the percent of patients diagnosed with sepsis that received the sepsis protocols in the Emergency Department. The baseline percentage was 0%, and the improved percentage was 5.4%, still far from meeting the target of 100%. The percentages are shown in Table 9.8.

The metrics before and after improvement are shown in Table 9.9. We have shown improvement in all of the metrics, even though we still need to ensure that the protocols are delivered promptly to improve the percent of patients that receive the protocols and to further reduce the sepsis mortality.

We redid the process capability for the improved process for the time from arrival to disposition (Fig. 9.7) and the time to treatment (Fig. 9.8). The processes are still not capable of meeting the desired specifications. Further improvement and adherence to the protocols is needed.

Table 9.8 Before and after percent of patients receiving sepsis protocol

Metric	Baseline	Improved (June to September)
Number receiving sepsis protocols	0	4
Number sepsis diagnosis	28	74
Percent patients receiving sepsis protocols	0%	5.4%
Statistically significant improvement?	Yes	P-value = 0.040

Table 9.9 Before and after metric improvement

Critical to satisfaction (CTS)	Metric	Baseline	Improved (June to September)	Target	Significant improvement/ P-value
Mortality rate of hospital sepsis patients	Mortality rate for septic patient	Mortality rate: 21.8%	Mortality rate: 12.2%	10%	Yes P-value = .032
Time to identify patients with sepsis Time to source control	Time to identify patients with sepsis: time from arrival to EDP disposition	Average = 245 min Standard deviation = 136 N = 111	Average = 197 Standard deviation = 86.8 N = 64	90 min	Yes P-value = .001
Time to treatment	Time to initial orders	Do not currently measure	Average = 33.6 min Standard Deviation = 28.1 N = 73	60 min	Met target
Patients receive sepsis protocols	Percent of patients receiving sepsis protocols	Percent receiving sepsis protocols: 0%	Percent receiving sepsis protocols: 5.4%	100%	Yes P-value = .040

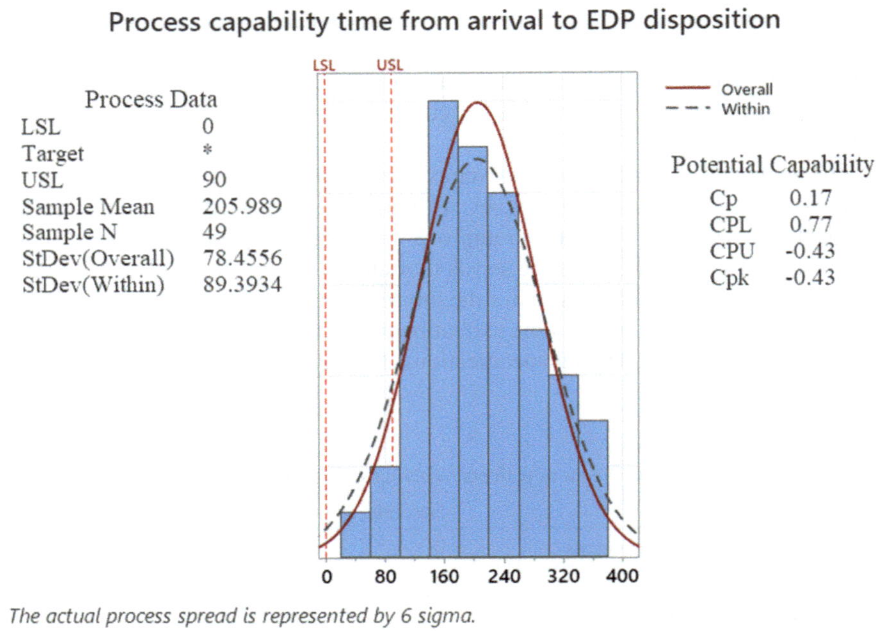

Fig. 9.7 Improved process capability of time to disposition

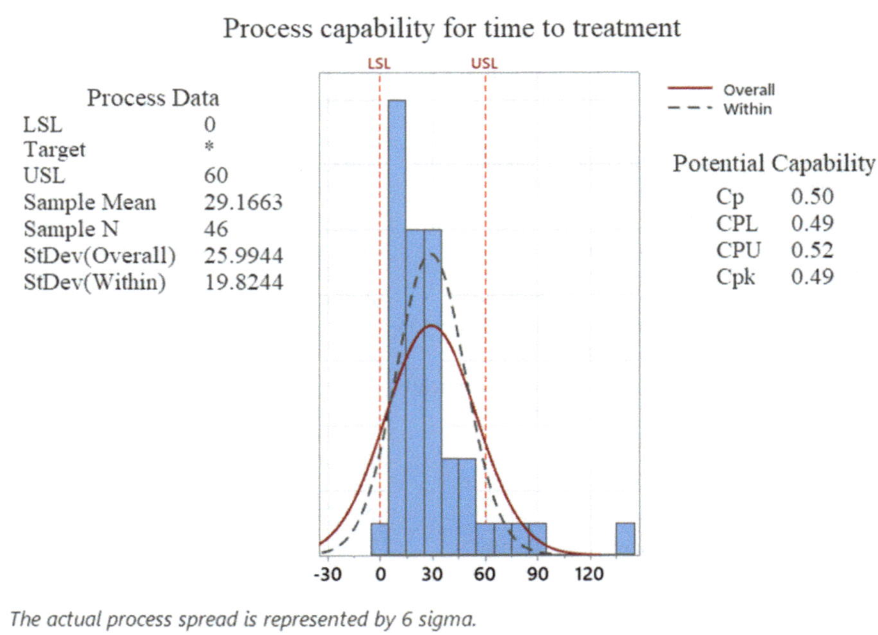

Fig. 9.8 Improved process capability of time to treatment

Report Scorecard Data and Create Process Control Plan

The control plan for the sepsis project will be to monitor the metrics on a monthly basis. The metrics dashboard is shown in Table 9.9.

Identify Replication Opportunities

The sepsis protocols can be implemented throughout the hospital. Most patients with sepsis are admitted through the Emergency Department, but these same protocols can be used and continued in the critical care units and other hospital departments as appropriate.

Develop Future Plans

We need to identify additional improvements so that the performance targets can be met and further statistical improvements can be attained for the remaining metrics.

Summary and Project Critical Success Factors

1. Engagement of the Emergency Department physicians, the intensivists, and sepsis specialists was critical to the success of the project.
2. Applying the Six Sigma methodology to assess the baseline and improved metrics and to apply the Six Sigma tools was also critical.
3. Facilitation and analysis by the Six Sigma Master Black Belt were critical to ensure that the value of Six Sigma was attained.

Engagement of ED staff, including the nurse manager, the charge nurses, and the ED nurses, is critical to ensure that the protocols are implemented in a timely manner.

References

Angus, D. C., Linde-Zwirble, W. T., Lidicker, J., et al. (2001). Epidemiology of severe sepsis in the United States: Analysis of incidence, outcome, and associated costs of care. *Critical Care Medicine, 29*(7), 1303–1310.

Annane, D., Aegerter, P., Jars-Guincestre, M. C., et al. (2003). Current epidemiology of septic shock: The CUB-Réa network. *American Journal of Respiratory Critical Care Medicine, 168*(2), 165–172.

Howell, M. D., & Shapiro, N. I. (2004). *MUST protocol.* Boston, Massachusetts: Beth Israel Deaconess Medical Center.

Kung, H. C., Hoyert, D. L., Xu, J., et al. (2008). Deaths: Final data for 2005. *National Vital Statistics Reports, 56*(10), 1–120.

Levy, M. M., Fink, M. P., Marshall, J. C., et al. (2003). 2001 SCCM/ESICM/ACCP/ATS/SIS international sepsis definitions conference. *Critical Care Medicine, 31*(4), 1250–1256.

Morrell, M. R., Micek, S. T., & Kollef, M. H. (2009). The management of severe sepsis and septic shock. *Infectious Disease Clinics of North America, 23*, 485–501 (Elsevier Inc.).

Sasse, K. C., Nauenberg, E., Long, A., et al. (1995). Long-term survival after intensive care unit admission with sepsis. *Critical Care Medicine, 23*(6), 1040–1047.

Chapter 10
Improving Satisfaction and Timeliness in Interventional Radiology Processes Through Lean Six Sigma

Abstract The Interventional Radiology department at an acute care hospital identified opportunities for improvement that improved throughput, patient and staff satisfaction, and provided a more robust and accurate operational data entry system to enhance continuous improvement. The Lean Six Sigma (LSS) team applied the DMAIC methodology with a heavy emphasis on statistical, graphical, and data analysis to baseline the current process. Root cause and lean analyses were critical in identifying root causes, inefficiencies, and wastes in the processes that were reduced and eliminated to improve the processes. The average on-time case start percentages increased 22% from the baseline period (19% to 41%) and achieved a high of 60% on-time starts during the improvement piloting period. The average on-time start times improved by 50%, and the percentage for the times in the holding area 30 min prior to the appointment time increased during the Q1 improvement period from 43 to 57%. The percentages in the procedure room, 10 min prior to the appointment time, increased during Q1 from 21 to 40%. In-room prep times improved by 5.5%, and the average procedure times were reduced by 24%, with a 30% standard deviation reduction. The end of exam to out of room times did increase during the improvement period, illustrating there was more improvement to be made.

Keywords Interventional radiology · Lean Six Sigma DMAIC methodology · Process improvement · Statistical analysis · Data-driven decision-making

Introduction

The Physician Director and Nurse Director of the Interventional Radiology (IR) department partnered with the Lean Six Sigma team made up of the clinical, registration, laboratory and tech areas to improve the patient and staff experience, as well as improve the on-time starts for the IR operating procedures, and the related process times. There was an additional focus in this project on reducing data entry errors related to recording the case start, end, and interim process times. The Lean Six Sigma (LSS) DMAIC methodology and tools were applied, as shown in Table 10.1.

S. Furterer, *Lean Six Sigma for the Healthcare Enterprise*,
https://doi.org/10.1007/978-1-4471-7617-6_10

Table 10.1 Six sigma tools applied

DMAIC phase	Tools applied
Define	• Project charter • Critical to satisfaction (CTS) • Stakeholder analysis • Project risk analysis • SIPOC
Measure	• Voice of the customer (VOC) • Data Collection Plan • Statistical, graphical, and data analysis • Value stream map • Process architecture maps
Analyze	• Cause-and-effect analysis—Why-why diagrams • Lean value analysis
Improve	• Improvement recommendations • Statistical, graphical, and data analysis • Hypothesis testing
Control	• Control plan • Statistical Process Control • Dashboard and alerts

Extensive statistical, graphical, and data analyses were used to understand the current process and assess improvements in the IR processes. Having ready access to the medical record system's data from the systems project manager along with her managing the improvement plan implementation was a critical success factor to the success of this project. Even though the team faced incredible challenges, including the pandemic occurring during the project, and having reduced staff in the IR department, the team rallied to improve the on-time starts from the third quarter in the baseline year to the first quarter in the following year, from 19 to 60%, resulting in a 41% improvement. There was a 50% improvement in the time that the procedure starts compared to the scheduled appointment time from quarter 1 of the baseline year to quarter 1 of the following year.

Define Phase

The LSS team developed, reviewed, and approved the project charter, shown in Table 10.2.

The Physician Director of the IR raised concerns related to procedure times' data entry errors, low patient and staff satisfaction, and on-time start percentages that were not meeting the desired targets. The Physician and Nurse Directors worked with a local university to form a LSS team of the clinical, registration, laboratory, and tech areas to optimize process flow for inpatients and outpatients who have IR

Table 10.2 Project charter

Project Overview, Problem Statement, and Goals
Project Overview: The project focuses on optimizing process flow of Interventional Radiology (IR) processes for outpatients (OP) and inpatients (IP) at the hospital. Problem Statement: IR processes at the hospital are not being carried out at an optimal level. IR Radiologists feel the process is less efficient due to the reasons like manual data entry points and lack of robust data analytics, communication issues during scheduling, multiple systems data sync issues, patient's lack of awareness of hospital location and their medical procedure. Goal(s): • Optimize process flow for IP and OP to reduce overall cycle times. • Reduce delays in a variety of areas. • Enhance financial performance through more efficient process flow. • Improve patient and staff satisfaction. • Provide robust and reliable data analytics. Project Scope: For outpatients: the process starts with procedure scheduling, and it ends when the patient leaves the hospital after the procedure. For inpatients: the process starts when the patient is being consulted for the procedure, and it ends when the patient returns to the room after the procedure. The scope includes the IR staff and patient satisfaction (current and future), patient check-in and check-out, and all processes related to the IR department. Benefits: Improved staff and patient satisfaction, robust data processes, and improved patient throughput time.

procedures, reduce delays, improve patient and staff satisfaction, and provide robust data analytics across the department.

The team identified four critical to satisfaction (CTS) measures (Table 10.3) that would be the thread throughout the project to make process improvements. The first CTS was patient throughput time, which included the process times for each step of the process, including registration, preparation in the surgical holding area, preparation in the room, the procedure, and post-procedure activities. Staff desired to serve patients through efficient processes with the least waiting times. Patients desired to have high-quality care. The robust data analytics included having all the data captured with the intended quality to be able to conduct data analysis for process improvements.

The next step in the Define phase was to identify the stakeholders that should be engaged in the LSS project. Table 10.4 shows the stakeholder analysis. The primary stakeholders included the inpatient schedulers, outpatient schedulers, advanced

Table 10.3 Critical to satisfaction (CTS) characteristics

	Critical to Satisfaction (CTS)	
	Title	**Description**
1	Patient throughput time	Outpatients process time: from the patient's arrival at the registration desk to departure from the hospital after the procedure. Inpatient process time: from the patient's arrival at the holding area to the return to the inpatient floor after the procedure.
2	Staff Satisfaction	Able to serve patients through efficient processes with the least waiting time and effective coordination.
3	Patient Satisfaction	High-quality care for the patient that meets the expectations and needs of the patient with less waiting time.
4	Robust Data Analytics	Having all the data captured with the intended quality to be able to conduct data analysis for process improvements.

patient treatment nurses, patient care techs (PCT), lab technicians, patients, anaesthesia, inpatient nurses, IR nurses, techs, and physicians. Secondary stakeholders included referring physicians, registration, and the transportation team. The primary role of each stakeholder group was defined, along with how each group is impacted by the project. The stakeholder group's initial receptivity to the project was assessed, and the need for the future receptivity was also documented.

Table 10.5 shows the project risk analysis for what could go wrong during the project that could prevent the project from completing successfully. The two critical risks were the inability to get all of the data needed, including lab and anaesthesia process times, and engaging the required departments. Identifying these risks at the start of the project can help to mitigate them so that they either don't occur, or if they do, to minimize the impact of the risk event occurring.

The SIPOC (Table 10.6) shows the high-level process steps included in the project, along with the inputs and outputs used in the process steps, along with the supplier providing the input, and the customer who receives the output of the process step. This tool helps to ensure that the project scope and the process to be improved is appropriate for the project. The process steps to be included start from the patient scheduling, registration, preparation in the pre-procedure holding area, preparation in the IR procedure room, the procedure, and the post-procedure activities. Preparation of the room is also included in the process to be improved. The suppliers and customers should also be included in the stakeholder analysis and provide a check and balance for the stakeholders and team members.

Table 10.4 Stakeholder analysis

Stakeholder analysis					
Identify the internal and external customer roles and concerns to ensure their expectations are addressed. Define roles, impacts, concerns, and receptivity					
Stakeholder	Type	Primary role	Potential impacts/concerns	Initial receptivity	Future receptivity
Referring physicians	**Secondary**	Treat patients that need a procedure in one of the following departments: Cath Lab, Endoscopy, Interventional Radiology (IR), Vascular Lab, Fluoroscopy, Pain Center, and Dental Services	Quality of treatment Process efficiency	**Neutral**	**Moderate support**
Inpatient schedulers	**Primary**	Schedules the inpatients for IR procedures	Right instructions to patients Process efficiency	**Neutral**	**Moderate support**
Outpatient schedulers	**Primary**	Schedules the outpatients for IR procedures	Right instructions to patients Process efficiency	**Strongly support**	**Strongly support**
Registration desk	**Secondary**	Oversees the registration, demographic, and financial agreement for outpatients	Correct registration Patient's waiting time	**Neutral**	**Neutral**
Advanced treatment area (ATA)—nurses and patient care techs (PCT)	**Primary**	Area where the patient is prepped before IR procedure and monitored after the IR procedure	Quality of care Patient throughput time	**Strongly support**	**Strongly support**
Laboratory technicians	**Primary**	Technicians who collect the samples and run the tests in the lab before the IR procedure	Turnaround time Patient's waiting time	**Neutral**	**Moderate support**

(continued)

Table 10.4 (continued)

Stakeholder analysis

Identify the internal and external customer roles and concerns to ensure their expectations are addressed. Define roles, impacts, concerns, and receptivity

Stakeholder	Type	Primary role	Potential impacts/concerns	Initial receptivity	Future receptivity
Transportation team	**Secondary**	Transports the patients from one department to another department for various treatments and procedures	Less waiting time Process efficiency	**Strongly support**	**Strongly support**
Patients	**Primary**	People who are being treated and IR procedures carried out on	Patient Satisfaction Quality of care	**Neutral**	**Neutral**
Anesthesia department	**Primary**	Gives anesthesia to the required patients during the IR procedure	On-time availability Quality of care	**Neutral**	**Moderate support**
Inpatient nurses	**Primary**	Attends the patients to prepare the patient for IR procedure	Quality of care Process efficiency	**Strongly support**	**Strongly support**
Interventional radiology (IR) staff—nurses and techs	**Primary**	Assists pre, post, and during the IR procedure	Quality of care Process efficiency	**Strongly support**	**Strongly support**
Physicians—interventional radiologists	**Primary**	Physicians (Radiologists) who conduct IR procedures for both inpatients and outpatients	Quality of treatment	**Strongly support**	**Strongly support**

Table 10.5 Project risk analysis

Project Risk Analysis			
Brainstorm the potential risks to a successful project, identify the likelihood that each will occur, the potential impact, and develop a mitigation strategy.			
Potential Risk to Successful Project	**Occurrence of Risk**	**Impact of Risk**	**Risk Mitigation Strategy**
Getting all data needed (lab and anesthesia)	Moderate	High	Ask other departments how the process happens and getting IT involved
Engaging departments	Moderate	High	Include all departments early in the project

Measure Phase

The Value Stream Map (Fig. 10.1) shows the patient arriving at the holding area, where holding preparation activities are performed. The patient is then moved to the IR procedure room and prepped in the room. The procedure is performed, the exam is recorded as complete, post-procedure activities are performed and then the patient is released and either moved to their inpatient room or discharged as an outpatient. The median time from holding to end of exam to out of the room is 89 min. There is additional time in the system for registration and pre-holding preparation prior to holding, and post-procedure activities, and transport to the inpatient room or discharge from the procedure.

A data collection plan (Table 10.7) identifies the CTS, metrics, operational definitions and data collection source. The metrics that were baselined are shown in the data collection plan.

The process maps provide an understanding of the IR processes. The process architecture cover page which summarizes the architecture elements of stakeholders, information, and technology used across the IR processes, is shown in Fig. 10.2.

The following process maps were documented:

1. Inpatient Schedule (Fig. 10.3).
2. Outpatient Schedule (Fig. 10.4).
3. Outpatient Registration (Fig. 10.5).
4. Inpatient—Holding, Procedure, and Post-Procedure (Fig. 10.6).
5. Outpatient—Procedure and Post-Procedure (Fig. 10.7).
6. Lab Department (Fig. 10.8).
7. Anaesthesia Department (Fig. 10.9).
8. Tech (Fig. 10.10).

Table 10.6 SIPOC

Supplier	Input	Process	Output	Customer
• Registration desk	• Patient demographics • Outpatient schedule	• Schedule and register patient	• Patient registration at the hospital (EPIC)	• Patient • Registration
• Registration • IR Holding Area	• Patient registration at the hospital (EPIC) • Physician order	• Prepare patient	• Patient information (EPIC) • Laboratory results	• Patient • Physician • Anaesthesia
• Transportation team	• Request to transport the patient	• Transport patient	• Transported to correct location	• Patient • Nurse • Physician
• Patients: IP and OP	• Patient information • Patient orders	• Prepare patient in holding	• Prepared patient • Consents for procedure	• Nurse • Physician
• IR Staff: Nurses and Techs	• Patient information from registration (EPIC) • Laboratory results (EPIC)	• Prepare room	• Current patient status (EPIC) • Patient ready for procedure • Patient in right location • IR room ready for procedure	• Patient • Physician • Anaesthesia
• Radiologists	• Patient info from registration (EPIC) • Laboratory results (EPIC) • Patient ready for procedure • Materials required for procedure	• Perform IR procedures and post-procedure activities	• Patient treated	• Patient

Voice of Process (VOP)

The VOP related to the baseline process metrics is discussed in this section. The percentage of on-time procedure starts is shown in Fig. 10.11. The first case of the day is the only case that is included in the calculation for the percentage of on-time starts. The first case is scheduled for an 8:00 am start time. The time that the procedure starts minus the appointment time (of 8:00 am) is calculated. If the time is less than or equal to 0 min, then the case is marked as starting on time or early. In Quarter 1, the baseline percentage of cases that started early was 30%, resulting in a negative value, since the start time is before the scheduled start time. Thirty-two percent (32%) of the cases started on time or early. Forty percent (40%) of the cases started within 5 min of the appointment, 48% started within 10 min of the appointment time, 61%

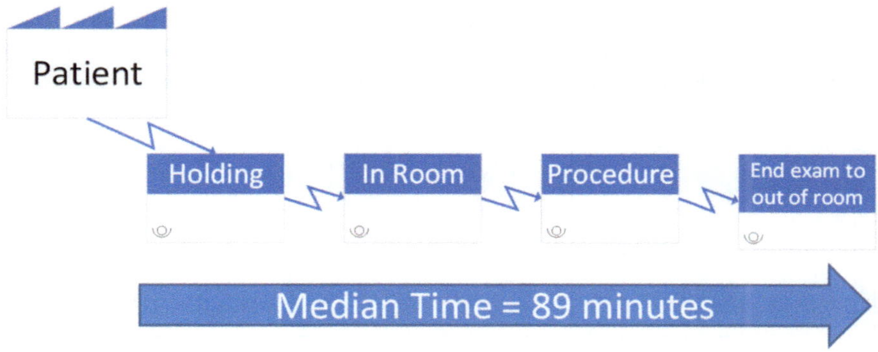

Fig. 10.1 Interventional radiology value stream map

Table 10.7 Data collection plan

Critical to satisfaction (CTS)	Metric (short title)	Operational definition (metric description)	Data collection source
Patient throughput time	Time spent by a patient in the system for IR procedures	• Outpatients: measure time from when the patient arrives at the registration desk to the time patient leaves the hospital after the procedure • Inpatients: measure time from when the patient arrives at holding area to the time when patient returns to the floor	EPIC system
	Percent on-time procedure starts	Time in-room minus appointment time ≤ 0 is an on-time start	EPIC system
	Activity times: • Holding time • In-room prep time • Procedure time • End of exam to out of room time	• Holding time: time from arriving in holding to move to procedure room • In-room prep time: time from arrival in room to begin procedure • Procedure time: time of procedure • End of exam to out of room: time from end of procedure to time out of room	EPIC system
Staff satisfaction	Staff survey	Send survey to employee after every payment	Google forms
Patient satisfaction	Patient survey	Send survey to patient after being discharged	Google forms
Robust data analytics	Data Collection	Having all the data captured with intended quality to be able to conduct data analysis for process improvements	EPIC system

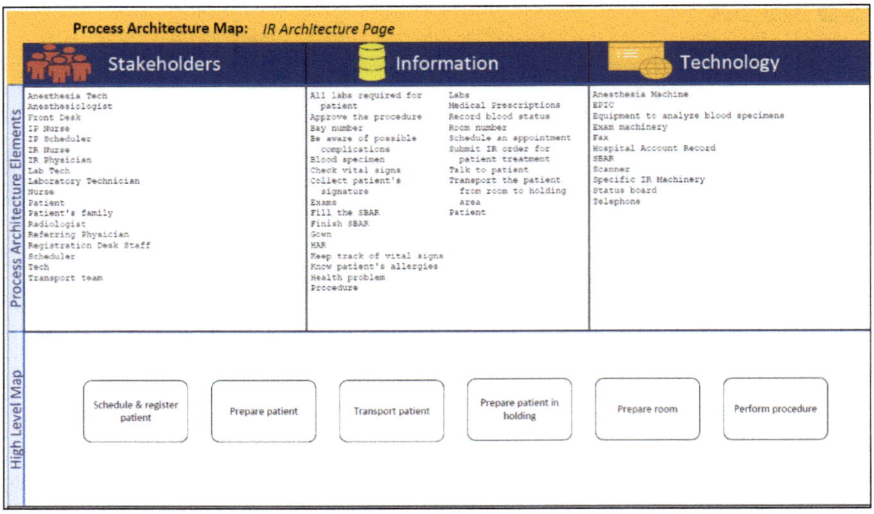

Fig. 10.2 Process architecture map (PAM) IR architecture page

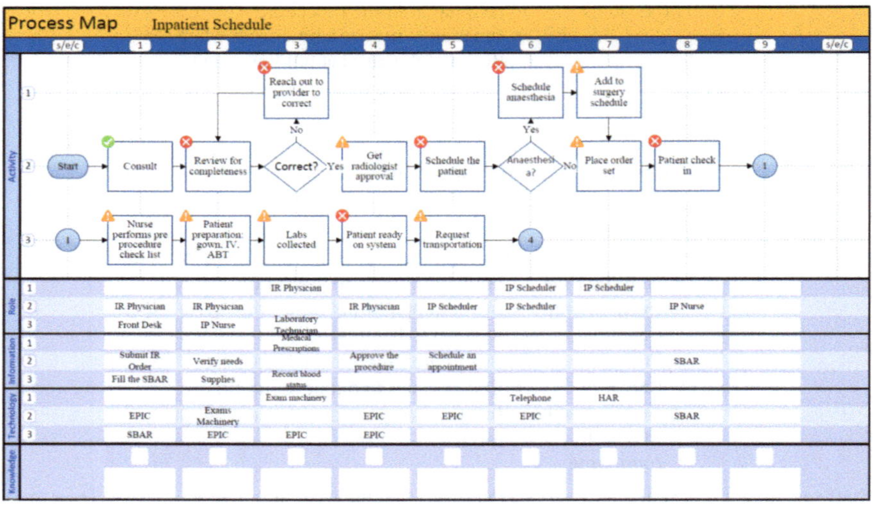

Fig. 10.3 Process map—inpatient schedule

started within 15 min, 70% within 20 min, and the remaining 30% started more than 20 min later than the scheduled start time. The percentage of on-time starts for the first case of the day is a common metric for operating room environments.

Table 10.8 illustrates the on-time starts and to follow case metrics for Q1 and Q3 of the baseline year. The data collection and project initially started in Q1 of the baseline year, but due to the pandemic, it was paused during Q2, and restarted in Q3,

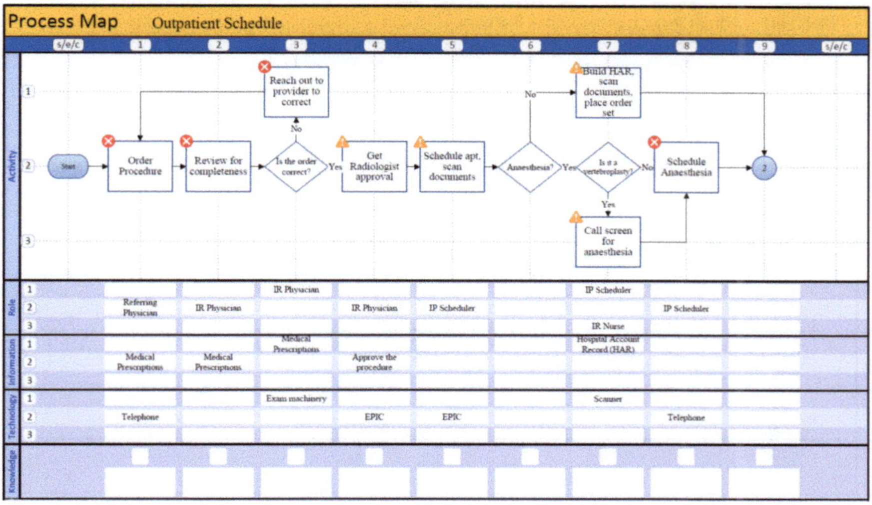

Fig. 10.4 Process map—outpatient schedule

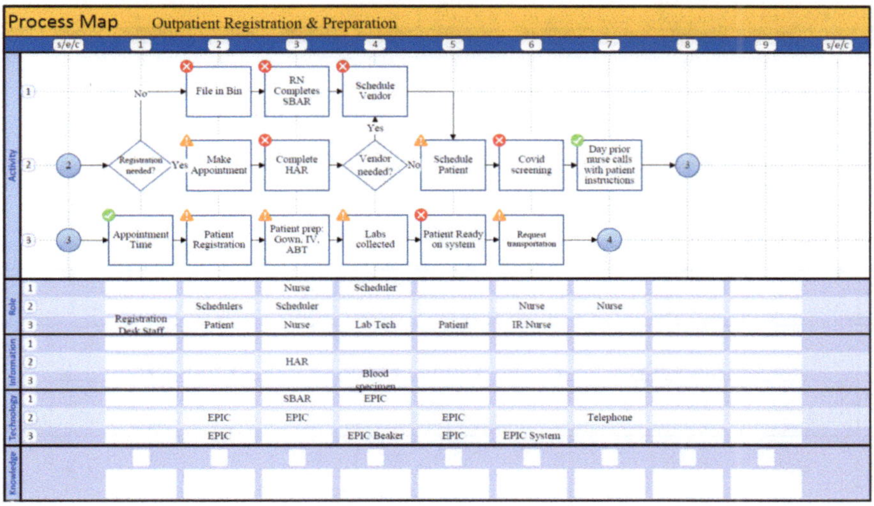

Fig. 10.5 Process map—outpatient registration and preparation

so both Q1 and Q3 baseline data are provided. Table 10.8 also shows the breakdown between inpatients and outpatients for the first cases and the to follow cases. The to follow cases are all of the remaining cases performed after the first case. The Q3 results were much worse than Q1 percentages, due mainly to understaffing across the hospital during the pandemic, as shown in Fig. 10.12.

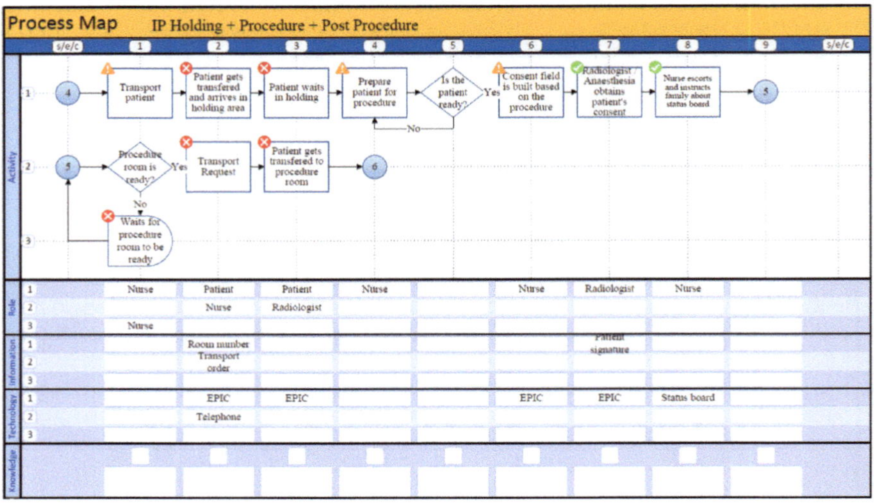

Fig. 10.6 Process map—IP holding, procedure, post-procedure

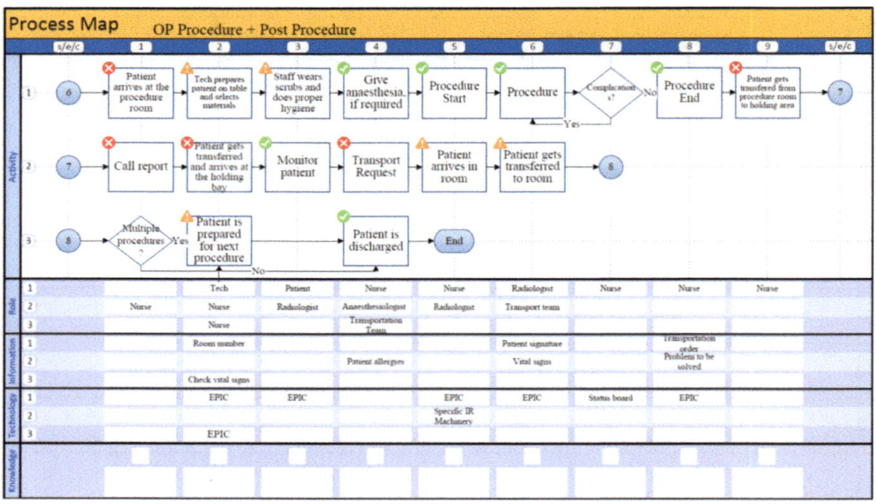

Fig. 10.7 Process map—OP procedure + post-procedure

Figure 10.12, the on-time starts for the baseline Q1 and Q3 are shown. Quarter 3 bars are 16% for % early, 19% for on time or early, 26% for within 5 min, 35% for within 10 min, 44% for within 15 min, 48% within 20 min, and 42% greater than 20 min.

The histogram distributions are shown in the next set of figures for.

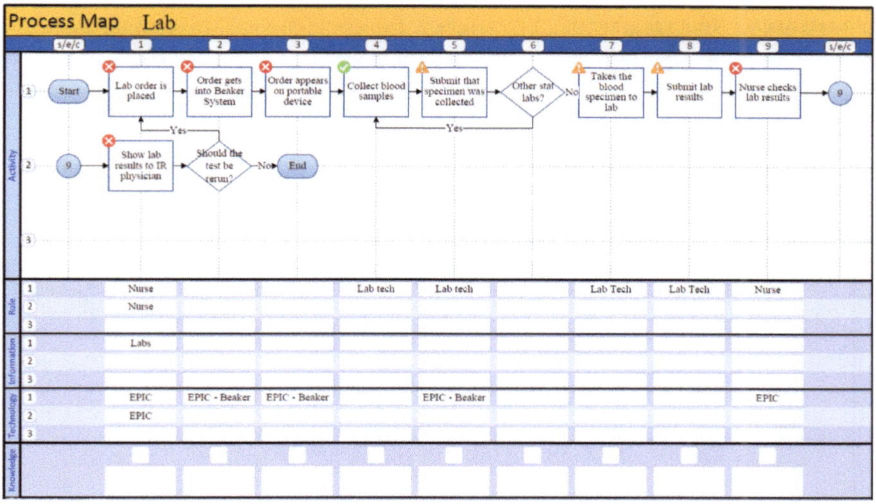

Fig. 10.8 Process map—lab

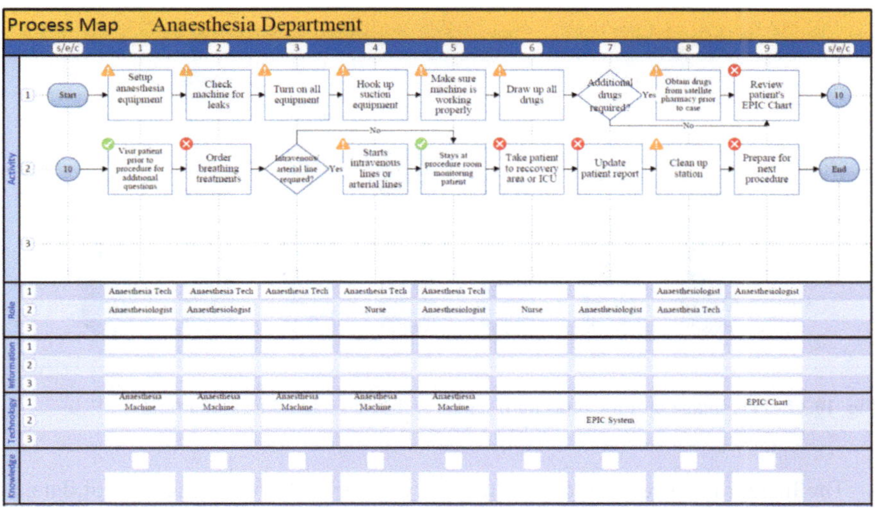

Fig. 10.9 Process map—anaesthesia department

- On-time starts time (Fig. 10.13).
- Holding time (Fig. 10.14).
- In-room prep time (Fig. 10.15).
- Procedure length (Fig. 10.16).
- End of exam to out of room time (Fig. 10.17).

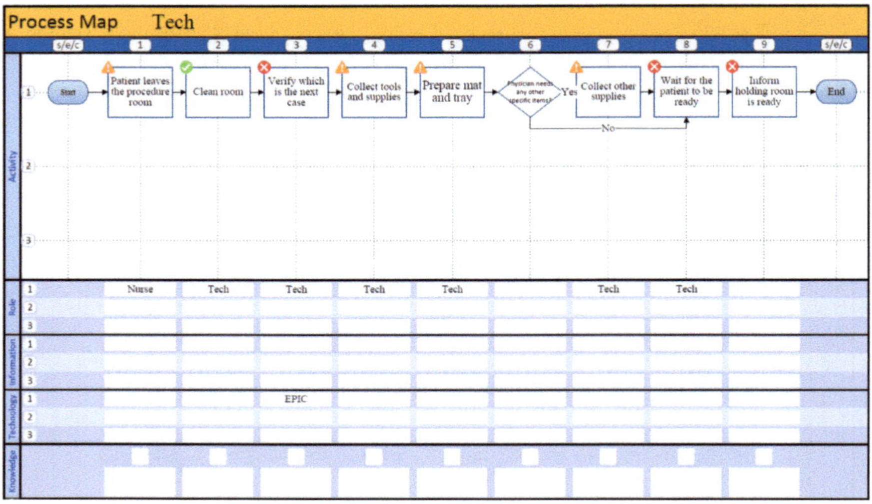

Fig. 10.10 Process map—tech

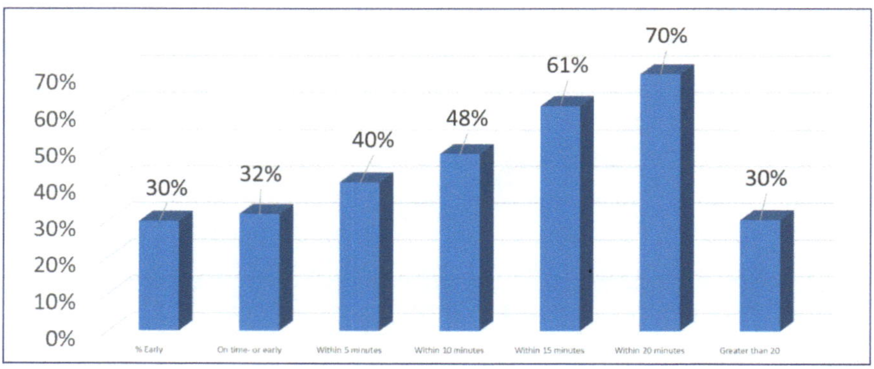

Fig. 10.11 Percentage on-time procedure starts for first case, baseline—quarter 1

The histograms show the central tendency or mean of the distribution of data, as well as the variation inherent in the process. Additionally, outliers are also evident within the distributions.

The on-time starts *time* is shown in Fig. 10.13 for Q1 baseline. The average on-time starts time, or time into room minus the appointment time for Q1 was 20.58 min with a standard deviation of 31.47 min.

The holding time for Q1 is shown in Fig. 10.14. The average was 35.85 min, with a standard deviation of 20.97 min.

The in-room prep time is 19.96 min with a standard deviation of 11.63 min for Q1, as shown in Fig. 10.15.

Table 10.8 Baseline data for first cases and to follow cases for inpatients and outpatients Q1, Q3

	Interventional radiology on-time starts (Q1, Q3 baseline year)							
	First cases				To follow cases			
	Inpatients		Outpatients		Inpatients		Outpatients	
	Q1 (%)	Q3 (%)	Q1 (%)	Q3 (%)	Q1 (%)	Q3 (%)	Q1 (%)	Q3 (%)
% Early	27	13	48	17	14	12	30	26
On time or early	29	13	51	20	15	13	32	27
Within 5 min	35	17	58	29	23	19	41	36
Within 10 min	48	25	65	38	31	24	50	44
Within 15 min	63	36	75	46	39	32	59	53
Within 20 min	73	53	82	59	47	38	67	58
Greater than 20	27	47	18	41	53	62	33	42

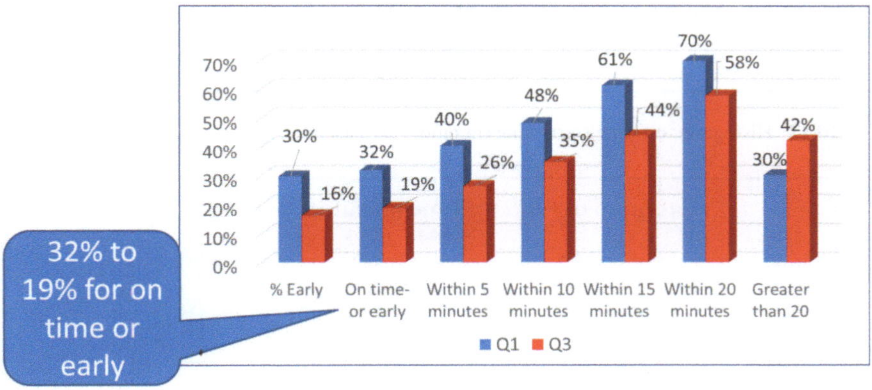

Fig. 10.12 On-time starts percentages Q1 to Q3 baselines

The average procedure length is 32.55 min, with a standard deviation of 36.18 min, as shown in Fig. 10.16.

The end of exam to out of room time for Q1 is shown in Fig. 10.17. The average time is 22.81 min with a standard deviation of 26.41 min.

Table 10.9 shows the missing data percentages for each activity metric for the baseline periods of Q1 and Q3. With such a high percentage of missing data elements, it becomes more difficult to have robust data for process improvement.

Voice of Customer (VOC)

A VOC survey of the IR staff was conducted. The following questions were asked in the survey:

(1) I feel that staffing is appropriate for my department (Fig. 10.18).
(2) I feel that leadership supports my department (Fig. 10.19).

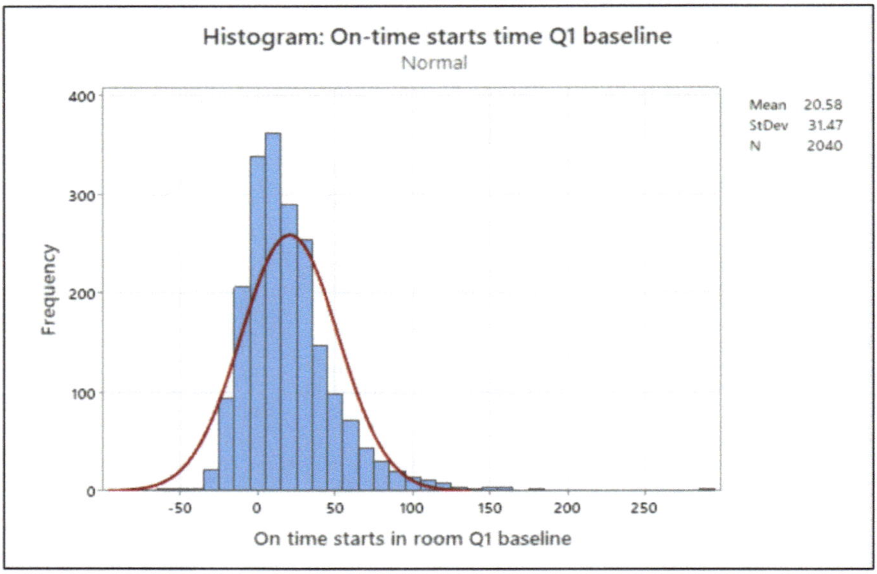

Fig. 10.13 Histogram: on-time starts time (minutes) Q1 baseline

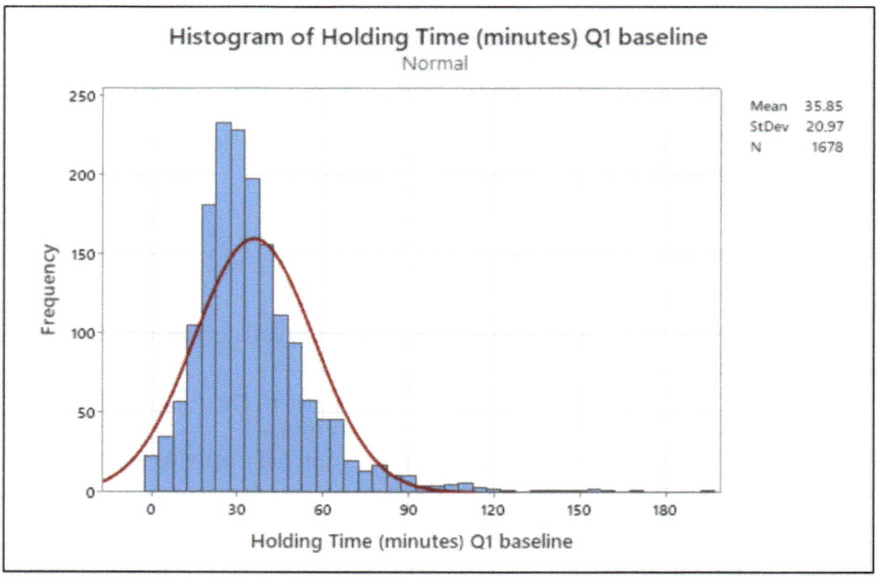

Fig. 10.14 Histogram: holding time (minutes) Q1 baseline

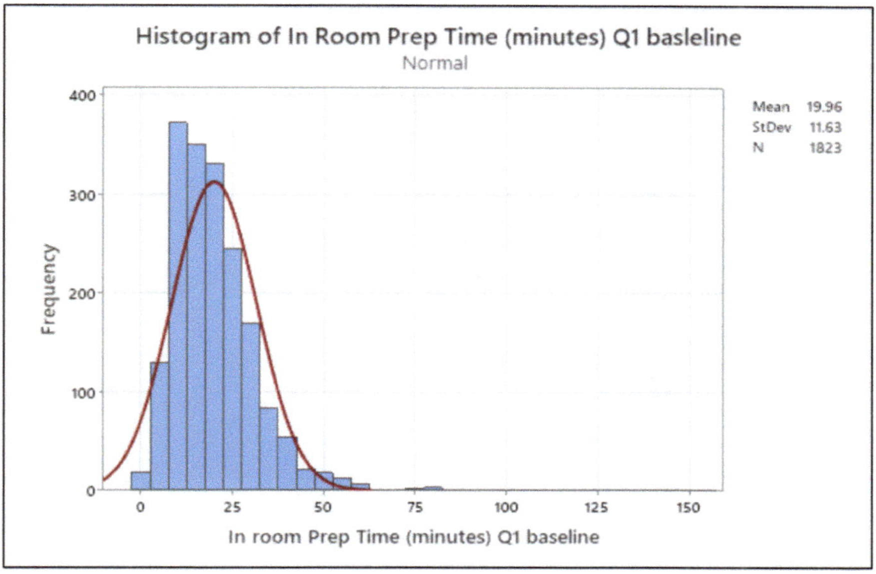

Fig. 10.15 Histogram: in-room prep time (minutes) Q1 baseline

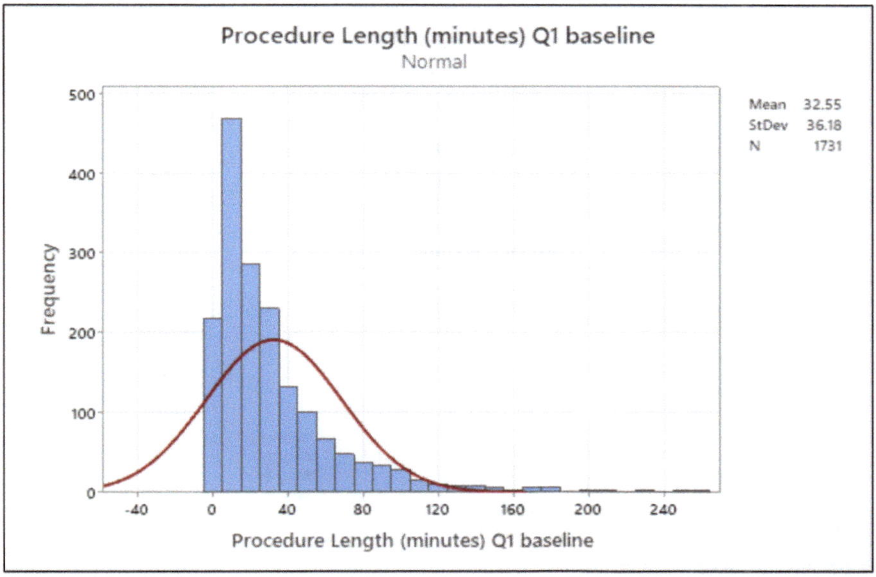

Fig. 10.16 Histogram: procedure length (minutes) Q1 baseline

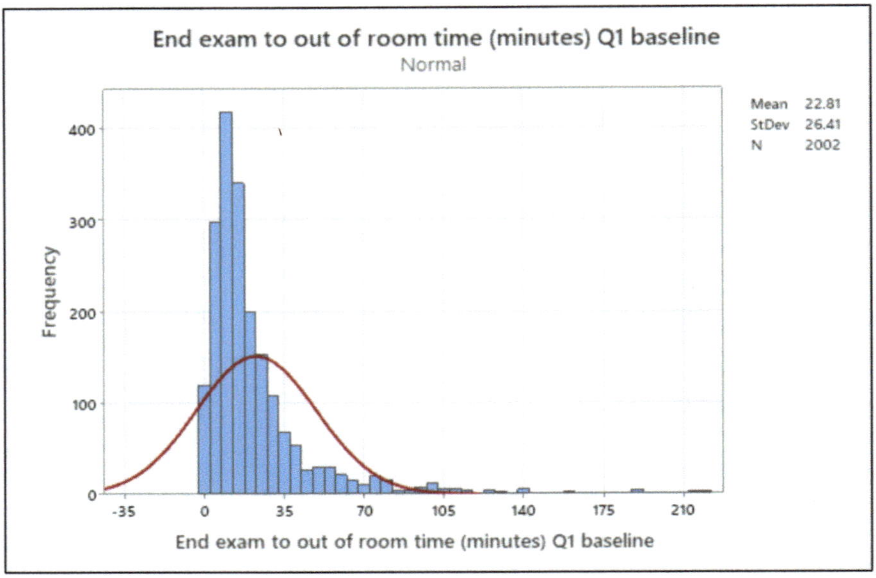

Fig. 10.17 Histogram: end of exam to out of room time (minutes) Q1 baseline

Table 10.9 Percent missing data baseline

Percent missing data		
Name of metric/activity	Baseline Q1 (%)	Baseline Q3 (%)
On-time starts	9	12
Wheels in to out	14	12
Procedure length	22	25
Holding time	25	31
In-room prep time	18	21
End exam to out of room	10	18

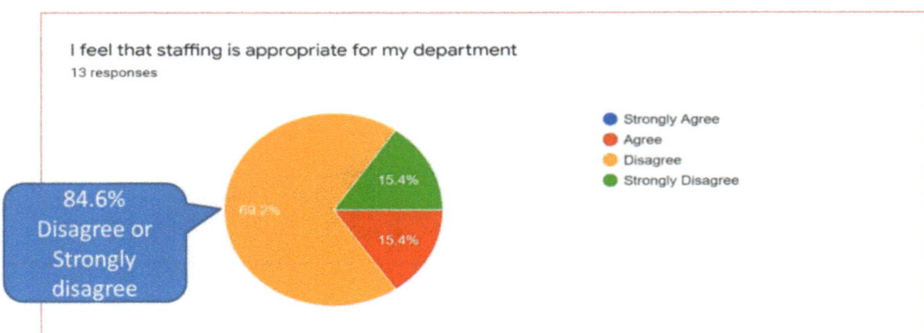

Fig. 10.18 Survey: I feel that staffing is appropriate for my department

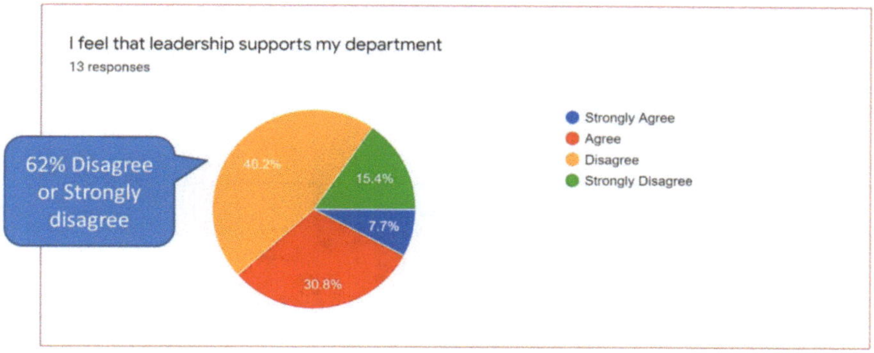

Fig. 10.19 Survey: I feel that leadership supports my department

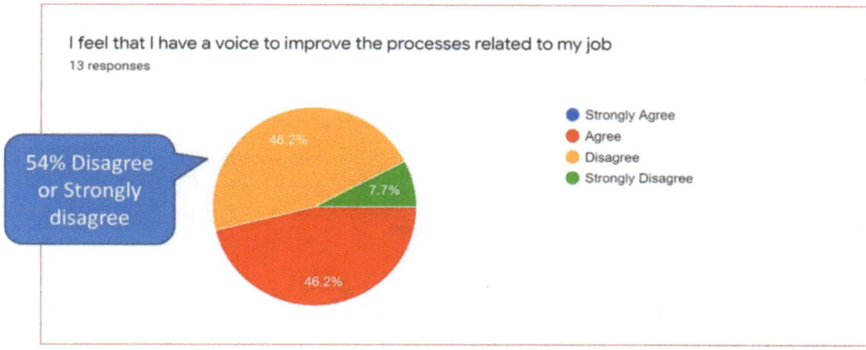

Fig. 10.20 Survey: I feel that I have a voice to improve the processes related to my job

(3) I feel that I have a voice to improve the processes related to my job (Fig. 10.20).

Almost eighty-five percent (84.6%) of the staff disagree or strongly disagree that staffing is appropriate in the IR department (Fig. 10.18).

Almost 62% of the staff disagree or strongly disagree that leadership supports the IR department (Fig. 10.19).

Fifty-four percent (54%) of the staff disagree or strongly disagree that they have a voice to improve the process related to their job (Fig. 10.20).

Analyze Phase

The team performed a cause-and-effect analysis using a why-why diagram to identify the root causes. The questions that were asked in the why-why diagrams were:

- Why is there low staff satisfaction? (Fig. 10.21)

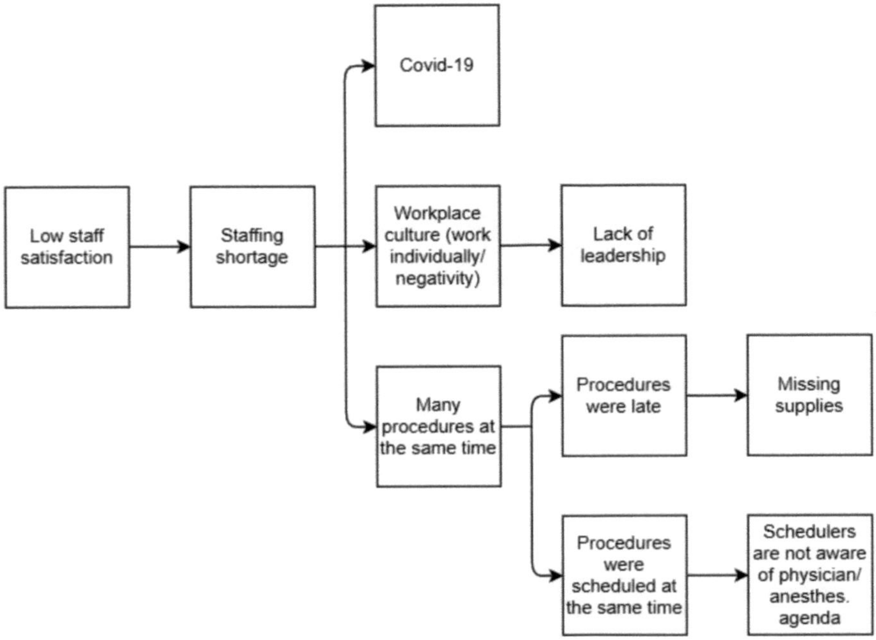

Fig. 10.21 Why-why diagram—Low staff satisfaction

- Why is there low patient satisfaction? (Fig. 10.22)
- Why is there wrong or missing data? (Fig. 10.23)

The Why-why diagram for "Why is there low staff satisfaction?" is shown in Fig. 10.21. The key root causes found were:

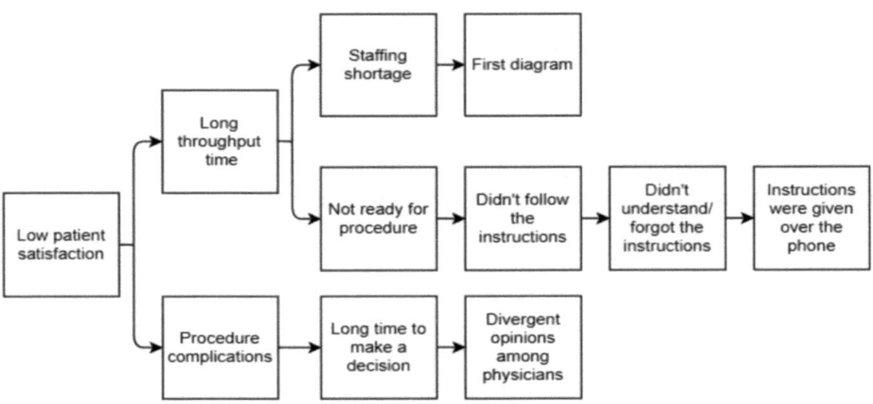

Fig. 10.22 Why-why diagram—Low patient satisfaction

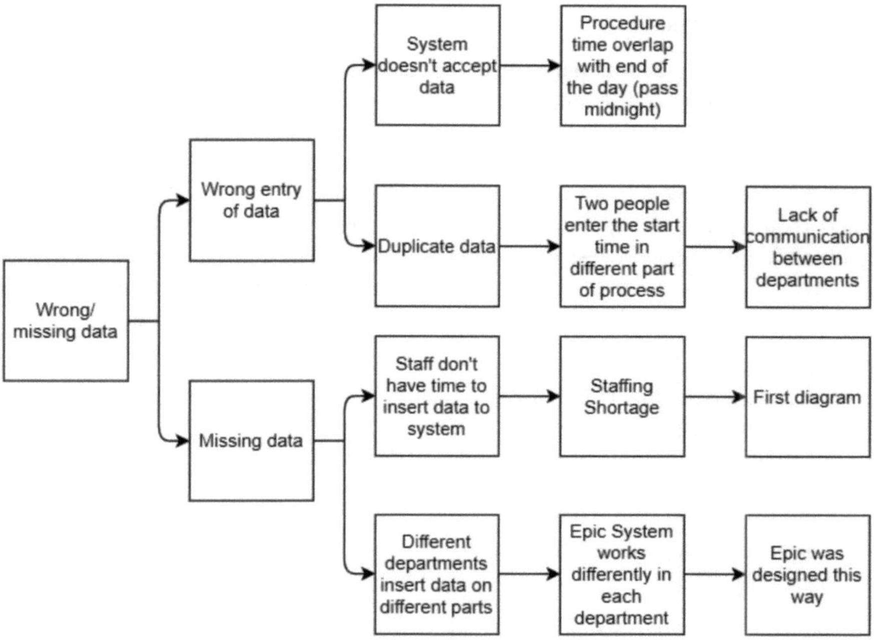

Fig. 10.23 Why-why diagram—wrong or missing data

- COVID -19 pandemic with staffing shortages.
- Lack of leadership.
- Missing supplies.
- Procedures scheduled same time.

The Why-why diagram for "Why is there low patient satisfaction?" is shown in Fig. 10.22. The key root causes found were:

- Staffing shortage.
- Instructions over phone.
- Procedure complications.

The Why-why diagram for "Why is there wrong or missing data?" is shown in Fig. 10.23. The key root causes found were:

- Staffing shortage.
- Poor user interface.
- Template not working.
- Lack of communication—2 people enter times.
- Duplicate fields.

Lean Value Analysis

A lean value analysis was performed for each activity in the process maps. Each activity was designated as either value-added, limited-value, or non-value-added. The value-added activities which create value for the patients are designated with a green check mark in the process maps and lean reports, shown in the next figures. A limited-value-added activity which doesn't add value from a customer perspective but is required due to regulatory or policy requirements is denoted with a yellow caution triangle in the process maps and lean reports. The non-value-added activities which do not add value to the patient are denoted with a red "X" in the process maps and lean reports. The following lean value analysis reports are provided:

1. Inpatient Schedule (Fig. 10.24).
2. Outpatient Schedule (Fig. 10.25).
3. Outpatient Registration (Fig. 10.26).
4. Inpatient—Holding, Procedure, and Post-Procedure (Fig. 10.27).
5. Outpatient—Procedure and Post-Procedure (Fig. 10.28).
6. Lab Department (Fig. 10.29).
7. Anaesthesia Department (Fig. 10.30).
8. Tech (Fig. 10.31).

The lean value analysis aids in the identification of areas of improvement when we move into the Improve phase. The percentage of value-added activities is 16%,

Lean Analysis Inpatient Schedule

#	Activity	Value Analysis	Trans-portation	Over Production	Motion	Defects	Delay	Inventory	Processing	People	Root Cause(s)
1	Reach out to provider to correct	X		X		X			X		
2	Schedule anesthesia	X							X		Not aware of anethesiologists' schedule
3	Add to surgery schedule	⚠							X		
4	Consult	✓									No available time to do the procedure (soon)
5	Review for completeness	X		X		X			X		Does the patient need another evaluation? Too many patients to check
6	Get radiologist approval	⚠									
7	Schedule the patient	X							X		Rework on paper/ computer
8	Place order set	⚠							X		
9	Patient check in	X		X							Entry information on paper and on computer
10	Nurse performs pre procedure check list	⚠				X					
11	Patient preparation: gown, IV, ABT	⚠							X		Duplicate data entry, checking the front desk information
12	Labs collected	⚠							X		Necessity to collect the blood specimen again
13	Patient ready on system	X							X		
14	Request transportation	⚠							X		
15											

Fig. 10.24 Inpatient schedule lean value analysis report

Lean Analysis Outpatient Schedule

#	Activity	Value Analysis	Trans-portation	Over Production	Motion	Defects	Delay	Inventory	Processing	People	Root Cause(s)
1	Reach out to provider to correct	X		X		X			X		
2	Build HAR, scan documents, place order set	⚠							X		
3	Order Procedure	X							X		
4	Review for completeness	X		X		X			X		
5	Get Radiologist approval	⚠									
6	Schedule apt, scan documents	⚠							X		Enter information manually
7	Schedule Anaesthesia	X		X					X		
8	Call screen for anaesthesia	⚠							X		
9											

Fig. 10.25 Outpatient schedule lean value analysis report

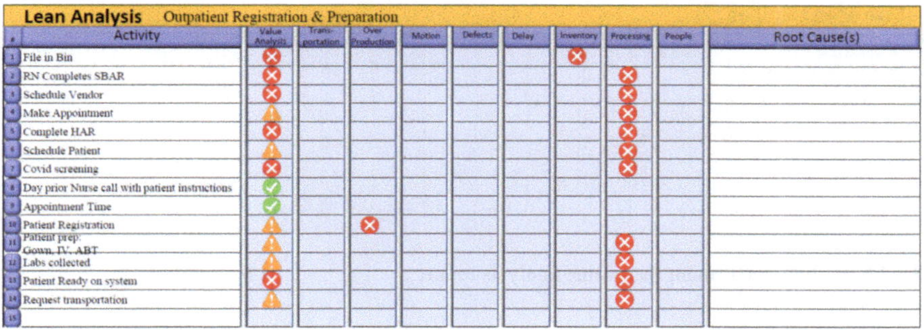

Fig. 10.26 Outpatient registration and preparation lean value analysis report

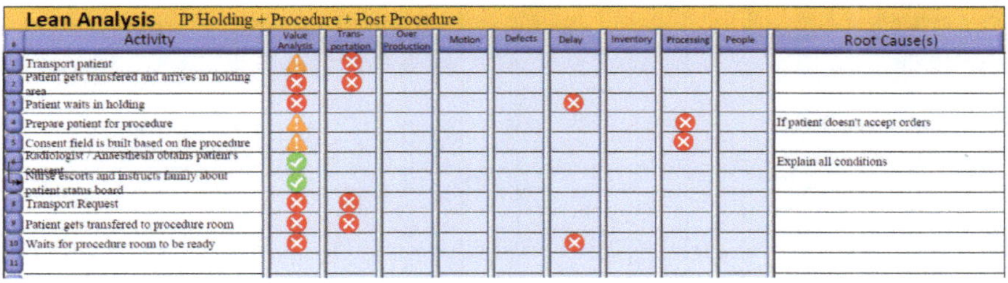

Fig. 10.27 Inpatient—holding + procedure + post-procedure lean value analysis report

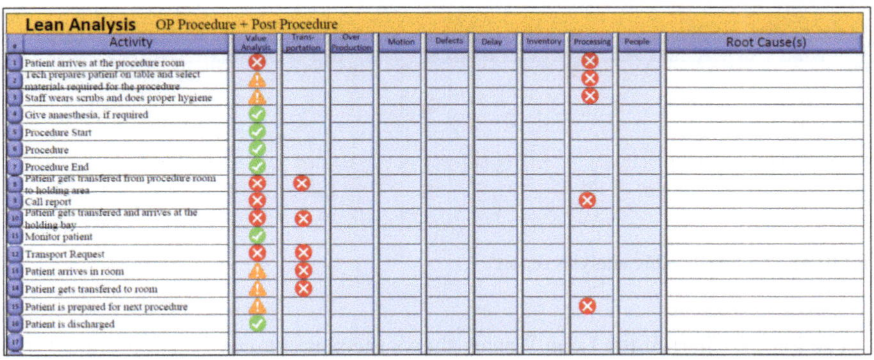

Fig. 10.28 Outpatient—procedure and post-procedure lean value analysis report

showing that there is a lot of room for improvement in the process to reduce and eliminate non-value-added (41%) and limited-value-added activities (43%), and design around the value-added activities. The wastes were also identified for each activity that was designated as limited or non-value-added activities. The most common wastes were processing, transportation, overproduction, and defects, as shown in Fig. 10.32.

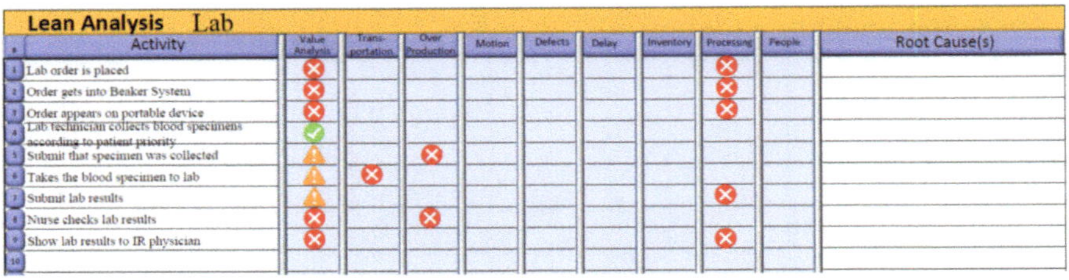

Lean Analysis Lab

#	Activity	Value Analysis	Trans-portation	Over Production	Motion	Defects	Delay	Inventory	Processing	People	Root Cause(s)
1	Lab order is placed	✗							✗		
2	Order gets into Beaker System	✗							✗		
3	Order appears on portable device	✗							✗		
4	Lab technician collects blood specimens according to patient priority	✓									
5	Submit that specimen was collected	⚠		✗							
6	Takes the blood specimen to lab	⚠	✗								
7	Submit lab results	⚠							✗		
8	Nurse checks lab results	✗		✗							
9	Show lab results to IR physician	✗							✗		
10											

Fig. 10.29 Lab department lean value analysis report

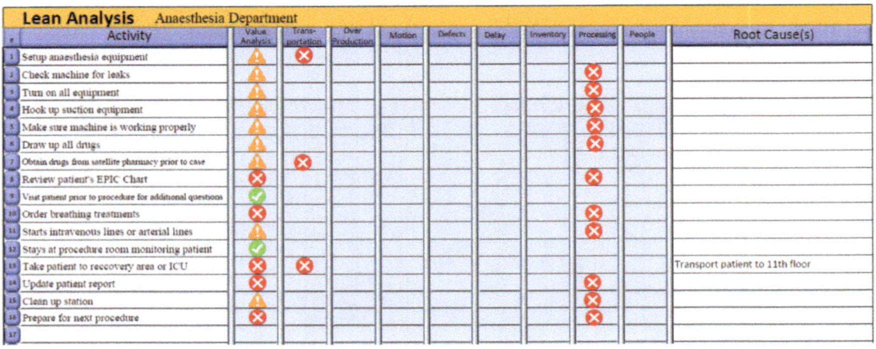

Lean Analysis Anaesthesia Department

#	Activity	Value Analysis	Trans-portation	Over Production	Motion	Defects	Delay	Inventory	Processing	People	Root Cause(s)
1	Setup anaesthesia equipment	⚠	✗								
2	Check machine for leaks	⚠							✗		
3	Turn on all equipment	⚠							✗		
4	Hook up suction equipment	⚠							✗		
5	Make sure machine is working properly	⚠							✗		
6	Draw up all drugs	⚠							✗		
7	Obtain drugs from satellite pharmacy prior to case	⚠	✗								
8	Review patient's EPIC Chart	✗							✗		
9	Visit patient prior to procedure for additional questions	✓									
10	Order breathing treatments	✗							✗		
11	Starts intravenous lines or arterial lines	⚠							✗		
12	Stays at procedure room monitoring patient	✓									
13	Take patient to recovery area or ICU	✗	✗								Transport patient to 11th floor
14	Update patient report	✗							✗		
15	Clean up station	⚠							✗		
16	Prepare for next procedure	✗							✗		
17											

Fig. 10.30 Anaesthesia department lean value analysis report

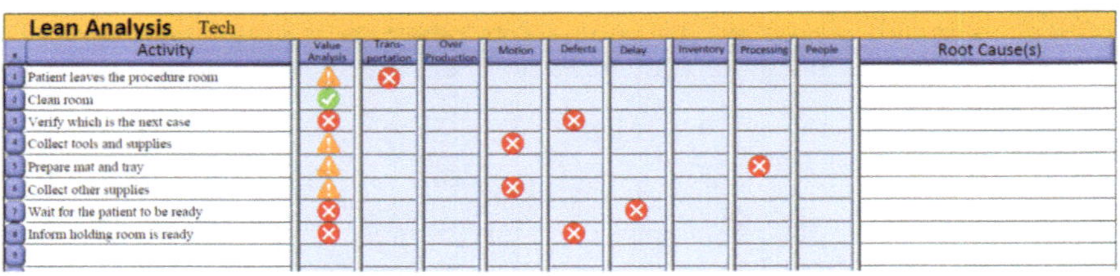

Lean Analysis Tech

#	Activity	Value Analysis	Trans-portation	Over Production	Motion	Defects	Delay	Inventory	Processing	People	Root Cause(s)
1	Patient leaves the procedure room	⚠	✗								
2	Clean room	✓									
3	Verify which is the next case	✗				✗					
4	Collect tools and supplies	⚠			✗						
5	Prepare mat and tray	⚠							✗		
6	Collect other supplies	⚠			✗						
7	Wait for the patient to be ready	✗					✗				
8	Inform holding room is ready	✗				✗					
9											

Fig. 10.31 Tech lean value analysis report

The Pareto Chart of the Lean Wastes shows the percentage of each of the 8 types of waste. Processing (58.6%), Transportation (16.1%), and Over Production (10.3%) wastes make up 85% of the waste types in the process. This can help us to target types of wastes in the Improve phase.

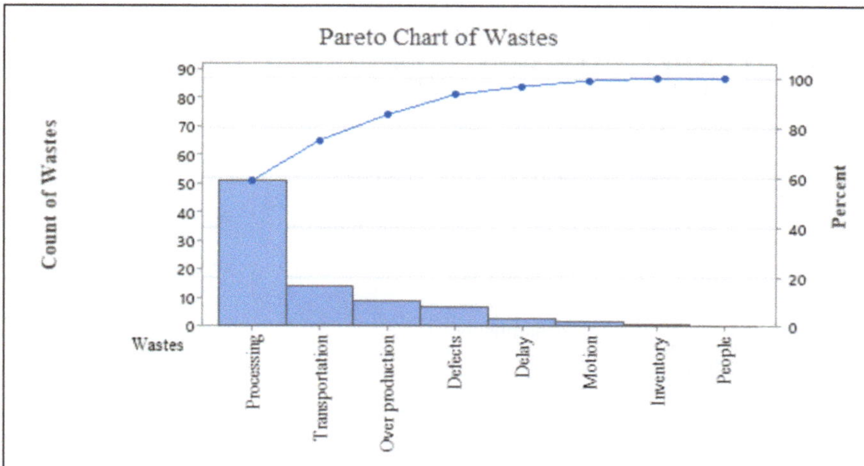

Wastes	Processing	Transportation	Over production	Defects	Delay	Motion	Inventory	People
Count of Wastes	51	14	9	7	3	2	1	0
Percent	58.6	16.1	10.3	8.0	3.4	2.3	1.1	0.0
Cum %	58.6	74.7	85.1	93.1	96.6	98.9	100.0	100.0

Fig. 10.32 Pareto chart of wastes

Improve Phase

The team generated improvement ideas that would align to the CTS and eliminate the root causes and wastes and reduce the limited-value and non-value-added activities, as shown in Table 10.10.

The improvements were piloted during the first quarter of the year following the baseline year.

CTS: Robust data analytics and staff satisfaction

To reduce the data entry errors or missing data, IR implemented standard data entry protocols. Previously there had been multiple templates within the EPIC system to enter the activities, which caused confusion and erroneous times to be entered. A cost–benefit analysis was also performed to justify hiring interns to help nurses with patient notes and data entry on the system. This would be a "win–win-win" for the patients, nurses, and interns, by allowing the nurses to spend more time with the patients and giving the interns experience in the IR department. Incorporating common notes checklists was also designed to reduce data entry errors. Additionally, a new EPIC tab would be introduced in the future, where patient vitals and data can be typed with the option of side notes and then imported to an official report.

CTS: Patient satisfaction

Some patients would go to the wrong hospital location, which were about 30 min apart for their procedure. To eliminate this issue, the IR department provided a brochure

Table 10.10 Improvement ideas and root causes

CTS	Problems/ inefficiencies/ wastes	Root causes	Improvement ideas	Category/ grouping	Priority
Robust data analytics	Duplicate data	• EPIC System allows multiple data entry • Lack of communication between departments	Make all departments use the same protocol to insert data	Processing/ delay	High
Staff satisfaction	Staffing shortage	• Lots of training • Low salary • Many procedures at the same time	Hire interns to help nurses with patient notes and data entry on system	People	High
	Many charts/ multiple notes	• Nurses do not have time to add all the information	Create common notes checklist	Processing/ delay	Medium
	All information is written on SBAR to be typed later	• No EPIC feature that allows sidenotes	Create an EPIC tab where patient vitals/ data (imported to an official report) can be typed with the option of side notes	Overproduction	High
Patient satisfaction	Patients heading to different hospitals	• Patient doesn't have information written/ typed • Patient doesn't understand the instructions • Other hospitals' options nearby	Send brochure with address (mail)/ email with location (link mode) with GPS	Motion/ processing	Medium

(continued)

Table 10.10 (continued)

CTS	Problems/ inefficiencies/ wastes	Root causes	Improvement ideas	Category/ grouping	Priority
	Patients not following prep instructions	• Patient doesn't have information written/ typed • Many items to remember • Patient doesn't understand the instructions	Send brochure with checklist (mail)/ email with checklist that can be edited	People/delay	High
Patient throughput time	Patients' treatment is usually not on time	• Schedulers aren't aware of physician/ anesthesiologist agenda • Staffing shortage • Each physician/ team have different procedure times	• Collect data and define how much time each physician spends in each procedure • Make easier for schedulers to access physicians/ anesthesiologist's agenda	Delay	High
	Procedure room doesn't have all supplies necessary for procedure	• Not enough space to stock all material	Put all material on the surgical cart	Inventory/ Transportation	Medium
	Lack of communication between departments	• Lack of communication between departments • Delays on information	Alerts on dashboards	Delay	High

with the address, an email with the location, and a link to the GPS directions. They also developed a checklist along with the brochure and email, so that the patients would bring everything that they would need for the procedure, such as medication lists. To enhance patient satisfaction the IR department implemented new targets for procedure times. Additionally, Information Technology made it easier for schedulers to access physicians' and anaesthesiologists' schedules to be able to schedule the procedures.

CTS: Patient throughput

Changes for pulling and organizing supplies for the surgical carts to reduce missing items were implemented. Dashboards with alerts were designed to identify when activity targets were in jeopardy of being met, to keep the patients moving through the processes. A sample dashboard is shown in Fig. 10.33. The status of each room and alerts for when the targeted activity times will be missed, along with color-coding showing the activities that have been successfully completed, are also shown on the dashboard.

Improvement metrics

The Q1 results of the year following the baseline year were greatly improved for all of the tracked metrics. The percentage of on-time starts through the first week of March improved an average of 22% from the Q3 baseline period. The percentage of on-time starts increased to 60% in the first year of March, a 41% improvement compared to the 19% in the Q3 baseline (Fig. 10.34).

The average for the improved Q1 (through the first week of March) was 41%, compared to 19% in the baseline year Q3, and 21% in the baseline Q1, as shown in Fig. 10.35. On-time start percentages improved from 19 to 41%, resulting in a 22% improvement, and achieved a high of 60% by the end of the pilot improvement period.

IR Alerts Dashboard

Room	Status	ATA	In holding	In room	Procedure start	Procedure end	Out of Room
IR-01							
IR-02	Waiting for lab results		>30 mins				
IR-03	Empty						
IR-04	Waiting for out of room						
IR-05	Empty						
IR-06	Empty						

Business Rules for Alerts:
1) Not in registration within 2 hours prior to appointment time
2) Not in holding 30 minutes from appointment time
3) Not in room 10 minutes from appointment time
4) Missing lab results
5) Room not ready 15 minutes after prior procedure out of room time

Fig. 10.33 IR alerts dashboard

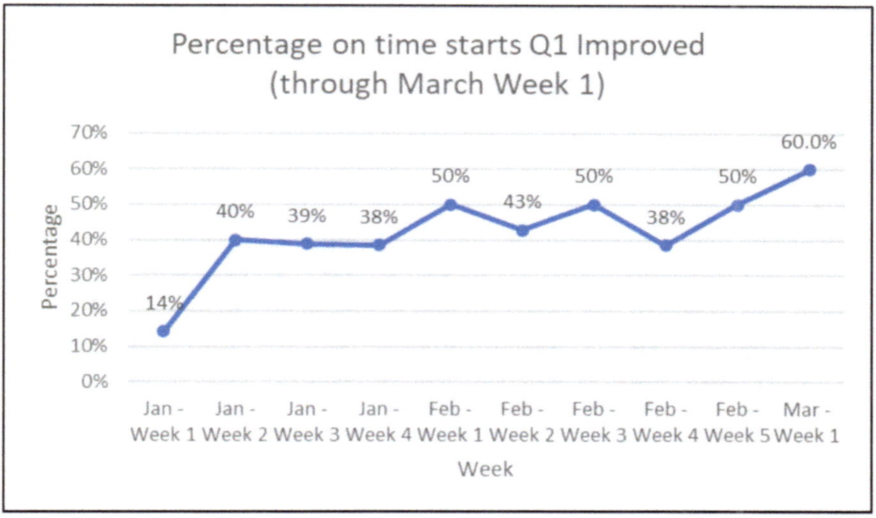

Fig. 10.34 Percent on-time starts Q1 improved

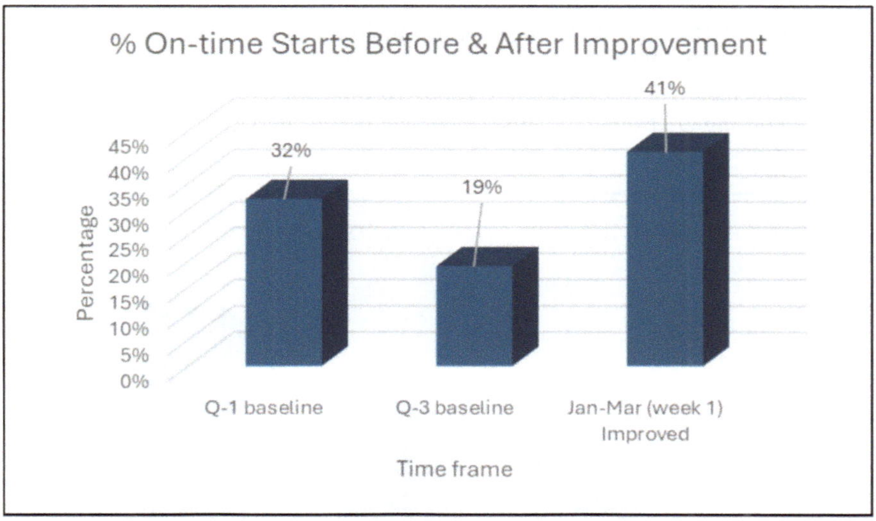

Fig. 10.35 On-time starts before and after improvement

The comparison bar chart between the on-time start percentages for the first cases comparing Q1 of the baseline year to Q1 of the improved timeframe is shown in Fig. 10.36, improving in every time category. The early cases were 30% to 38%, on time or early was 32% to 40%, within 5 min from 40 to 50%, within 10 min from 48

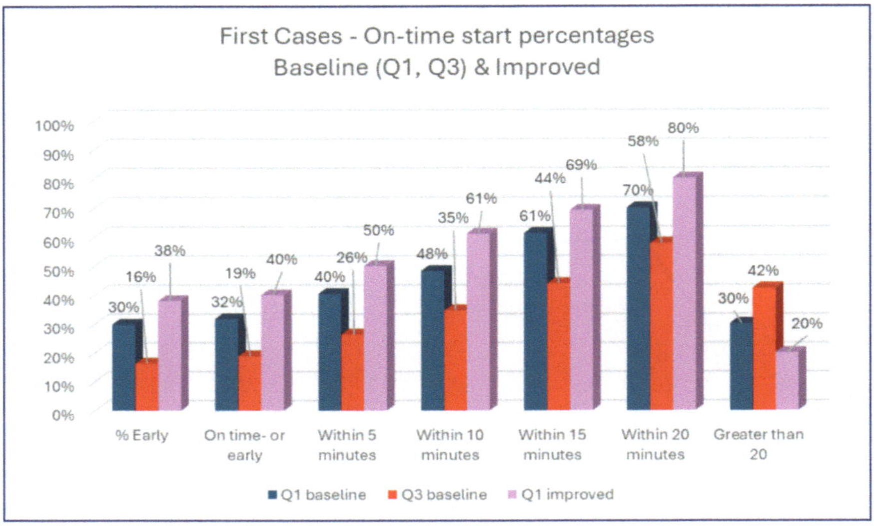

Fig. 10.36 First cases on-time start percentages baseline (Q1, Q3) and improved

to 61%, within 15 min from 61 to 69%, within 20 min from 70 to 80%, and greater than 20 min 30% to 20%.

The percentages in the procedure room 10 min prior to the appointment time increased during Q1 from 21 to 40%, as shown in Fig. 10.37.

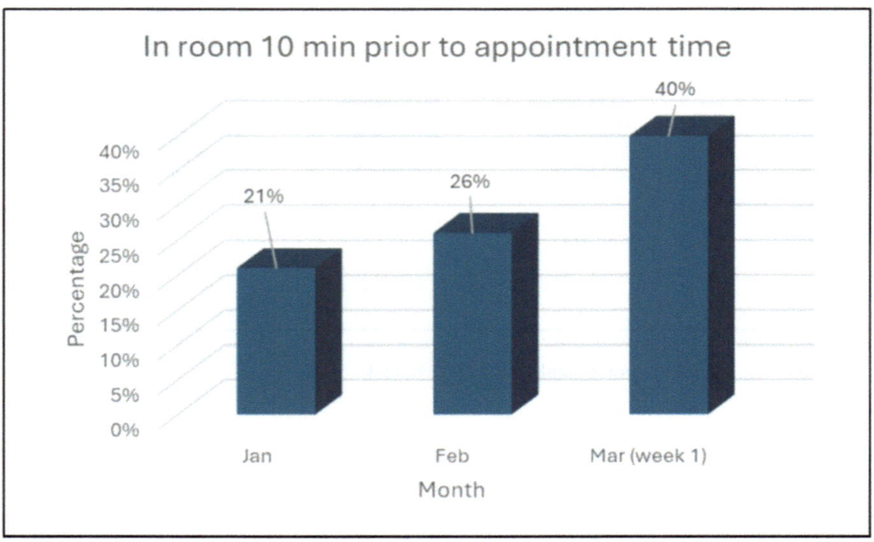

Fig. 10.37 In room 10 min prior to appointment time percentages

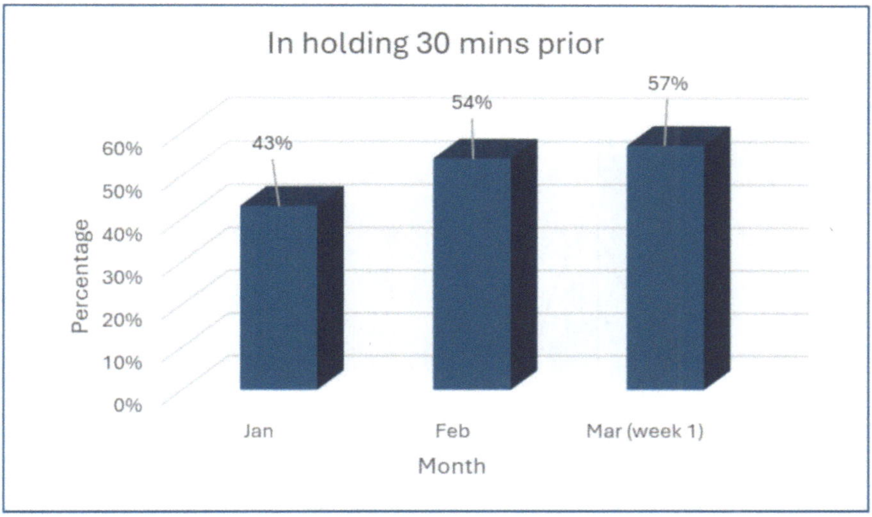

Fig. 10.38 In holding 30 min prior to appointment times

The percentages for the times in the holding area 30 min prior to the appointment time increased during the Q1 improvement period from 43 to 57%, as shown in Fig. 10.38.

The average on-time start times for the baseline Q1 (20.58 min) in Fig. 10.39 and improved times Q1 (10.49 min) are shown in Fig. 10.40. The two-sample t-test with a p-value of 0.000 showed a statistically significant difference in the average times, at 50% improvement.

The average holding times stayed about the same, 35.85–37.12 min from Q1 baseline (Fig. 10.41) to Q1 improved time frame (Fig. 10.42). For the two-sample t-test comparing the average times, with equal variances, the p-value was 0.227, demonstrating that there was not a statistically significant difference for holding time before and after improvement.

The histograms for in-room prep times before (19.96 min) and after improvement (18.87 min) are shown in Figs. 10.43 and 10.44. The in-room prep time was reduced by 5.5%, with a p-value of 0.048, showing a statistically significant difference before and after improvement.

The average procedure times before (32.57 min) versus after improvement (24.72 min) (Figs. 10.45 and 10.46) were statistically significant with a p-value of 0.000 and a 24% improvement. The standard deviation also improved (was reduced) from 36.18 to 25.46 min, with a p-value of 0.000, showing a statistically significant difference in the variances as well, and a 30% reduction in standard deviation.

The end of exam to out of room times increased between the baseline Q1 (22.8 min) and the improved Q1 time frames (35.05 min), shown in Figs. 10.47 and 10.48. The end of exam to out of room times increases significantly (p-value of 0.000). This would need to be focused on to improve in the future.

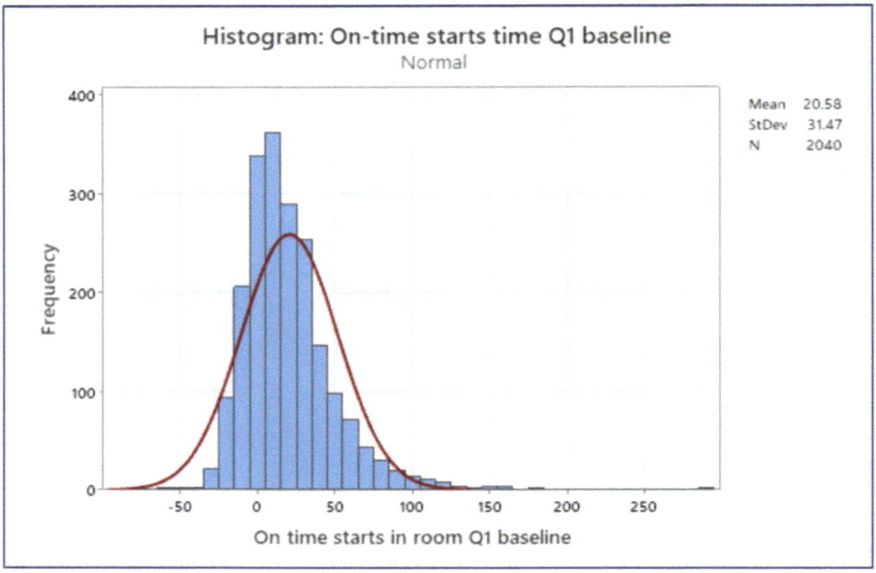

Fig. 10.39 On-time start times for baseline Q1

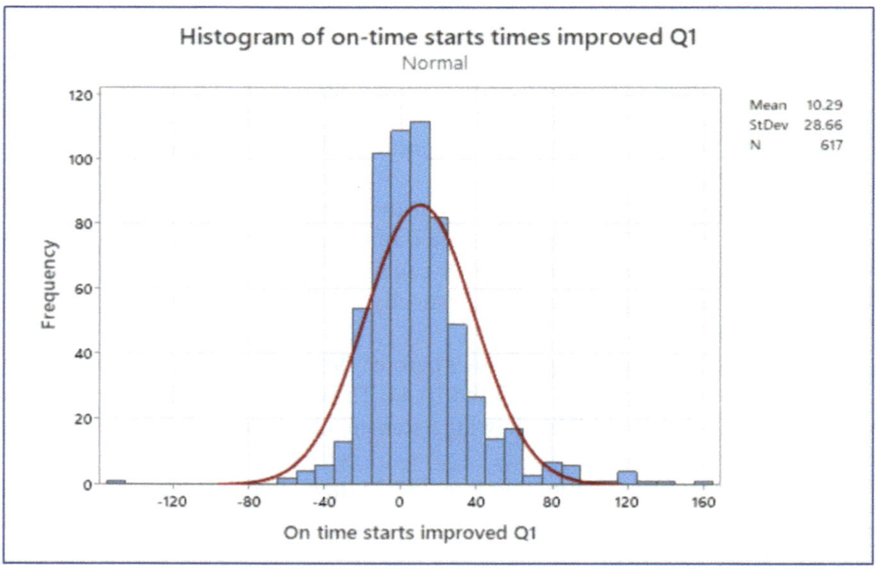

Fig. 10.40 On-time start times for improved times Q1

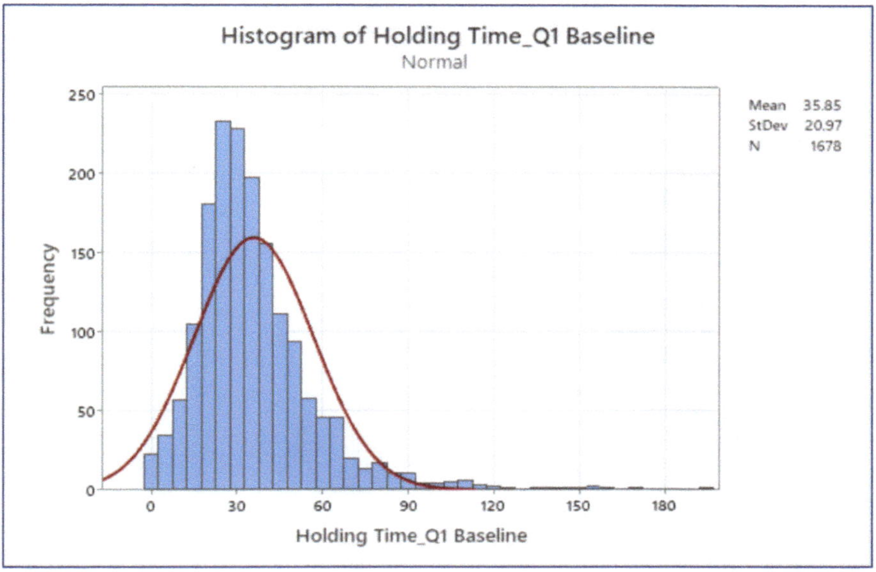

Fig. 10.41 Histogram of holding time, baseline Q1

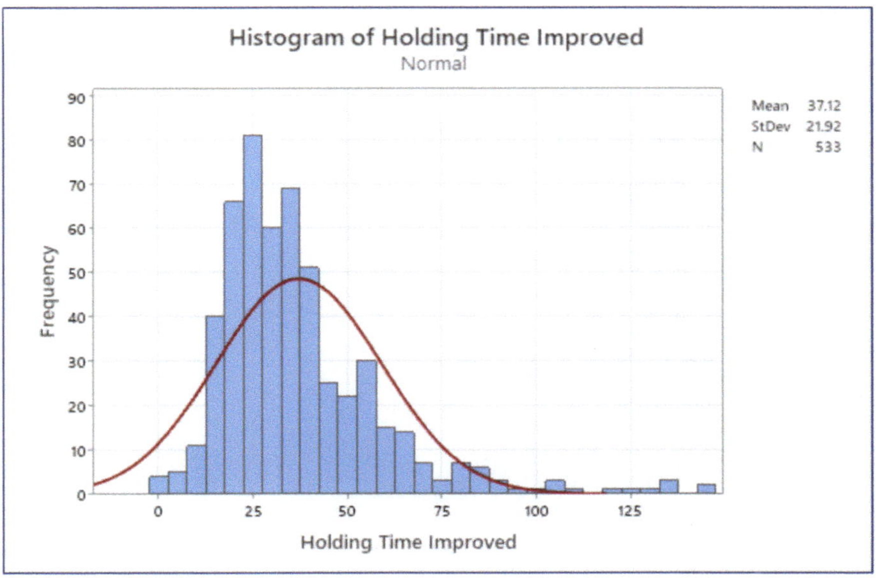

Fig. 10.42 Histograms of holding time improved Q1

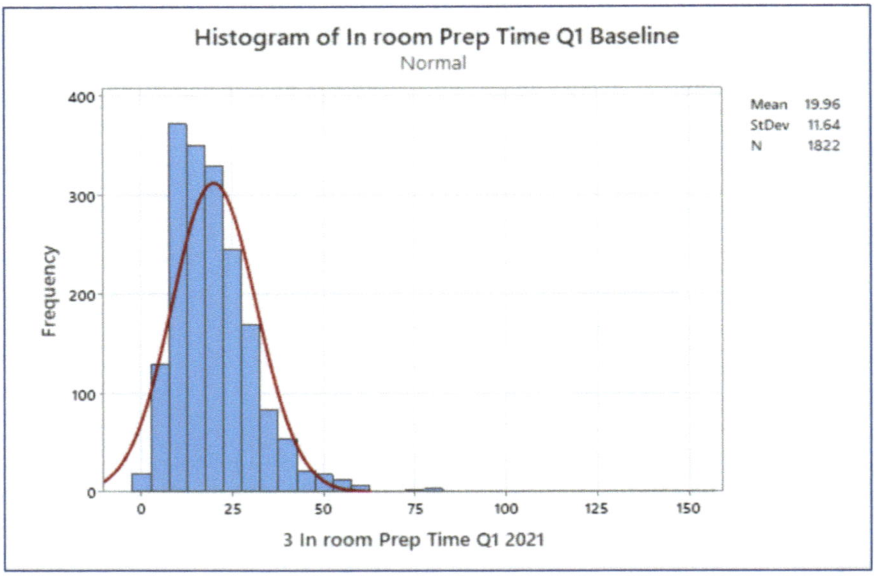

Fig. 10.43 Histograms of in-room prep time improved Q1

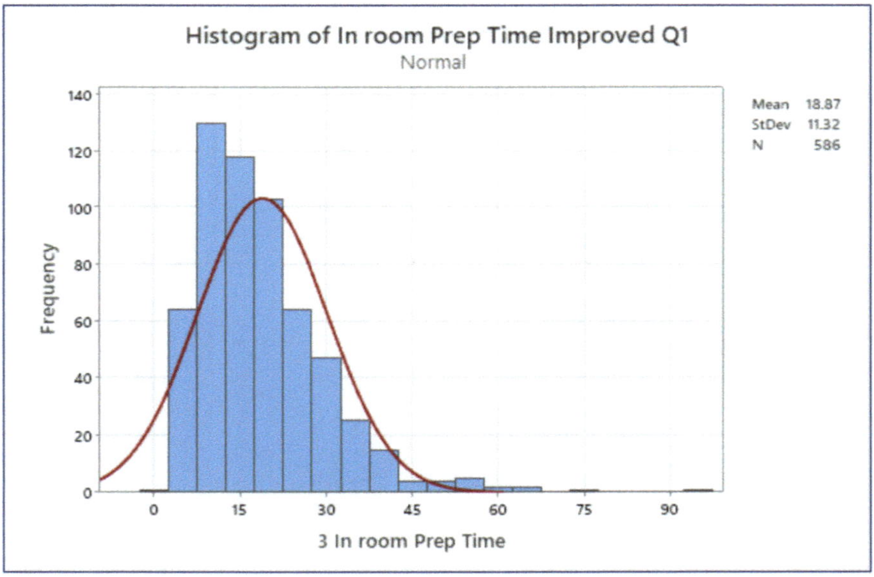

Fig. 10.44 Histograms of in-room prep time improved Q1

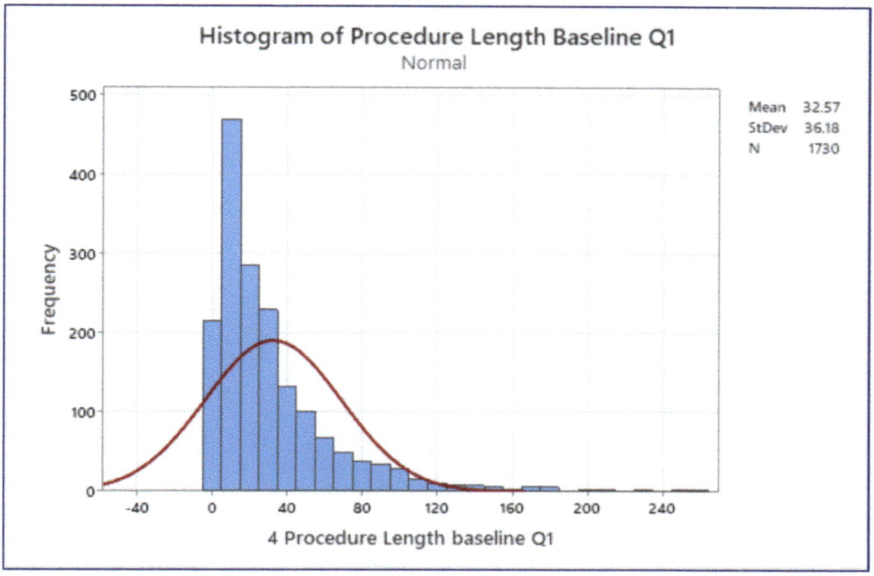

Fig. 10.45 Histograms of procedure times baseline Q1

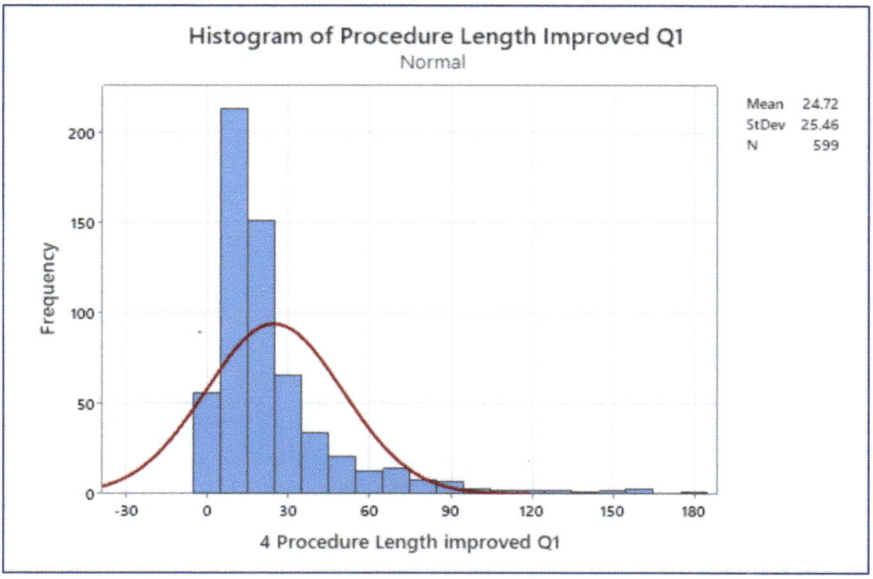

Fig. 10.46 Histograms of procedure times improved Q1

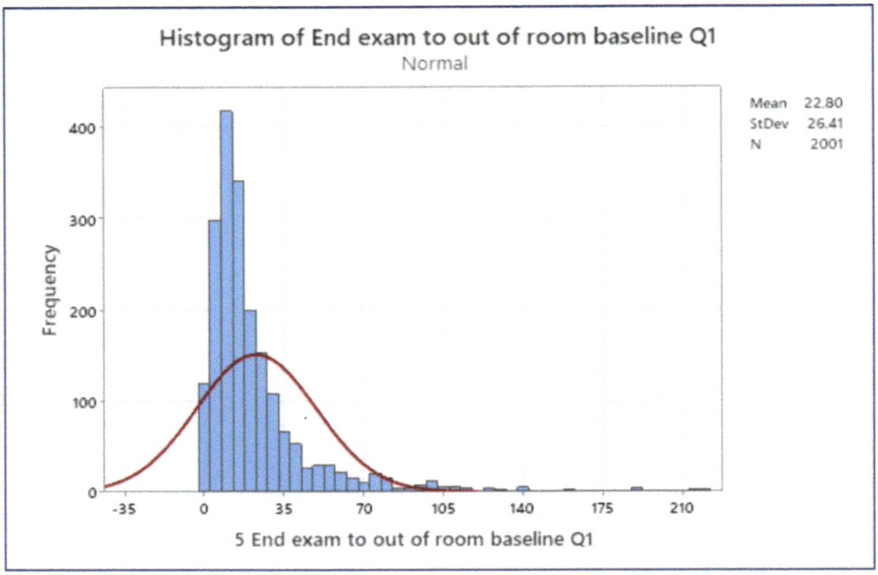

Fig. 10.47 Histogram of end exam to out of room baseline Q1

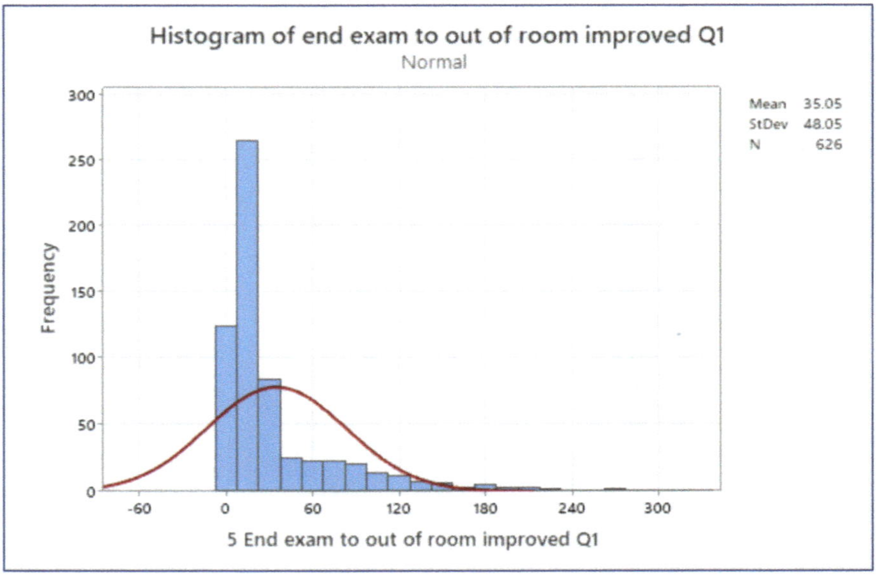

Fig. 10.48 Histogram of end exam to out of room improved Q1

The process times for the IR activities before and after improvement are shown in Table 10.11, along with the p-values for the 2-sample t-tests. The on-time start times and the procedure times improved significantly, 50% for on-time starts and 24% for the procedure times. The end of exam to out of room increased significantly (54%).

The two-sample t-test for the on-time starts baseline compared to improved times is shown following.

Two-Sample T-Test and CI: On-time starts baseline Q1, On-time starts improved Q1

Method

μ_1: population mean of On-time starts baseline Q1
μ_2: population mean of On-time starts improved Q1
Difference: $\mu_1 - \mu_2$

Equal variances are assumed for this analysis

Descriptive Statistics

Sample	N	Mean	StDev	SE mean
On-time starts baseline Q1	2039	20.6	31.5	0.70
On-time starts improved Q1	617	10.3	28.7	1.2

Estimation for Difference

Difference	Pooled StDev	95% CI for difference
10.28	30.84	(7.50, 13.05)

Test

Table 10.11 Hypothesis test results

Process activity	Average baseline Q1	Average improved Q2	Statistically significant improvement?/ P-value	% improvement
On-time start times	20.57	10.29	Yes/0.000	50%
Holding time	35.85	37.12	No/0.227	Not applicable
In-room prep time	19.96	18.87	Yes/0.048	5.5%
Procedure length	32.57	24.72	Yes/0.000	24%
End of exam to out of room	22.8	35.05	Yes/0.000	54% increase

Table 10.12 Percentage of missing data by IR activity

Percent missing data			
Name of metric	Baseline Q1 (%)	Baseline Q3 (%)	Improved Q1 (%)
On-time starts	9	12	7
Wheels in to out	14	12	7
Procedure length	22	25	11
Holding time	25	31	24
In-room prep time	18	21	13
End exam to out of room	10	18	6

Null hypothesis	H_0: $\mu_1 - \mu_2 = 0$
Alternative hypothesis	H_1: $\mu_1 - \mu_2 \neq 0$

T-value	DF	P-value
7.25	2654	0.000

The percentage of missing data also improved for each IR activity, as shown in Table 10.12 and Fig. 10.49. The missing data improved for on-time starts from 9% in baseline Q1 and Q3 of 12% to 7% in Q1. The Wheels in to out for baseline Q1 of 14% and Q3 of 12% reduced to 7% in Q1 improved. The Procedure length was baseline Q1 of 22% and Q3 of 25% improved (reduced) to 11% in Q1 improved. The holding time was 25% in Q1 baseline and 31% Q3 baseline improved in Q1 to 24%. The in-room prep time missing data was 18% in Q1 baseline and 21% in Q3 baseline and improved to 13% in Q1 improved. The end exam to out of room in Q1 baseline was 10%, and 18% in Q3 baseline and improved to 6% in Q1 improved time period.

Control Phase

There was significant improvement as illustrated in the Improve phase across most of the operational IR activities, as well as a reduction in missing data by IR activity. In the Control phase, the improvement control plan was used to track the improvements for the process metrics as shown in Table 10.13. Standard work and metrics were incorporated, especially around collection of the data to reduce the percentage of missing data. Control charts were proposed to track the IR activities times. An example of an individual and moving range control chart is shown in Fig. 10.50. The control charts help to monitor when the process becomes unstable, and an assignable cause occurs in the process, which must be investigated to bring the process back into control. The control charts were a future improvement that wasn't yet implemented at the end of the piloting period. A common notes checklist and common

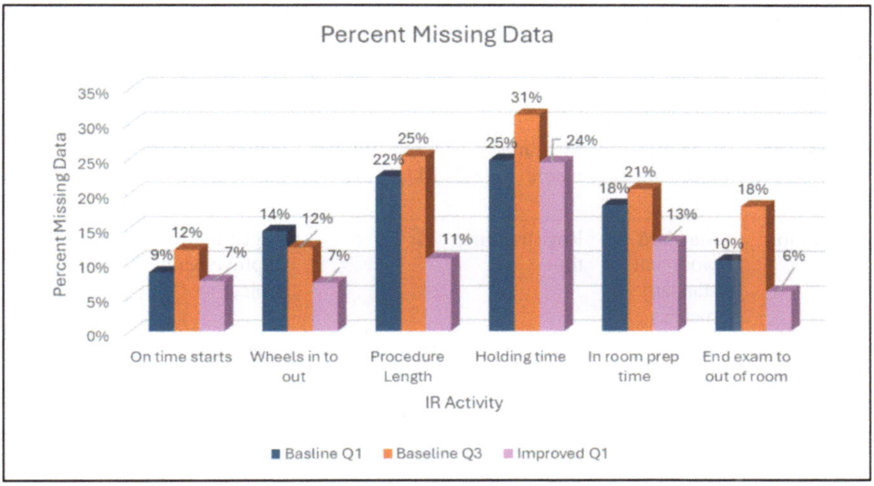

Fig. 10.49 Percentage of missing data by IR activity—baseline (Q1, Q3) and improved Q1

data entry templates were also incorporated as control mechanisms for the process. Performance metric scorecards were also incorporated to control the process and ensure improvements continued. The new dashboard and alerts were also proposed to be implemented, as shown in Fig. 10.33, within the Improve phase. The alerts would alert staff when there was a delay in the process that needed to be addressed.

There were many successes in this project as well as challenges.

The challenges included:

- Pandemic with incredible staffing challenges.
- Code Yellow identifying a lack of resources available in the IR department during Q3, which supported the poor performance metric results in Q3.
- The team was still waiting on major dashboard alerts by the end of the piloting period.
- Due to staffing shortages and the pandemic, there was an inability to work with the floors to improve patient readiness.

The incredible successes included:

- Engagement of nurse manager, Physician and Nurse Directors in the IR and IR process owners.
- The project conveyed the importance of having a robust data collection system to enhance continuous improvement.
- The project helped to transfer knowledge and the value of Lean Six Sigma.

Future work:

The future work within the IR department included:

Table 10.13 Improvement control plan

Improvement control plan					
The control plan is a detailed assessment and guide for maintaining all of the positive changes that you have implemented					
Process steps	Control mechanism	Measure/ metric	Criticality (H M L)	Action taken if problems occur	Owner
Registration	Standard work and standard metrics	Registration time	**M**	Registration improvement project	Registration director
Holding	Control charts	Time from holding to in-room	**H**	Investigate assignable causes	IR nurse director
In room	Standard work and standard metrics	In-room time On-time starts	**H**	Investigate assignable causes	IR nurse director
Procedure	Common notes checklist Procedure times	Procedure times	**M**	Investigate assignable causes	IR physician director
End exam to out of room	Dashboard alerts	End exam to out of room times	**L**	Investigate assignable causes	IR nurse director
Time from holding to out of room	Scorecards	On-time starts Percent missing data	**H**	Investigate assignable causes	IR nurse director IR physician director

- Re-assessing improvement metrics when they implement the dashboards and alerts.
- Continuing to reduce staffing shortages and train staff on the standard processes.
- Restart and re-energize the continuous improvement outside of the pilot period and implement improvements more widely.

Conclusions

This chapter described the use of the Lean Six Sigma DMAIC methodology within the Interventional Radiology department. Engaged IR leadership, including Physician and Nurse Directors, along with key process owners, helped implement significant improvements within the IR department to reduce the times of most IR activities, and increase the percentage of cases that started on-time. When the first cases of

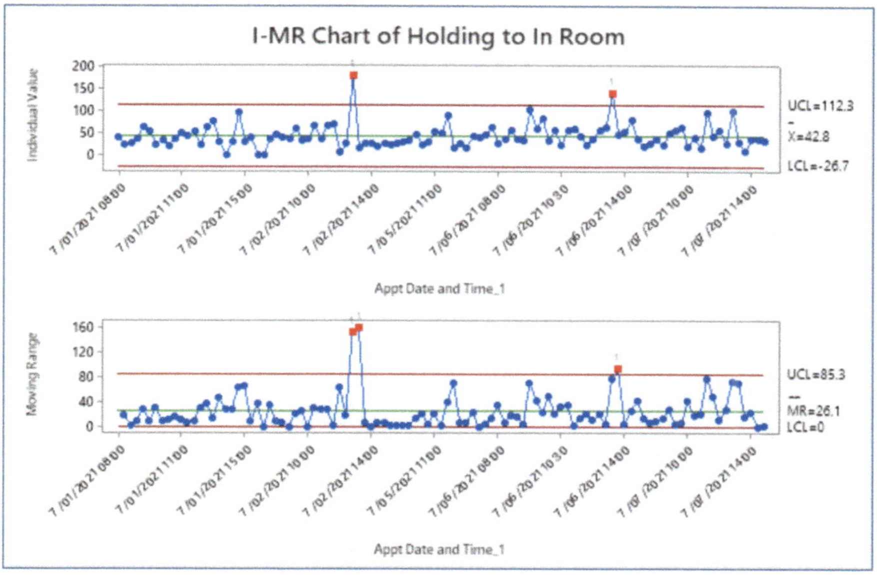

Fig. 10.50 Individual and moving range charts for holding time

the day start on-time, overall daily operations run more smoothly, enhancing both patient and staff satisfaction, and reducing overtime. The focus of this project was to enhance the data system to have more robust and accurate performance metrics and a data analytics system is essential for continuous process improvement.

Chapter 11
Innovating the Patient Care Operating Model: Integrated Value Care

Abstract The healthcare industry and hospitals in general have been providing care from an operating model perspective in much the same way since hospitals began. Yes, the medical technology has become much more advanced: imaging technology providing an ever-advancing and amazing view inside the body without a single incision; surgical techniques with cameras and robots enabling surgeons to perform surgeries with smaller incisions and quicker recoveries are just two examples. But the basic operating model has not changed. We are still managing the flow of patients through the healthcare system with a focus on the processes and care providers within a department providing care to each patient, but largely ignoring the flow and progression of a patient to the next step in their care. Patients sit in emergency rooms, with long delays, because the model doesn't incorporate pulling the patient through their emergency care by understanding where the patient needs to be in their next step of care in a proactive manner. The patient arrives in the emergency room, signs in, and sits down to wait. The triage nurse triages the patient and returns them to the lobby to wait, with a vague priority rating of the rough order that they will be called to an emergency room. The patient is finally placed in a room and waits for the next care provider to pick them out of a list of their other patients, in an unknown priority, to assess their medical status, and push them to the next step of the process. Each care step is provided in a somewhat random order, not based on the series of activities that must be performed to make the patient well, but on them sitting in multiple queues for separate departments such as imaging, transportation, registration, scheduling, admitting, emergency physician, nurse, admitting physician, consulting physician, etc. Another example is in the inpatient operating model. The patient comes to the hospital in one of several input channels, being admitted through the emergency room, being a direct admit to registration based on a physician's orders, or being admitted based on either a planned or emergency surgery. Once the patient is admitted to an inpatient floor, which could be after a long wait in one of the input channels, the patient sits and waits again for their initial assessment. "Normal" service level requirements, typically based upon a regulatory requirement from the government, require a patient's initial inpatient assessment to be performed within 24 h of admission. This seems like a long time to identify an initial diagnosis, let alone design a potential path of care for the patient. The patient then waits to be

S. Furterer, *Lean Six Sigma for the Healthcare Enterprise*,
https://doi.org/10.1007/978-1-4471-7617-6_11

"roughly" scheduled for diagnostic tests that provide more insight into their maladies. The "roughly" term is used because a diagnostic test is ordered, but there is really no inpatient schedule. The patient is put on a list for the required department but has little or no insight into when the test will be run, other than within a rough 24-h period timeframe. Wouldn't it be patient-centered to provide the patient with a schedule each morning of the care, and tests that will be provided and performed for that day? In this way the family members can easily be a part of the patient's care, without waiting for days not knowing when the patient will get their tests. Wouldn't it also be novel to know when the patient is expected to be discharged based on standardized patient care protocols and inpatient schedules of patient care activities? The existing operating model in a hospital is inefficient and causes immense delays in patient care, when what could be minutes and hours of delays become days and weeks of delays. Is there a better way? The author believes there is a better way to manage the flow of care for a patient through the myriad activities, tests, transports, and handoffs in a hospital, based on the principles of Lean, Six Sigma, Systems Engineering, and Business Architecture. This chapter will provide a proposal for a new and innovative value care operating model for providing patient care in a hospital setting. It is a proposal for future research and application for those brave care providers, in particular management engineering pioneers who seek to provide a better way.

Keywords Staffing level · Patient flow · Spiritual care · Order entry · Patient demand

Define Phase

In the Define Phase we will define the scope of the project and the project goals. We will identify the key stakeholders for the project and identify their initial Voice of the Customer, their concerns, and how the project impacts them.

The following steps were executed in the Define Phase:

1. Develop project charter.
2. Perform stakeholder analysis.
3. Perform initial Voice of Customer (VOC) and Identify Critical to Satisfaction (CTS).

Develop Project Charter

The project charter is shown in Table 11.1 and discussed next.

Table 11.1 Project charter

Project Overview, Problem Statement, & Goals
Provides a description of the problem to be solved. Be as descriptive as possible. Include improvement goals that are SMART (Specific, Measurable, Attainable, Realistic, and Time-Based).
Project Name: Innovating the Patient Care Operating Model: Value Centric Care Problem Statement: The existing operating model in a hospital is inefficient and causes immense delays in patient care, when what could be minutes and hours delays become days and weeks of delays. Goal(s): This project will provide a proposal for a new and innovative value centric care operating model for providing patient care in a hospital setting. Scope Statement: The scope of the project includes the inpatient care activities including: register patient, assign room, prepare room, transport patient, admit patient, schedule patient, collect specimens, provide therapy, provide health care, provide diagnostic tests & results, manage patient flow, manage patient care, provide spiritual care. Projected Financial Benefits: Reduction of the length of stay in the hospital. Reduction in severity of illness, reduction of cost of care.

Project Charter

The existing operating model in a hospital is inefficient and causes immense delays in patient care, when what could be minutes and hours of delays become days and weeks of delays. The patient flow is managed within each siloed department, not across the continuum of care. The time that the patient spends in the hospital is impacted immensely by the delays in providing care and moving the patient through each step in their care, and in the handoffs between each department. If a patient stays over the weekend, their length of stay can be increased by almost 2 days, because physicians may not be easily available to write orders, perform consultations, and discharge the patient. Ancillary staff may not be available to perform diagnostic tests and therapy, due to not scheduling staff on the weekend, due to higher pay differentials on the weekend. Patients are typically, in Lean terminology, "pushed" through the process based on physician orders, not "pulled" based on a plan of care, and an estimated and desired length of stay.

The primary stakeholders are the patients and their families, physicians, nurses, Lab, Imaging, therapists, Scheduling, Registration, Transportation, Bed Assignment, Environmental Services, Case Management, Administration, and Spiritual Care. The goal of the project is to provide a proposal for a new and innovative value centric care operating model for providing patient care in a hospital setting. The scope of the project includes the inpatient care activities including register patient, assign room, prepare room, transport patient, admit patient, schedule patient, collect specimens, provide therapy, provide healthcare, provide diagnostic tests and results, manage patient flow, manage patient care, provide spiritual care. The CTS criteria are Patient Length of Stay; Process times; Delays in patient flow.

Perform Stakeholder Analysis

The stakeholders are patients who are admitted to the hospital as inpatients. The Physicians provide care to the patients and are concerned with the quality of the patient's care; patient length of stay; appropriate diagnosis; access to tests and lab results; reduced patient delays in getting tests and results. The nurses provide care to the patient. They are concerned with the quality of their patient's care; patient length of stay; and nursing staffing levels being appropriate to provide quality of care. The Lab staff obtains lab specimens from the patient. They are concerned with the quality of care and lab staffing levels being appropriate to collect specimens in a timely and safe manner. The Imaging staff performs imaging tests for inpatients and is concerned with quality of the patient's care, imaging staffing levels, scheduling of tests, and number of reschedules of tests.

Therapists provide physician and occupational therapy care to the patients. They are concerned with quality of care, therapist staffing levels, and being incorporated into the patient care continuum. Scheduling schedules the patients for exams. They are concerned with the complexity of scheduling patients for tests, having appropriate clearances, and insurance authorizations. The Registration department registers the patients as an inpatient. They are concerned with timely patient registration; reduced number of times the patient needs to provide the same information for multiple registrations. The Transportation department is concerned with patient transport turnaround time, and transportation staffing levels. The Bed Assignment staff assigns inpatient beds to the patient. They are concerned with the patient being assigned to the right bed in a timely manner, and knowledge of clean, available beds. Environmental Services clean the room and prepare them for the next patient. They are concerned with room cleaning turnaround time and environmental services staffing levels. Case Management personnel manage the patients' cases, monitor their length of stay, and connect the patient to social services. They are concerned with sufficient staffing levels to provide case management services. Hospital Administration manages the business aspects of the hospital. They are concerned with productivity of personnel, the number of Full Time Equivalents (FTEs), the cost of providing services, patient experience, and the quality of the patient care. Spiritual Care staff provides spiritual and emotional support for patients and their families. They are concerned about being included in the patient care continuum.

The stakeholder analysis is shown in Table 11.2.

Table 11.2 Stakeholder analysis

Stakeholder analysis			
Stakeholder	Type	Primary role	Impacts/concerns
Patients and their families	Primary	Patients who are admitted to the hospital and receive care, and their families	Quality of care; reduced delays; knowledge of scheduled exams; knowledge of continuum of care
Physicians	Primary	Physicians who provide patient care and diagnostics	Quality of care; patient length of stay; appropriate diagnosis; access to tests and lab results; reduced patient delays in getting tests and results
Nurses	Primary	Nurses who provide patient care	Quality of care; patient length of stay; nursing staffing levels
Lab staff	Primary	Lab staff, who enter orders, and collect lab specimens from patients	Quality of care; lab staffing levels
Imaging staff	Primary	Imaging staff who enter orders and perform diagnostic tests	Quality of care; imaging staffing levels; incorporated into the patient care continuum
Therapists: physical and occupational	Secondary	Physical Therapists, Occupational Therapists who provide patient therapy	Quality of care; therapist staffing levels; incorporated into the patient care continuum
Scheduling	Secondary	Schedulers who schedule tests	Complexity of scheduling patients for tests; having appropriate clearances and insurance authorizations
Registration	Secondary	Registrars who register the patient into the hospital	Timely patient registration: reduced number of times patient needs to provide same information for multiple registrations
Transportation	Primary	Transportation staff who transport the patients for admission, discharge, tests and patient care	Patient transport turnaround times; transportation staffing levels
Bed assignment	Secondary	Nurses who assign inpatient hospital beds from direct admits, emergency room, and surgical patients admitted to the hospital	Patient is assigned the right bed in a timely manner; knowledge of clean, available beds
Environmental services	Primary	Environmental services who clean and prepare the patients' rooms	Room cleaning turnaround time; environmental services staffing levels

(continued)

Table 11.2 (continued)

Stakeholder analysis			
Stakeholder	Type	Primary role	Impacts/concerns
Case management	Secondary	Case management personnel who manage the patients' cases, monitor their length of stay, and connect the patient to social services	Sufficient staffing levels to provide case management services
Administration	Secondary	Hospital administration who manage the business aspects of the hospital	Productivity; number FTEs; cost of providing services; patient experience; quality of care
Spiritual care	Secondary	Staff who provide spiritual and emotional support for patients and their families	That they are included in the patient care continuum

SIPOC (Suppliers-Inputs-Processes-Outputs-Customers)

The SIPOC is used to ensure the scope of the project is well understood and is shown in Table. 11.3. It is also a check and balance for the customer and stakeholder analysis. Any customers and suppliers should also be identified as a project stakeholder.

The high-level process steps are the register patient, assign room, prepare room, transport patient, admit patient, schedule patient for tests, collect specimens for lab tests, provide therapy, provide healthcare, provide diagnostic tests and results, manage patient flow, manage patient care, provide spiritual care, and discharge patient. The patients are admitted to the hospital in one of three ways: direct admission from physician's orders, from the emergency room, or from the surgical suite. Admitting registers the patient. The Bed Assignment nurse assesses the type of bed and availability and assigns a bed once the desired bed type becomes available. Environmental Services receives an order to clean the room, and then cleans the room, after a potential delay because of other beds in the queue to be cleaned. The patient is transported to their room by a transportation tech. Again, there may be a delay waiting for a transport. The patient's initial admission assessment is performed by the nurse within 24 h of arrival to the room. The physician sees the patient and provides orders for tests, therapy, and care. The tests are scheduled, performed, and the results are provided to the physician. The patient care is managed based on Case Management reviewing the patient's cases where there are excessively long lengths of stay compared to the average expected length of stay for a patient with the primary diagnosis. The patient is provided with therapy and healthcare based upon the physician's orders. Spiritual care is provided at the request of the patient. The patient flow is not proactively measured, other than Case Management reviewing their case if their length of stay exceeds the established averages. The patient is discharged and transported out of the hospital.

Table 11.3 SIPOC

SIPOC				
Suppliers	Inputs	Process	Outputs	Customers
Physicians patients	Physician orders Patient information	Register patient	Patient registration	Nurses transportation
Physician	Physician orders	Assign room	Room assignment	Environmental services transportation
Physician	Room clean order	Prepare room	Cleaned room	Patient nurse
Bed assignment	Room assignment transportation order	Transport patient	Transport patient	Patient
Patient physician nurse	Physician orders registration	Admit patient	Admission assessment	Patient nurses physicians
Patient physician	Physician orders	Schedule patient for tests	Scheduled tests	Patient imaging
Lab patient	Lab orders	Collect specimens for tests	Specimens	Lab
Patient physician	Therapy orders	Provide therapy	Therapy results	Patient
Patient physician nurse	Physician orders	Provide healthcare	Healthcare	Patient
Patient physician	Imaging orders	Provide diagnostic tests and results	Test results	Patient physician
Patient physician nurse	Physician orders physician diagnostics	Manage patient flow	Patient care	Patient nurse
Patient physician	Physician orders physician diagnostics	Manage patient care	Patient care	Patient
Patient spiritual care	Request for care	Provide spiritual care	Spiritual care	Patient
Patient physician	Physician orders	Discharge patient	Discharged patient	Patient

Perform Initial Voice of Customer and Identify Critical to Satisfaction

The Critical to Satisfaction Criteria for this project are:

- Patient length of stay as an inpatient. This is measured from the time that the patient is admitted to when they are discharged.

- Process Time is the time it takes to process different activities within the patient flow.
- Delay in patient flow: this is measured by tracking the time it takes to receive diagnostic tests, therapy, and care from when the physician order is requested until when the care is received.

Measure Phase

The Measure Phase included the following activities:

1. Define the current process.
2. Define detailed Voice of Customer (VOC).
3. Define the Voice of Process (VOP) and current performance.

Define the Current Process

The first step in the Measure phase was to understand the current processes. Several techniques were applied: value chain and functional decomposition; value stream mapping; and process mapping. Each technique will be discussed.

A value chain consists of the high-level activities that provide value to the customer, or in this case, the inpatient. The value chain is made up of "value activities," those activities that are direct patient facing and bring value to the patient to help them receive care. The value chain is also made up of "supporting activities," those activities that are necessary to be performed, such as registration and billing, but doesn't "touch" the customer or directly provide care. The inpatient value chain is shown in Fig. 11.1. The functional decomposition provides an inventory of the activities performed in the value chain. It helps us to understand the activities performed to provide inpatient services. The functions can be decomposed or broken into their next more detailed level of activities. This will typically continue for about three levels, or until a natural sequencing of the activities begins to form. At this point, it is time to create the process maps, which shows the flow, or sequence of activities to be performed to achieve the process objective.

Value stream mapping can be used to understand the high level of activities and the time it takes to complete the activities. It is typically used to represent information or material flow through the system. In our case, it will show how the patient moves through our inpatient system. A value stream map for the inpatient services is shown in Fig. 11.2. It incorporates the high-level activities, shows the inventory of patients waiting before each activity, and can show the processing times for standard diagnoses.

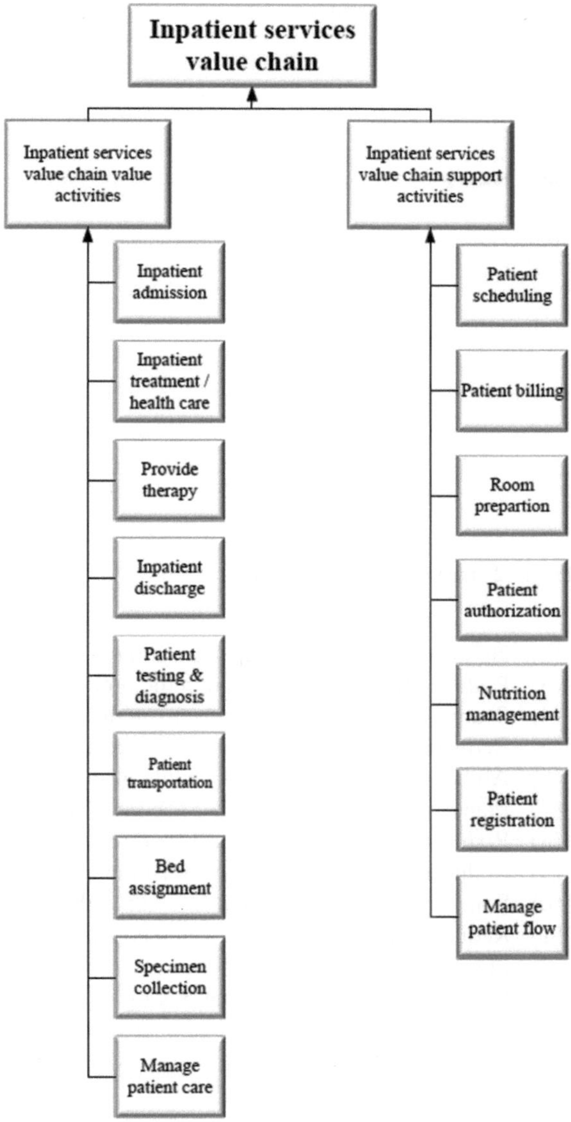

Fig. 11.1 Inpatient value chains and functional decomposition

Define Detailed Voice of Customer

The Voice of Customer was defined through identifying inefficiencies and pain points in the existing processes. The following customer requirements were defined:

(1) Register once in the hospital, whether it is in the Emergency Room, Surgical Lobby, or Admitting Lobby.

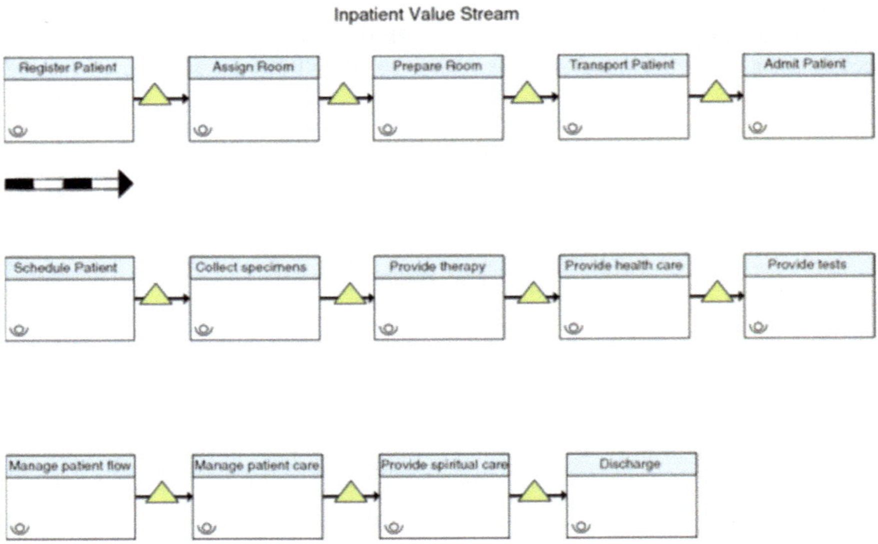

Fig. 11.2 Inpatient value stream map

(2) No or little wait for an inpatient room.
(3) Reduced time between steps in the process.
(4) Seamless handoffs between steps in the process.
(5) Reduced time for admission assessment.
(6) Ability to have a planned schedule of inpatient tests and diagnostics.
(7) Plan of care, with estimated discharge dates.
(8) Reduced delays in getting tests and results.
(9) Reduced delays in physician consultant referrals.
(10) Someone to manage patient flow through the healthcare system.
(11) Reduced avoidable delays, which equals reduced length of stay.
(12) Standard treatment protocols.
(13) Someone to manage the plan of care.
(14) Updated, online, real-time bed board and room status.
(15) No holding beds when the patient is discharged, not updating the bed status immediately.
(16) Reduced delays in waiting for transport.
(17) Reduced delays in waiting for rooms to be cleaned.

We developed a Quality Function Deployment House of Quality, shown in Fig. 11.3. We developed technical requirements that would potentially fulfill the customer requirements. These requirements could be process and/or information technology requirements. We rated a strength of relationship between the customer and technical requirements, by assigning a 9 to a high strength of relationship, a 3 to a medium strength, a 1 to a low strength relationship, and a blank to no relationship. The

Quality Function Deployment-House of Quality 1

Technical Requirements

Customer Requirements	Importance (15 = high, 1 = low)	Registration system and data integrated	Integrated Value Centric Care Model	Pull Patients	Service levels of process times	Patient Flow Managers	Inpatient Scheduling	Standard Treatment Protocols of Care	Patient Care Managers	Order Entry Process Improvement	Bed board status updates, real time	Integrated Care Providers Managing across value stream
1) Register once in the hospital, whether it is in the Emergency Room, Surgical Lobby or Admitting Lobby.	1	9										
2) No or little wait for an inpatient room.	6		9		9	9		1	1		3	9
3) Reduced time between steps in the process.	4		9		9	9	3	3	3	1		9
4) Seamless hand offs between steps in the process.	8		9			9	3	3	3	1		9
5) Reduced time for admission assessment.	10	3	3		9	9			3			3
6) Ability to have a planned schedule of inpatient tests and diagnostics.	14		3			3	9	3	3	9		3
7) Plan of care, with estimated discharge dates.	15		9		9	3	3	9	9	3		1
8) Reduced delays in getting tests and results.	7		9		9	9		9	9	9		9
9) Reduced delays in physician consultant referrals.	5		3		9	9		9	9	9		3
10) Someone to manage patient flow through the healthcare system	13			9	9	9	3	1		3		9
11) Reduced avoidable delays, which equals reduced length of stay	9			9	9	9	3	9		9		9
12) Standard treatment protocols	12				9	3	3	9	9	3		1
13) Someone to manage the plan of care.	11						3	9	9	3		9
14) Updated, on-line, real time bed board and room status	3									3	9	9
15) No holding beds when the patient is discharged, not updating the bed status immediately	2									9	9	9
Weighting		30	447	198	729	681	342	628	564	507	63	681
Ranking		11	7	9	1	2	8	4	5	6	10	2

Fig. 11.3 QFD house of quality

strength of relationship identifies how much the technical requirement helps to meet the customer requirements. We ranked the importance of each customer requirement, based on our assessment of importance, with a 15 being the highest importance to a 1 being the lowest importance. We then multiplied the importance times each strength of relationship to derive a relative weighting across the technical requirements. The absolute weighting is the ranking across the technical requirements to determine the priority when implementing improvements to process and technology.

The highest technical requirement in enhancing the inpatient customer experience is to incorporate service levels for process times. Many times, we don't measure how long it takes the patient to receive services and move through the process, so it is extremely important to measure the processes so that we know if we improve. The next two most important technical requirements are to incorporate patient flow managers and integrate the care providers across the value chain. These two requirements will ensure that personnel are engaged and focused on managing the flow and pulling the patient through the care processes based on a care treatment plan, instead of pushing them through based mainly on physician orders. To help to reduce the delays in the process, and be able to estimate the length of stay, implementing standard care protocols is the next most important technical requirement. Incorporating Patient Care Managers who manage the care of the patients throughout their inpatient stay and ensure that the patient is moving through their care plan is the next ranked technical requirement.

The next technical requirement is to redesign the order entry processes to reduce the time to enter orders so that the patients will receive their diagnostic tests and results more quickly. The next technical requirement is to design the integrated value centric care model, focusing on the value activities that provide value to the inpatient as they move through the process. This model will pull the patient through the steps of the processes and reduce delays by providing better coordination between steps and handoffs between care providers. Implementing inpatient scheduling, which would provide a daily schedule for the patient with a listing of their scheduled tests, therapy, and care, would help to provide a better patient experience, as well as reduce delays to receive care.

The requirement to pull the patient will help to focus the Patient Care Managers on pulling the patient through their standard care protocol. A requirement that is an enabler of getting patients admitted to beds quicker is to incorporate online, real-time updates, and status of the bed assignment system. This will ensure showing a bed as dirty, clean, available, or not available, which can help with bed assignments. The last requirement is to integrate the multiple registration systems to consolidate the data into the registrations that are required for billing purposes. The patient shouldn't need to provide the same information each time they move to a different service in the hospital, whether it is emergency services, inpatient services, surgical services, or outpatient services. This information should automatically transfer to a single registration, and the billing system can pick up the appropriate information with the registration type related to the types of services being provided. Many medical information systems are designed first to handle patient billing, not with a focus on ease of use, and simplification of the patient flow. The House of Quality provides a great way to ensure that the Voice of the Customer is collected and is aligned to the technical process and information system requirements.

Define the Voice of Process and Current Performance

The VOP and current performance includes mapping out the detailed processes related to providing inpatient care, as defined in the Value Chain and Functional Decomposition. Sample process maps are shown in Figs. 11.4, 11.5, and 11.6 for the inpatient admission process. A Process Architecture Map cover page shows the inventory of the stakeholders, information, and technology used in the admissions process. The detailed process maps consist of a new patient notification and report, and patient arrival and admission assessment.

Multiple process maps were developed for providing the inpatient care, including:

- Admission.
- Doctor calls on the floor.
- Doctor interaction.
- Daily huddles.
- Administering medication.
- Transport.
- Nursing morning activities.
- On call consultations.
- Order entry flow.

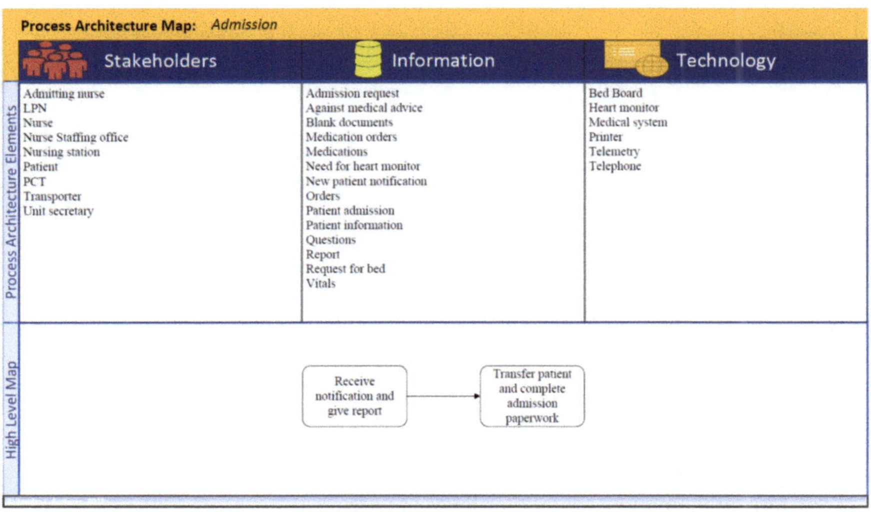

Fig. 11.4 Process architecture map—admission

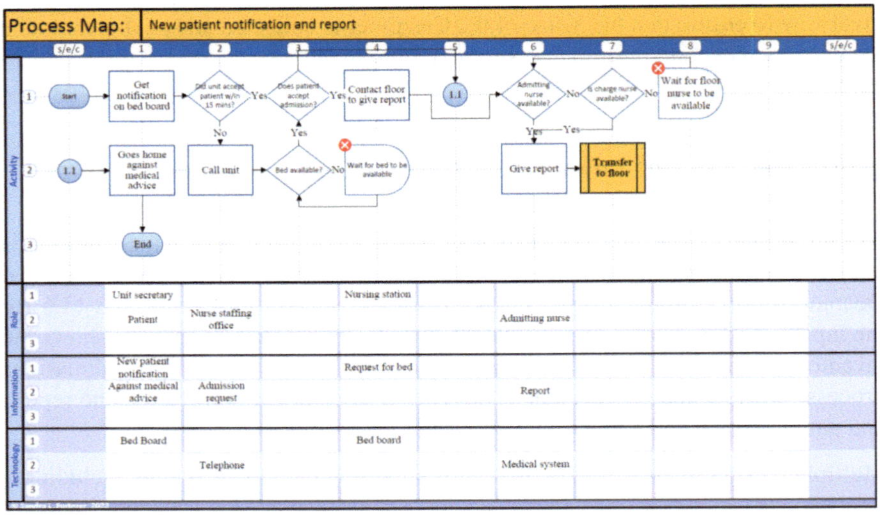

Fig. 11.5 Process map—new patient notification and report

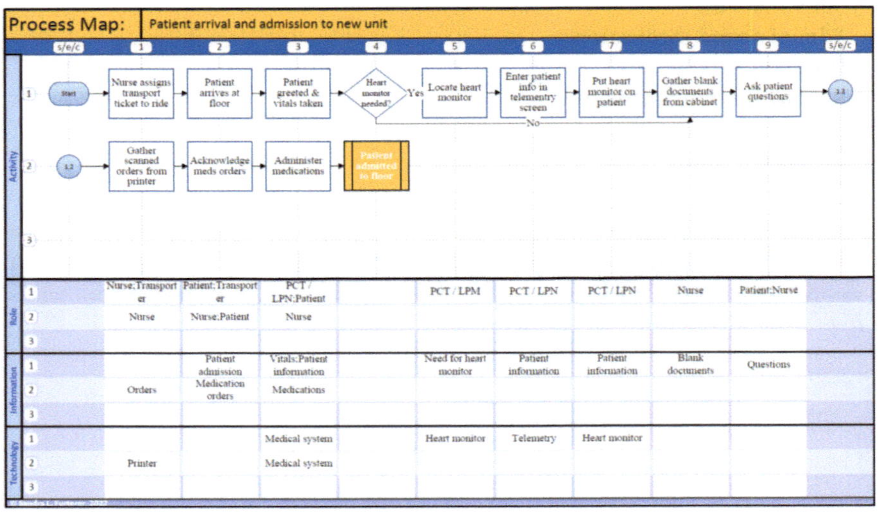

Fig. 11.6 Process map—patient arrival and admission to new unit

- Recurring tasks.
- Location supplies.
- Taking and giving report.
- Transferring the patient to another level of care.

Analyze Phase

The following activities were performed for the Analyze Phase to identify and validate the root causes.

1. Develop cause-and-effect relationships.
2. Determine and validate root causes.

Develop Cause-and-Effect Relationships and Determine and Validate Root Causes

Cause-and-Effect Analysis

A cause-and-effect analysis using the 5 Whys and Why-Why Diagrams was generated, shown in Figs. 11.7, 11.8, 11.9, 11.10, 11.11, 11.12, 11.13, 11.14, and 11.15. The following question was the starting point for understanding the root causes, and then we asked Why at least 5 times until the root causes were understood. The following potential root causes were identified.

Why are There Delays in the Inpatient Processes?

- Staff are not available to provide services.
  - Productivity goals are defined by department, not by value chain.

    Lack of knowledge of how to set the goals better.
  - Productivity goals are not set using real data, but arbitrary national averages.

    Lack of knowledge of how to set the goals better.
    Easier to set them without time, resources, and cost to collect real data.
    Always done it that way.
    Culture doesn't support questioning the productivity standards.
  - Nurses call out sick at last minute.

    Lax absence policies.
    Nurse is sick.
  - Staff not aligned to patient demand.

    No process to align staff to patient demand.
    Manage work by departmental key performance indicators, not across value chain.
- Lack of visibility of orders for care, diagnostic tests, and therapy.

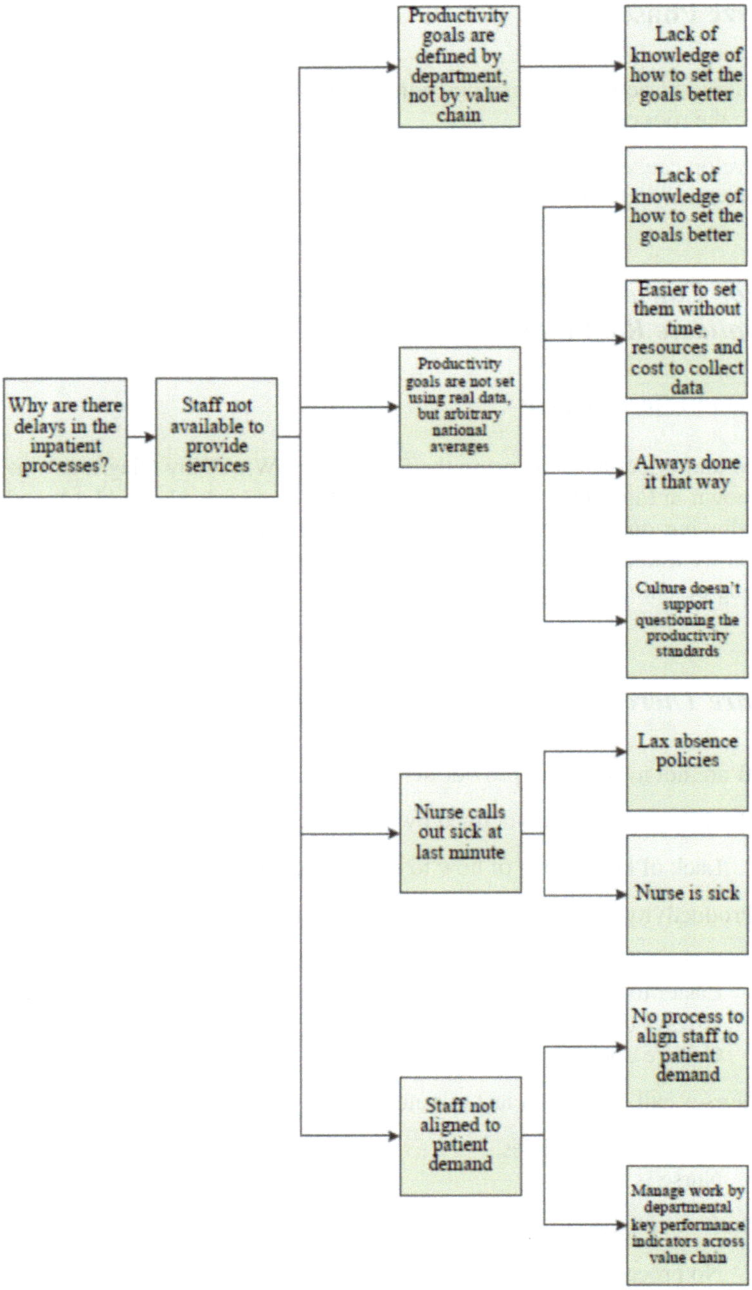

Fig. 11.7 Why-why diagram 1

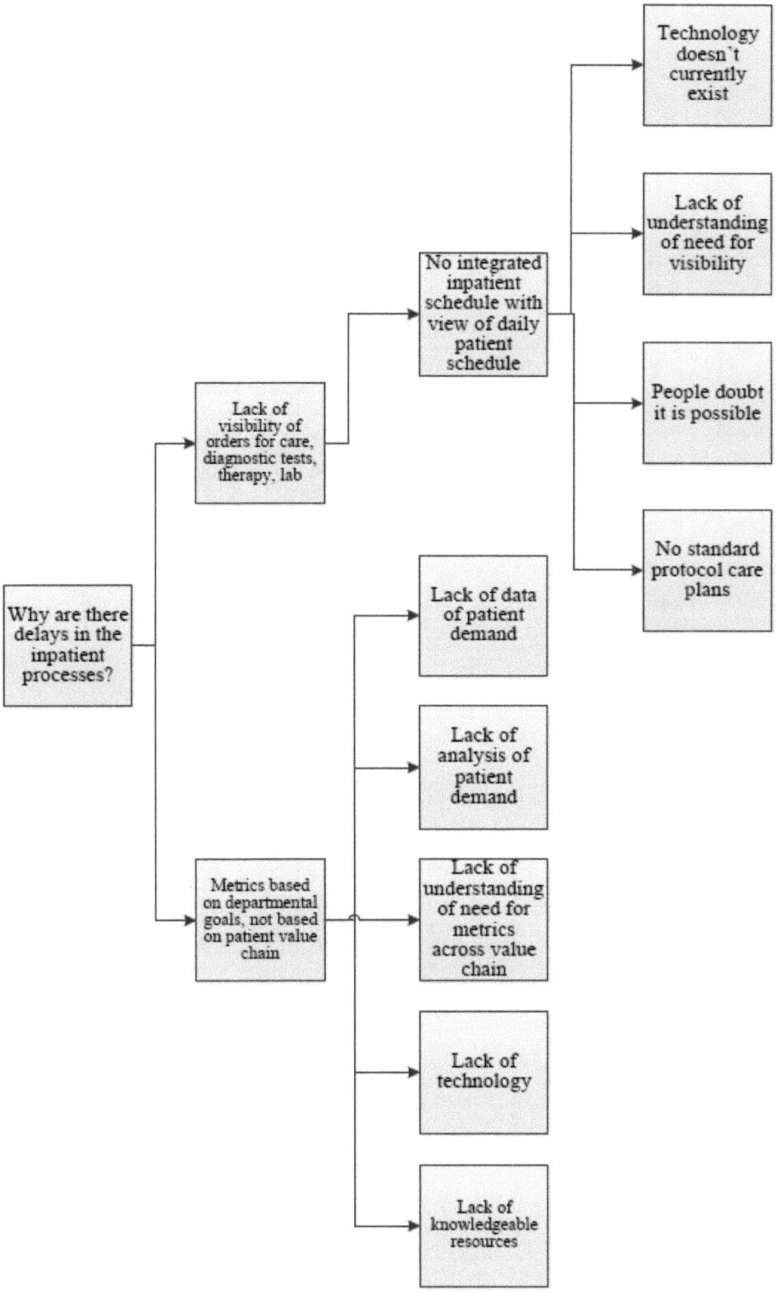

Fig. 11.8 Why-why diagram 2

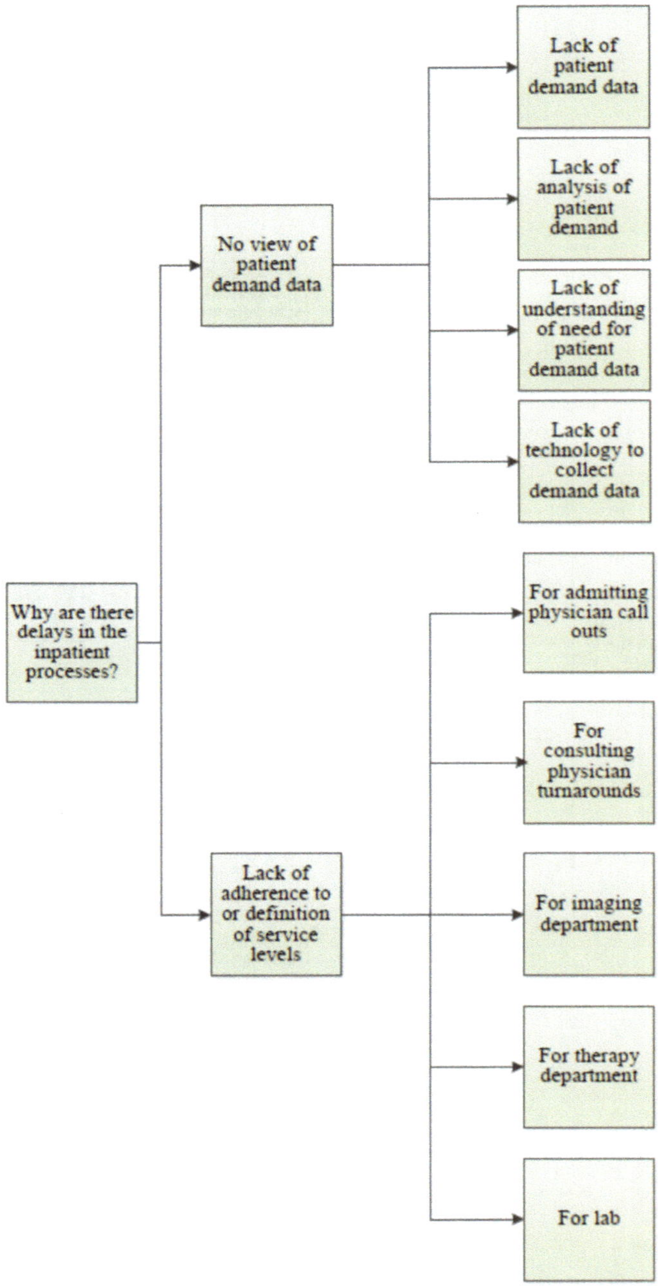

Fig. 11.9 Why-why diagram 3

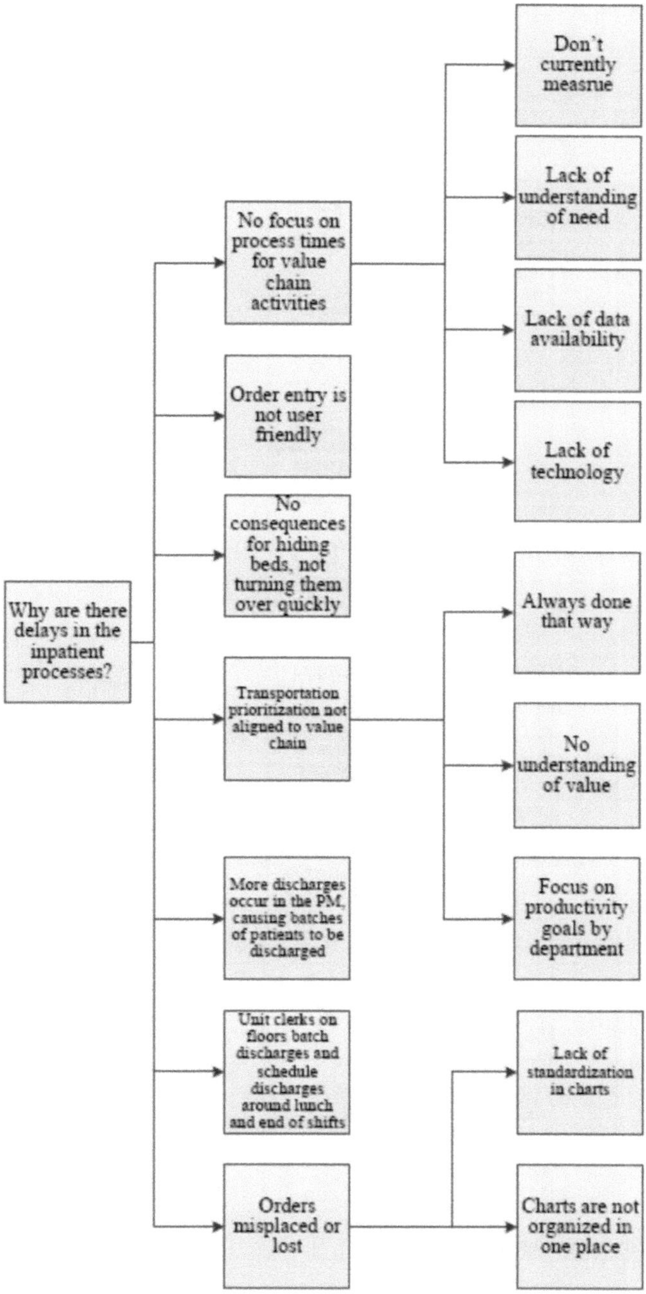

Fig. 11.10 Why-why diagram 4

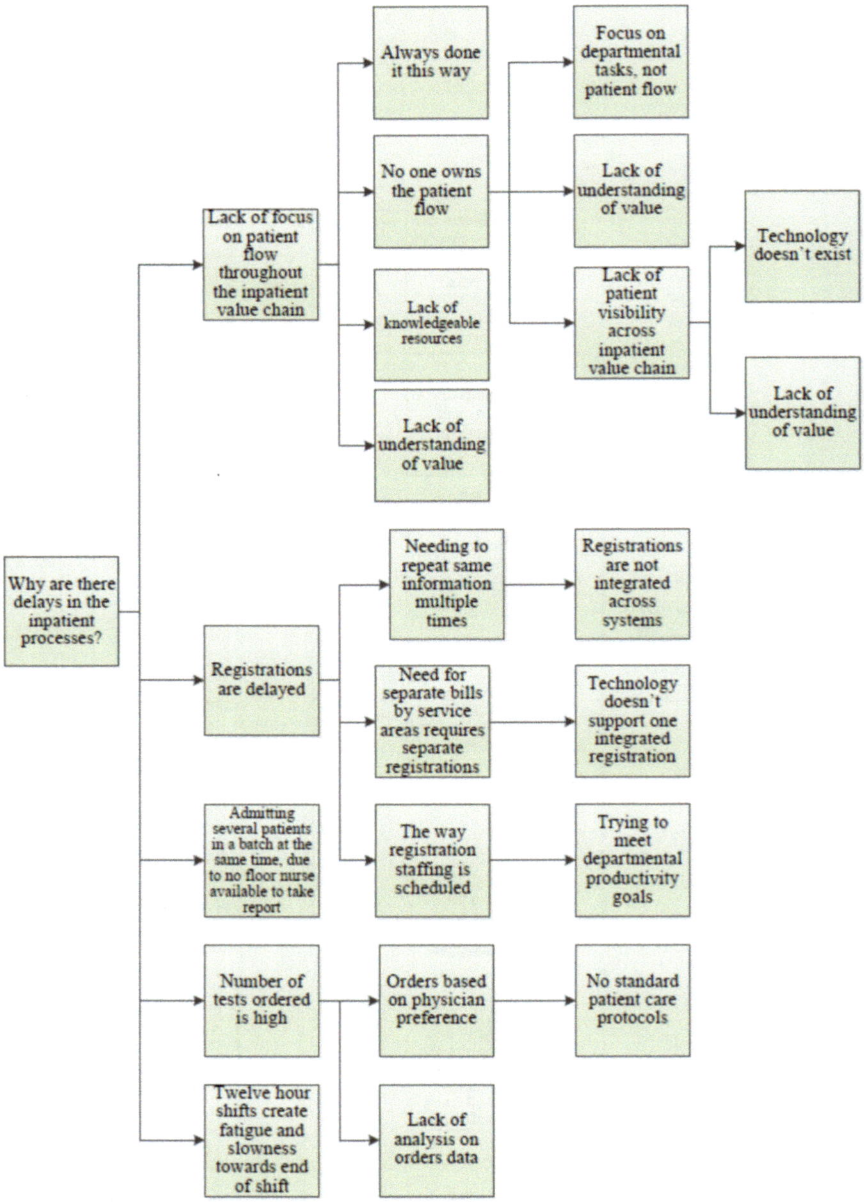

Fig. 11.11 Why-why diagram 5

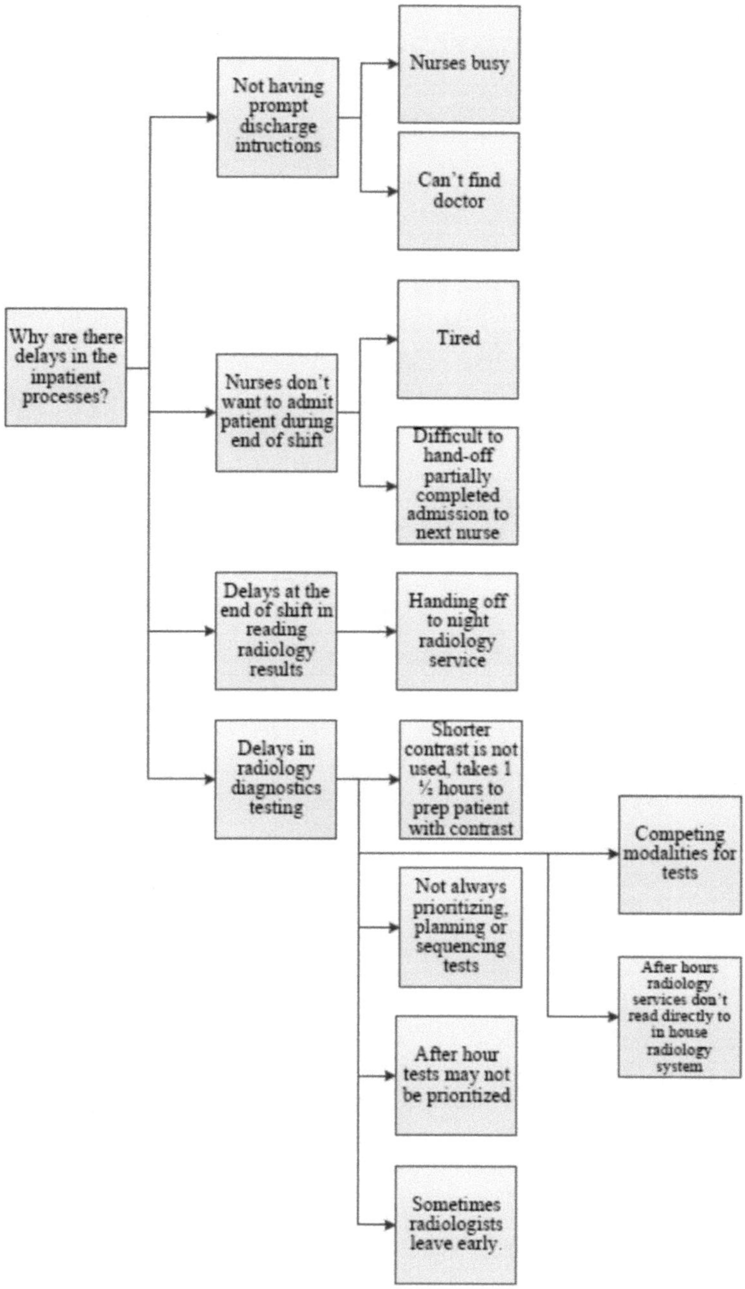

Fig. 11.12 Why-why diagram 6

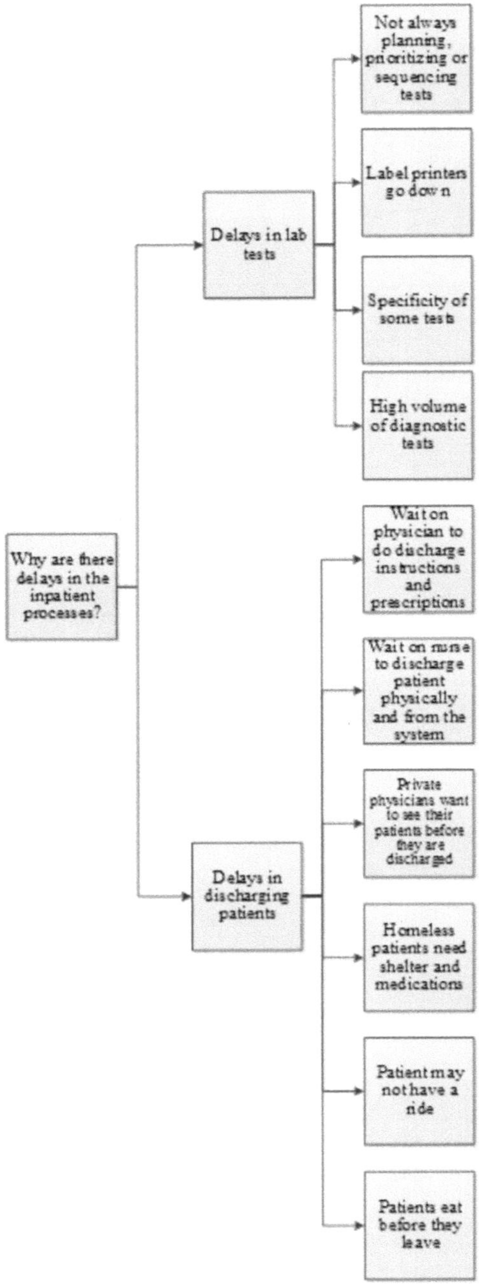

Fig. 11.13 Why-why diagram 7

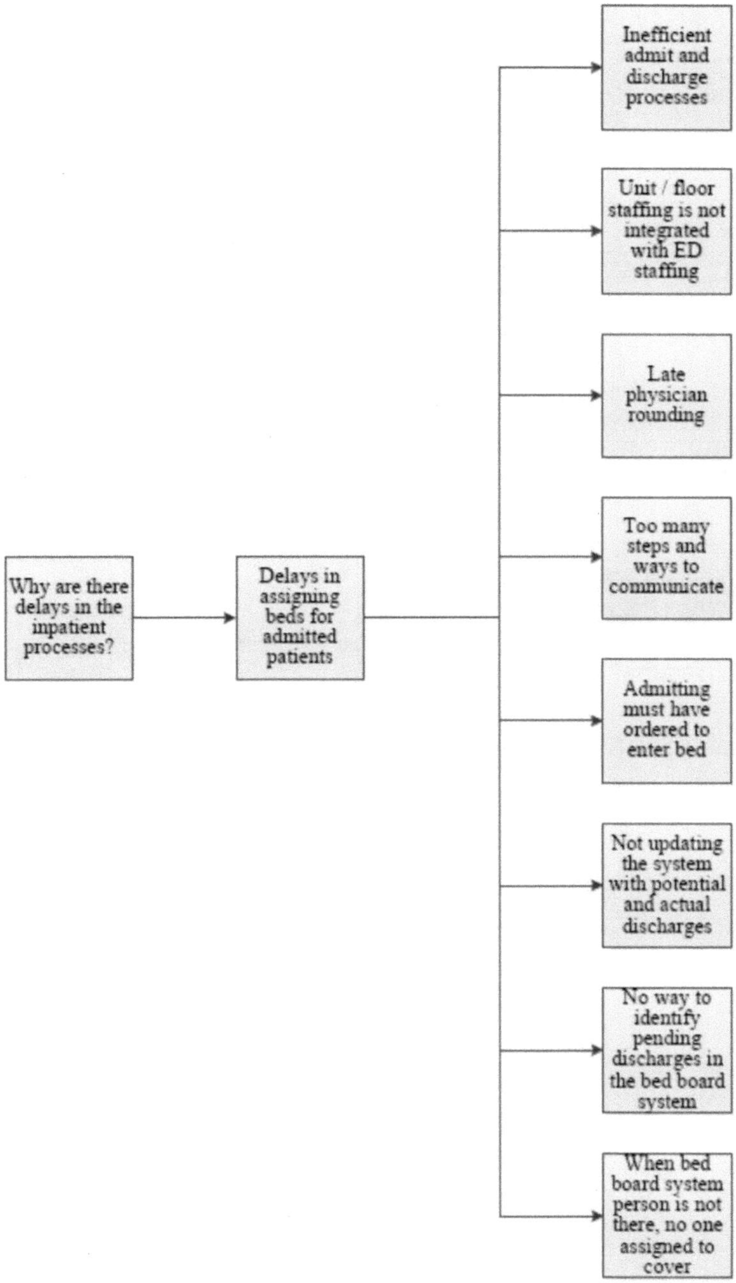

Fig. 11.14 Why-why diagram 8

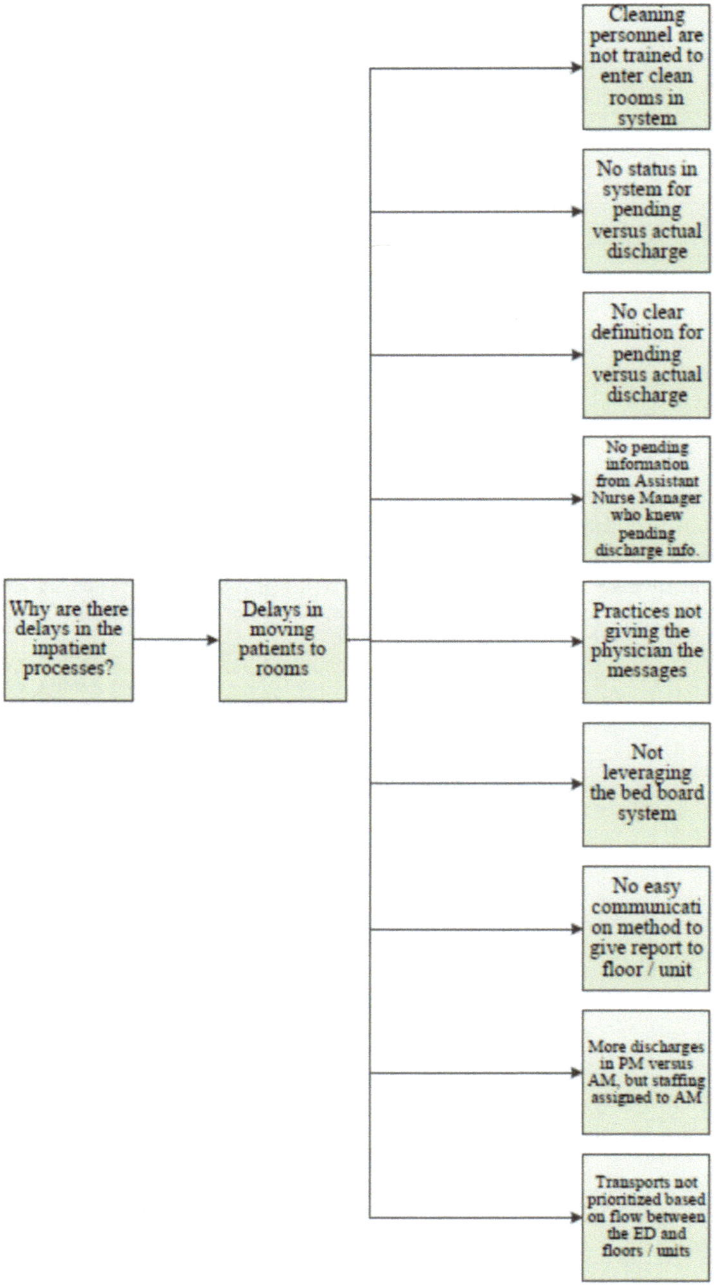

Fig. 11.15 Why-why diagram 9

  - No integrated inpatient schedule with view of daily patient schedule.

    Technology doesn't currently exist.
    Lack of understanding of need for visibility.
    People doubt it is possible.
    Poor process.

- No standard protocol care plans.
- Metrics based on departmental goals, not based on patient value chain.
  - Lack of data of patient demand.
  - Lack of analysis of patient demand.
  - Lack of understanding of need for metrics across value chain.
  - Lack of technology.
  - Lack of knowledgeable resources.
- No view of patient demand data.
  - Lack of patient demand data.
  - Lack of analysis of patient demand.
  - Lack of understanding of need for patient demand data.
  - Lack of technology to collect demand data.
- Lack of adherence to or definition of Service Levels.
  - For admitting physician call outs.
  - For consulting physician turnaround time.
  - For imaging department.
  - For therapy department.
  - For lab.
- No focus on process times for value chain activities.
  - Don't currently measure.
  - Lack of understanding of value.
  - Lack of data availability.
  - Lack of technology.
- More discharges occur in the PM, causing batches of patients to be discharged.
- Unit clerks on the floors batch discharges and schedule discharge order entry around lunch and end of shifts, which clogs the hospital rooms.
- Orders misplaced or lost.
  - Lack of standardization in charts.
  - Charts are not organized in one place.
- Order entry is not user friendly.
- No consequences for hiding beds, not turning them over quickly.
- Lack of focus on patient flow throughout the inpatient value chain.
  - Always done it this way.

  - Lack of understanding of value.
  - Lack of knowledgeable resources.
  - No one owns the patient flow.

    Focus on departmental tasks, not patient flow.
    Lack of understanding of value.
    Lack of patient visibility across inpatient value chain.

      Technology doesn't exist.
      Lack of understanding of value.

- Registrations are delayed.
  - Needing to repeat same information multiple times.

    Registrations are not integrated across systems.

  - Need for separate bills by service area requires separate registrations.
  - Technology doesn't support one integrated registration.
  - The way registration staffing is scheduled.
  - Trying to meet departmental productivity goals.
- Transportation prioritization not always aligned to value chain.
  - Always done that way.
  - No understanding of value.
  - Focus on productivity goals by department.
- Admitting several patients in a batch at the same time, due to no floor nurse available to take report.
- Number of tests ordered is high.
  - Orders based on physician preferences.

    No standard patient care protocols.

  - Lack of analysis on orders data.
- Not having prompt discharge instructions.
  - Nurses busy.
  - Can't find doctor.
- Twelve-hour shifts create fatigue and slowness toward end of shift.
- Nurses don't want to admit patient during end of shift.
  - Tired
  - Difficult to handoff partially completed admission to next nurse.
- Delays at the end of shift in reading radiology results.
  - Handing off to night radiology service.
- Delays in radiology diagnostic testing.

  - Shorter contrast is not used; it takes 1 ½ hours to prep patient with contrast.
  - Not always prioritizing, planning, or sequencing tests.
  - Competing modalities for tests.
  - After hours tests may not be prioritized.
  - After hours radiology services don't read directly on in-house radiology system.
  - Sometimes radiologists leave early.

- Delays in lab tests.
  - Not always planning, prioritizing, or sequencing tests.
  - Label printers go down.
  - Specificity of some tests.
  - High volume of diagnostic tests.
- Delays in discharging patients.
  - Wait on physician to do discharge instructions and prescriptions.
  - Wait on nurse to discharge patient physically and from the system.
  - Private physicians want to see their patients before they are discharged.
  - Homeless patients need shelter and medications.
  - Patient may not have a ride.
  - Patients eat before they leave.
- Delays in assigning beds for admitted patients.
  - Inefficient admission and discharge processes.
  - Unit/floor staffing is not integrated with ED staffing.
  - Late physician rounding.
  - Too many steps and ways to communicate.
  - Admission must have an order to enter the bed.
  - Not updating the system with potential and actual discharges.
  - No way to identify pending discharges in the bed board system.
  - When the bed board system person is not there, no one is assigned to cover.
- Delays in moving patients to rooms.
  - Cleaning personnel are not trained to enter cleaned rooms directly into the system.
  - No status in system for pending versus actual discharges.
  - No clear definition for pending versus potential discharges.
  - No pending information from Assistant Nurse Manager who know pending discharge information.
  - Practices not giving the physician the messages.
  - Not leveraging the bed board system.
  - No easy communication method to give report to floor/unit.
  - More discharges in PM versus AM, but staffing assigned to AM.
  - Transports not prioritized based on flow between the ED and floors/units.

The next step in the cause-and-effect analysis was to prioritize them based on the Critical to Satisfaction Criteria of Length of Stay, process times, and reducing delays in the processes. We first ranked the importance of the CTSs on a scale of 1–10, with 10 being high and 1 being low. We rated delays as the most important with a 10, reducing process times as a 5, and because if delays and process times are reduced, it will have a positive impact on delays, we ranked that the lowest, with a 1. There are also many other types of improvement projects that can also effectively impact length of stay related to clinical pathways and protocols.

We then applied an impact rating for each potential cause for how much the root cause impacted the CTS, using a scale of 9 as high, 3 as medium, 1 as low, and blank as none.

One of the Cause-and-Effect matrices is shown in Table 11.4.

We achieved the overall weightings by multiplying each CTS importance rating by the strength of relationship weighting. Ranking the root cause weighting gives us a focus on the high-priority improvement areas as we move into the Improve phase of the project. The high-priority improvement areas are shown in Tables 11.5, 11.6 and 11.7.

Improve Phase

The following steps were executed in the Improve Phase:

1. Identify improvement opportunities and plans.
2. Design future state.
3. Establish performance targets and project scorecard.

Identify Improvement Opportunities and Plans, and Design Future State

Improvement plans were developed based on eliminating the identified root causes and improving flow across the inpatient value chain. Improvements were categorized into the following themes:

- Value Chain integration.
- Enhance productivity measurement.
- Inpatient scheduling.
- Predictive bed demand.
- Order entry and diagnostics.
- Information Technology.
- Patient Transport.
- Discharge planning.
- Value Chain Process Metrics.

Table 11.4 Cause-and-effect analysis prioritization by CTS—page 1

Cause-and-effect analysis prioritization by CTS					
					Low = 0–50
		CTS			Med = 51 to 100
Possible root causes	Patient length of stay	Process time	Reduce delays		High = Over 100
	3	**5**	**10**	**Priority**	**Ranking**
Why are there delays in the inpatient processes?					
Staff not available to provide services	9	9	9	162	**High**
Productivity goals are defined by department, not by value chain	1	9	9	138	**High**
Lack of knowledge of how to set the goals better	1	3	3	48	Low
Productivity goals are not set using real data, but arbitrary national averages	1	3	9	108	**High**
Lack of knowledge of how to set the goals better	1	3	9	108	**High**
Easier to set them without time, resources and cost to collect real data		1	3	35	Low
Always done it that way		1	1	15	Low
Culture doesn't support questioning the productivity standards	1	3	9	108	**High**
Nurses call out sick at last minute	1	3	3	48	Low
Lax absence policies	1	3	3	48	Low
Nurse is sick	1	3	3	48	Low
Staff not aligned to patient demand	9	9	9	162	**High**

(continued)

Table 11.4 (continued)

Cause-and-effect analysis prioritization by CTS					
					Low = 0–50
No process to align staff to patient demand	9	9	9	162	**High**
Manage work by departmental key performance indicators, not across value chain	3	9	9	144	**High**

Value Chain Integration

This improvement idea is to integrate the process activities and resources that perform these activities across the inpatient value chain. We suggest integrating the key resources that manage these activities including:

- Nurse patient care coordinators.
- Case Management.
- Transportation.
- Bed Assignment.
- Room Cleaning—Environmental Services.
- Physicians.
- Inpatient Scheduling.
- Imaging Diagnostics.
- Lab.

Some best practices in the healthcare industry encourage this type of care coordination where nurse patient care coordinators in a physically centralized location, which we will call the Central Care Hub, work with nurse managers or patient care coordinators in the units and floors to coordinate patient care, using standard practice protocols based on the patient's primary diagnosis while in the hospital. One such example is CareLogistics®, (CareLogistics 2013). They have a patient care model augmented by their software that manages the patient flow through every step of the process, including scheduling of tests and diagnostics.

For our model, the central location is where the bed assignments are managed with the centralized nurse care coordinators. They coordinate with environmental services to ensure that the rooms are cleaned and that the most recent status of the beds (dirty or clean) is updated in the bed board system. An environmental services manager works out of the Central Care Hub to align with the needs of bed assignment. A Transportation manager works in the Central Care Hub to align with the needs of transporting patients to inpatient rooms from whatever location in the hospital they are admitted from. The Central Care Hub Patient Care Coordinator works with the floor and unit nurse managers, the Case managers, and the physicians to coordinate

Table 11.5 Highest priority root causes

Cause-and-effect analysis prioritization by CTS					
		CTS			
Possible root causes	Patient length of stay	Process time	Reduce delays		
	3	**5**	**10**	**Priority**	**Ranking**
Why are there delays in the inpatient processes?					
Staff not available to provide services	9	9	9	162	**High**
Productivity goals are defined by department, not by value chain	1	9	9	138	**High**
Productivity goals are not set using real data, but arbitrary national averages	1	3	9	108	**High**
Lack of knowledge of how to set the goals better	1	3	9	108	**High**
Culture doesn't support questioning the productivity standards	1	3	9	108	**High**
Staff not aligned to patient demand	9	9	9	162	**High**
No process to align staff to patient demand	9	9	9	162	**High**
Manage work by departmental key performance indicators, not across value chain	3	9	9	144	**High**
Lack of visibility of orders for care, diagnostic tests, therapy	9	9	9	162	**High**
No integrated inpatient schedule with view of daily patient schedule	9	9	9	162	**High**
Technology doesn't currently exist	9	9	9	162	**High**
People doubt it is possible	1	3	9	108	**High**

(continued)

Table 11.5 (continued)

Cause-and-effect analysis prioritization by CTS					
		CTS			
Metrics based on departmental goals, not based on patient value chain	9	9	9	162	**High**
Lack of data of patient demand	3	9	9	144	**High**

and execute the standard care protocol plans. Inpatient Scheduling schedules Imaging Diagnostics and the Lab orders within the Central Care Hub. The Central Care Hub acts as a command center where all patient flow activities are managed to reduce delays in the flow of the patient through the inpatient value chain activities. The key ingredients for a Central Care Hub are shown in Fig. 11.16.

Enhance Productivity Measurement

A huge drain on productivity and an inherent cause of delays in the inpatient flow is the way that many hospitals measure productivity of their personnel. Since the human resource is a large "controllable" cost of providing hospital services, most hospitals attempt to closely manage personnel and align resources with patient demand. One inherent weakness is the lack of sound data on which to base the productivity standards, as well as attempting to manage too closely to a forecasted demand, that is always going to have errors. Some days the hospital will over schedule and have too much staff, and end up sending staff home, or the inverse, under staff the units, floors, and departments, and need to call in staff. This can cause morale challenges when staff are called in and then sent home without working a full shift. Unrealistic productivity standards that are not based on the actual work performed and the variables of the particular hospital can result in staff being required to take vacation days when the hospital dictates, and end up with no or few vacation days for the staff to take at their own discretion. This can also lead to morale challenges and a lack of employee engagement.

A fuller description of the flaws and best practices of labor management in healthcare is provided in Chap. 12.

Inpatient Scheduling

Inpatient scheduling is an innovative concept that virtually no hospitals that we found were doing. The concept is to provide a daily schedule for inpatients of their diagnostic tests including the following modalities: radiology, ultrasound, CT scan, MRI, nuclear medicine, interventional radiology, and dialysis. A second phase of

Table 11.6 Highest priority root causes (continued)

Cause-and-effect analysis prioritization by CTS					
		CTS			
Possible root causes	Patient length of stay	Process time	Reduce delays		
Lack of understanding of need for metrics across value chain	1	3	9	108	**High**
Lack of technology	1	9	9	138	**High**
Lack of knowledgeable resources	3	9	9	144	**High**
No view of patient demand data	1	3	9	108	**High**
Lack of patient demand data	1	3	9	108	**High**
Lack of technology to collect demand data		3	9	105	**High**
For consulting physician turnaround time	3	3	9	114	**High**
For imaging department	1	3	9	108	**High**
Don't currently measure	3	9	9	144	**High**
Lack of focus on patient flow throughout the inpatient value chain	9	9	9	162	**High**
Lack of understanding of value	9	9	9	162	**High**
No one owns the patient flow	9	9	9	162	**High**
Focus on departmental tasks, not patient flow	9	9	9	162	**High**
Lack of patient visibility across inpatient value chain	9	9	9	162	**High**
Technology doesn't exist	9	9	9	162	**High**
Lack of understanding of value	9	9	9	162	**High**
Technology doesn't support one integrated registration	1	9	9	138	**High**
The way registration staffing is scheduled	1	9	9	138	**High**
Technology doesn't exist	9	9	9	162	**High**

(continued)

Table 11.6 (continued)

Cause-and-effect analysis prioritization by CTS					
		CTS			
Lack of understanding of value	9	9	9	162	**High**
Technology doesn't support one integrated registration	1	9	9	138	**High**

inpatient scheduling could incorporate schedules for patients' physical and occupational therapy, but this endeavor is fraught with many challenges, due to the complex nature of scheduling appropriate therapy type and skill level of the therapists.

The current state of inpatient scheduling is extremely manual. There is not extensive technology for the healthcare industry to enable easy centralized inpatient scheduling. There are many nuances in the requirements for providing an inpatient schedule that is different from outpatient scheduling. Key inpatient scheduling requirements are as follows:

(1) Provide a detailed inpatient schedule on a daily basis, available in the morning of the patient's daily appointments for diagnostic tests.
(2) Enable streamlined order entry from the floors and inputs.
(3) Incorporate resource availability of the delivering departments into the scheduling technology.
(4) Incorporate technician skill set of the delivering departments into the scheduling technology.
(5) Provide the ability to seamlessly order transportation for when the patient must travel to their appointment throughout the hospital.
(6) Provide the ability to change appointments and schedules when necessary based on patient or service delivering department and notify the nurse and patient of the revised patient schedule.
(7) Provide automated reminders for patient preparation.
(8) Provide prioritization of appointments by priority, modality, preparation time, recovery time, etc.
(9) Provide the ability to schedule stat orders.
(10) Provide metrics for schedule adherence and timeliness of service delivery.
(11) Provide historical demand data.

A future state process map is provided in Fig. 11.17, proposing just one possible way to design an inpatient scheduling process.

Predictive Bed Demand

It can be very valuable for the Central Care Hub to perform statistical analysis on the number of Adjusted Occupied Beds. The Adjusted Occupied Beds is a statistic that

Table 11.7 Highest priority root causes (continued)

Cause-and-effect analysis prioritization by CTS					
		CTS			
Possible root causes	Patient length of stay	Process time	Reduce delays		
The way registration staffing is scheduled	1	9	9	138	**High**
Trying to meet departmental productivity goals	1	9	9	138	**High**
Transportation prioritization not always aligned to value chain	1	9	9	138	**High**
Focus on productivity goals by department	1	9	9	138	**High**
Admitting several patients in a batch at the same time, due to no floor nurse available to take report		3	9	105	**High**
Number of tests ordered is high	9	9	9	162	**High**
No standard patient care protocols	9	3	9	132	**High**
Nurses busy	1	3	9	108	**High**
Can't find doctor	1	3	9	108	**High**
Delays in radiology diagnostic testing	1	3	9	108	**High**
Delays in lab tests	1	3	9	108	**High**
High volume of diagnostic tests	9	3	9	132	**High**
Delays in assigning beds for admitted patients	3	3	9	114	**High**
Late physician rounding	9	3	9	132	**High**
Delays in moving patients to rooms	9	3	9	132	**High**
Not leveraging the bed board system	9	9	9	162	**High**
More discharges in PM versus AM, but staffing assigned to AM	9	3	9	132	**High**
Transports not prioritized based on flow between the ED and floors/units	9	3	9	132	**High**

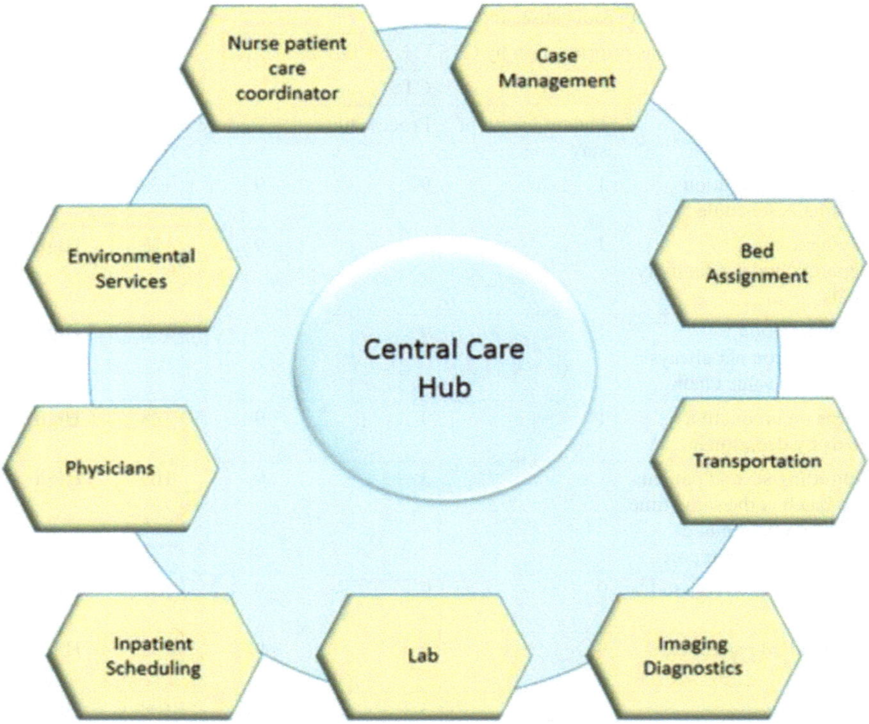

Fig. 11.16 Central care hub

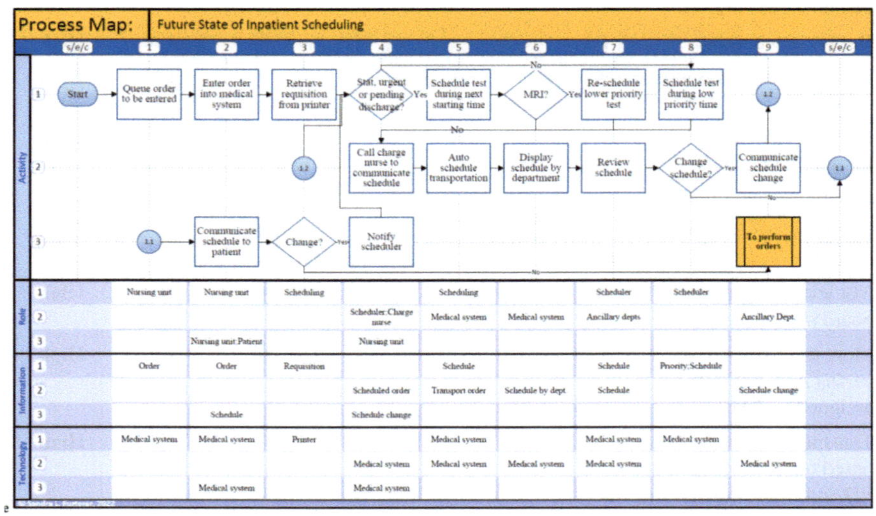

Fig. 11.17 Inpatient future state process map

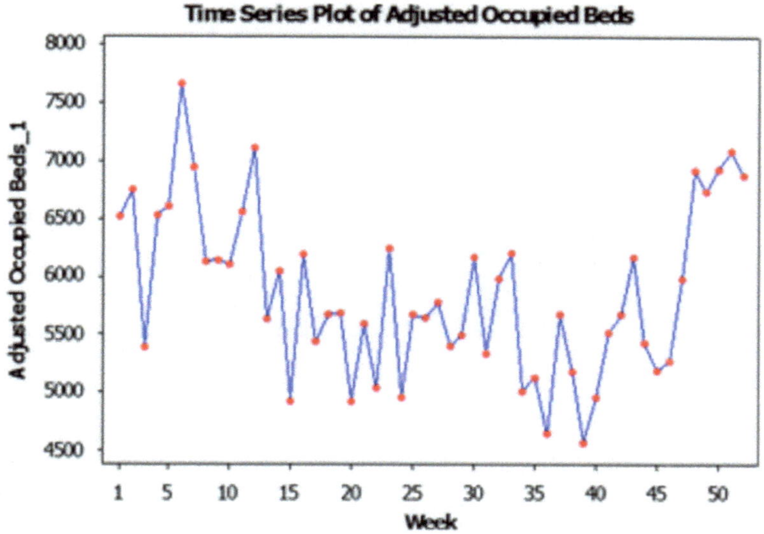

Fig. 11.18 Patient demand profile time series plot

measures the number of patients in inpatient beds on a weekly basis. This number is used to staff and set productivity targets. An example analysis showing a time series chart in Fig. 11.18, and the related histogram is shown in Fig. 11.19. This statistical analysis can help us plan for staffing. This same analysis can be performed to understand trends on a daily, weekly, monthly, and seasonal basis. There are hospitals that are located in warm vacation climates that are heavily impacted by the tourist season, so this seasonality would want to be considered in the statistical analysis.

Order Entry and Diagnostics

It is important to reduce the delays in entering orders into the medical information system for diagnostic tests. Some of the root causes of the delays around order entry include batching of orders around the lunch hour and change of shift; orders being misplaced or lost due to a lack of standardization and organization in patient charts. Additionally, the number of tests ordered is high due to a lack of standard patient care protocols. The order entry system is also not user friendly and takes extra time to enter orders. A process improvement project should be initiated to reduce delays in getting orders in the system; sometimes this can cause an additional one-half to a day of delays in the patients receiving their diagnostic tests.

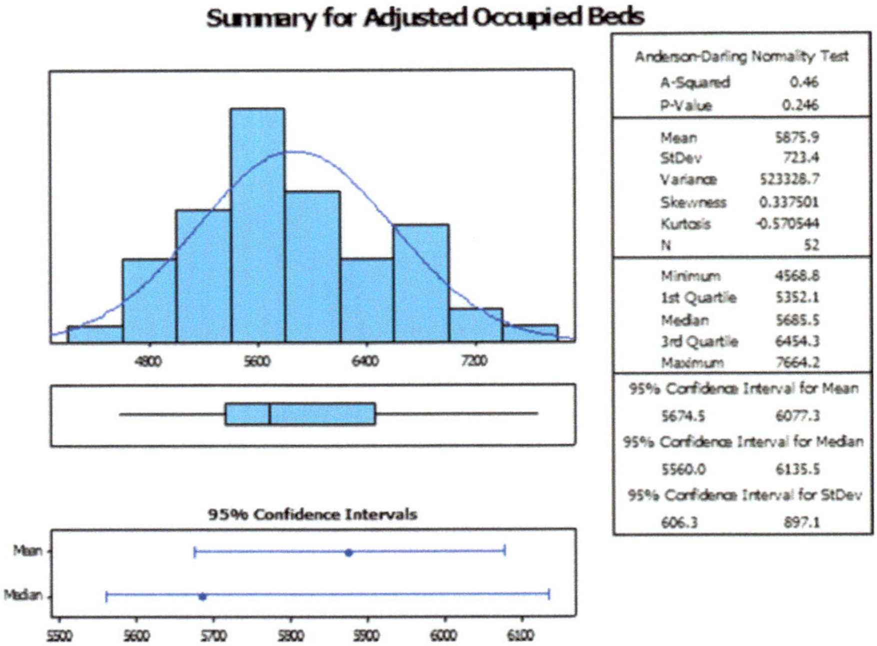

Fig. 11.19 Patient demand profile histogram

Information Technology

The information technology to support providing an inpatient schedule so that the patient and their family know roughly when the patient will receive tests, care, and therapy was not currently available. Some software organizations are now creating enterprise systems that can help to know where the patient is at any moment, and schedule tests, care, and therapy. One particular healthcare organization was attempting to adapt the outpatient scheduling system to be used for inpatient scheduling. However, this was not going to be an easy task, nor did it currently have the ability to meet all of the defined requirements listed in the inpatient scheduling section of this chapter. As more healthcare organizations begin to think innovatively to be more patient-focused and responsive, they will begin asking for this type of technology that can schedule and track patients' movements throughout the hospital via radio frequency identification and similar technologies. Patient tracking technologies can help with patient transportation, entering orders for transport to and from tests.

Using simulation software applications can be very effective to simulate the end-to-end inpatient value chain activities. The trouble, or challenge, is to get the data that is required to identify appropriate statistical distributions and assumptions to ensure simulation model validity. But just because it is challenging, does not mean that it's not valuable. Hiring an industrial or management engineer with this skill set would enhance the hospital's ability to be able to analyze process times, better predict patient demand, and perform staffing and resource capacity analysis.

Enhancing the bed assignment technology to be able to integrate the bed assignment and knowledge of open beds throughout the hospital, understanding where the demand for beds is coming from (emergency department, surgery, direct admit), integrating with order entry and scheduling, as well as transportation orders, will enhance the ability to manage inpatient flow. Again, there are starting to be enterprise patient tracking, bed assignment, and transportation systems that enable managing the inpatient value chain. Some of these information systems are still in their infancy from a user friendliness to provide metrics that help to manage patient flow, but as hospitals demand this type of detailed process level data, the software vendors will most likely respond. Business Process Management Systems (BPMS) can also be extremely helpful to enhance managing the inpatient value chain and flow of the patient throughout the hospital. Many of the current information systems are built to satisfy billing requirements instead of patient flow and process improvement, but again as the healthcare industry demands innovation, these systems may become more popular as well.

There is a very real need for integrating registration information across the many places where patients are registered, including the emergency department, surgical services, outpatient services, and inpatient services. Gone should be the day where patients need to keep providing the same information every time they move to a new service provider in the healthcare system. The Ohio State University medical system does an excellent job at integrated electronic medical record. Once a patient provides their registration information, it is accessible to any OSU medical provider, whether it is within a hospital setting, a physician's office, an outpatient facility, etc.

Patient Transport

Transporting the patient to and from tests, care, and therapy throughout the hospital needs to be more tightly integrated with the ordering and scheduling of these physician orders. We discussed the need for integrating the technology of order entry, scheduling, and patient locating ability. The processes also need to be integrated within the Central Care Hub.

Discharge Planning

Another area that can help reduce delays in the inpatient activities is to improve the discharge planning. The idea for discharge planning is to understand discharge patterns throughout the day, and focus on understanding root causes of delays, so that patients can be discharged earlier in the day. Reducing discharge delays can make room for the patients that need rooms as they come into the hospital through the emergency room, surgery, and direct admits. A sample discharge time in minutes by month is shown in Fig. 11.20.

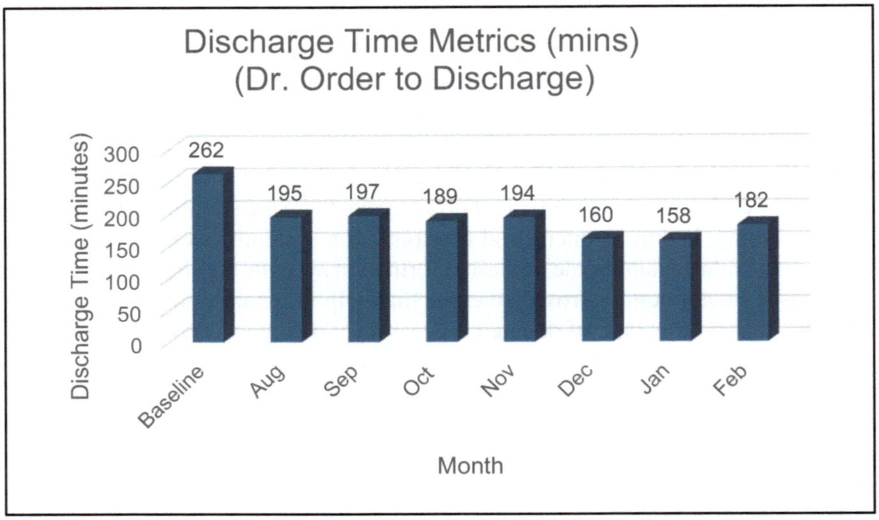

Fig. 11.20 Discharge statistics

Process improvement projects can be executed to reduce delays in the inpatient value chain activities and provide a better patient experience.

Value Chain Process Metrics

When the goal is to improve processes, we must have process metrics that indicate that we have improved. Many organizations do not collect data at a process level that can provide insight into how our processes are performing. A BPMS which collects data on how long it takes to complete tasks within a workflow can be used to collect process times on the value chain activities. A process measurement system should be put in place that can measure key metrics including:

- Process times.
- Process defects.
- Cost of processes.

Implementing a value chain process measurement system is a huge undertaking, but it can be prioritized and focused to collect key metrics in our existing information systems and then migrate to a more robust BPMS. The potential inpatient value chain scorecard (Table 11.8) can be used as a guide for key metrics to collect in the inpatient value chain.

Table 11.8 Potential inpatient value chain scorecard

Value chain activities	Target average process time (minutes)	Actual average process time (minutes)	Variance %
Register patient	30	60	*−100%*
Assign room	30	45	**−50%**
Prepare room	30	42	**−40%**
Transport patient	15	14	***7%***
Admit patient	60	188	**−213%**
Schedule patient for tests	30	175	**−483%**
Collect specimens for lab tests	60	52	***13%***
Provide therapy	60	110	**−83%**
Provide healthcare	60	130	**−117%**
Provide diagnostic tests and results	120	115	***4%***
Discharge patient	60	85	*−42%*
True length of stay (LOS) (time from admission to discharge)	4.8 days	5	*−4.2%*

Control Phase

The purpose of the Control Phase is to implement control mechanisms and procedures to ensure that the process maintains the levels of improvements attained during the pilot activities.

The following activities were executed in the Control Phase:

1. Create scorecard data framework and create process control plan.
2. Identify replication opportunities.
3. Develop future plans.

Measure Results and Manage Change

Since this chapter is a conceptual design of an innovative care model, we did not yet implement this model and will develop how we might control the processes once they are implemented.

Create Scorecard Data Framework and Create Process Control Plan

The control plan is a detailed guide for maintaining all of the positive changes that you have implemented. We developed the control plan defining the metric and control mechanisms for each step in the value chain. The control plan scorecard is shown in Table 11.8.

We encourage the designation of a value chain owner that would be responsible for implementing the process improvements, the enhanced or new technology, and monitoring the processes post implementation to ensure that improvement in the metrics is achieved.

Conclusions

We challenge the reader to embrace the principles of systems engineering and process improvement that are core elements of this case study and attempt to design and implement the Patient Care Operating Model of Integrated Value Care.

Reference

Care Coordination Model. (2013). Website: http://www.carelogistics.com/solution/care-coordination-model.aspx, CareLogistics®.

Chapter 12
Labor Management in Healthcare: Flaws and Best Practices

Abstract Labor Management is critical in healthcare and especially so in acute care hospitals, where 50–70% of the costs are attributed to labor costs. According to the American Hospital Association labor represents more than half the cost of business in a healthcare facility (aha.org). The purpose of this article is to describe best practices of labor management, and how they can be applied to effectively manage an acute care hospital.

Keywords Actual work · Productivity standard · Benchmark database · Nonproductive time · Unavoidable delay

Introduction to Labor Standards

Labor management and labor standards were first widely developed and applied starting in the peak of the industrial revolution in the late 1800s. Frederick Taylor, commonly referred to as the father of Scientific Management, was an extremely successful Industrial Engineer. He furthered Adam Smith's division of labor principle (Smith, 1776), where each person completes a specialized number of small tasks. Frederick Taylor incorporated work measurement including time studies to understand work elements and assess how long each element takes to perform. The methodology of developing labor standards, or how long it should take to perform work, and therefore how much work the operators are expected to complete in a certain time period, is developed by performing the following direct observation time study steps (Smith, 1978):

(1) Step 1: State the objective

   a. For direct observation time study, the objective is usually "to determine the time which should be allowed to perform the observed tasks."

(2) Step 2: Notify the appropriate persons

S. Furterer, *Lean Six Sigma for the Healthcare Enterprise*,
https://doi.org/10.1007/978-1-4471-7617-6_12

a. The department manager and the operator who is performing the work being measured should be notified. This is to ensure that the process is stable, that the conditions are representative of the normal working environment, and that the operator has the reasonable skills to perform the work.

(3) Step 3: Determine the scope

a. The time frame to collect the data and ensuring the appropriate number of observations is developed in the scope.

(4) Step 4: Select the operator

a. The operator should use the agreed upon and appropriate method to perform the work during the study.

(5) Step 5: Establish activity elements

a. The elemental activities that make up the work should be identified and then timed as activity elements. Those elements that are paced and/or performed by a machine should be separated from operator-based activities. The elements should make sense. They should be easily recognizable from a starting and end point perspective. They should be of a reasonable length to observe and capture the timing of them.

(6) Step 6: Conduct the study

a. When conducting the study, the job condition should be recorded, and then the time to perform the tasks.
b. The analyst should rate the operator's performance. Normal pace is assumed to be 100%. A person working faster than normal receives a performance rating higher than 100%, and one who is working at a slower pace than normal would receive a performance rating below 100%. The performance factor will increase fast (shorter times) and reduce slow (longer) times to the normal time.

(7) Step 7: Specify the allowances

a. Allowances are adjustments to the normal time to account for nonproductive portions of the workday.
b. Personal time includes breaks and lunch.
c. Loss of production due to fatigue.
d. Delays, usually unavoidable that are not within the control of the operator.

The underlying philosophy of work measurement and setting a productivity standard through measuring the actual work performed should enable the operator to make a target volume of quantity by working at a normal pace. The magnitude of the allowance should be a matter of bilateral agreement between the operators and management. Negotiated allowances are used because only the unavoidable delays are considered to be measurable with standard technology. The unavoidable delays are commonly measured using work sampling techniques. Work sampling defines the activity elements that make up the work, has an analyst observe what tasks are being performed at set intervals, to determine the percent of the day that is spent on the different activities.

There are *critical elements* that must be applied to develop an appropriate labor standard:

(1) The actual work performed should be observed and measured to develop the work standard.
(2) The unavoidable delays should be identified through observation, and an allowance should be applied to the performance factor, so as not to penalize the operator for delays beyond his or her control, which are systemic to the work design. Only through application of process improvement efforts can these unavoidable delays be removed and therefore enable a tighter standard.
(3) The work standard should be agreed upon between management and the operator.

This practice of measuring work has over 130 years of application defining labor standards and has been widely accepted as a sound methodology.

Healthcare Development of Labor Standards

Healthcare as an industry has adopted vastly different methods to develop labor standards, which will now be described.

Standards Based on Organization Structure, Not Work Performed

Healthcare develops standards typically based on organizational structure and management reporting departments, such as operating room, lab, rehab therapy, endoscopy, critical care floor, emergency department, and so on. While these organizational reporting units can represent a group of similar work, the actual work activities can vary widely. For instance, the operating room can include several different work functions, such as:

(1) *Anaesthesia activities*: Room prep, patient assessment, administer anaesthesia, monitor patient.
(2) *Orderly activities*: Retrieving and setting up or equipment; helping to position patients; ordering surgical supplies; stocking surgical supplies, cleaning, and turning over surgical suites after surgery.
(3) *Scrub tech activities*: Open and set up cases, including instruments and supplies; assist in surgery by providing needed supplies and instruments; help remove used instruments, trays, and supplies.

(4) *Circulator nurses*: Interview and assess patient' readiness for surgery; document patient care, treatment, and status; transfer patient care to post anaesthesia care, and record specimens.
(5) *Department managers*: Manage staff; define staffing schedules; problem solve; manage crises; monitor time reporting; apply discipline; manage surgeon, anaesthesia, and vendor relations; administer performance evaluations, rewards, and recognition.
(6) *Surgical coordinators*: Manage cases; update preference cards; interface with surgeons.

Similar vast differences exist in the lab related to just one of the departments, pathology:

(1) *Pathologist activities*: Perform gross cut of specimens; documentation; assess pathology of specimens; dictate results and reports.
(2) *Lab technicians*: Enter orders, process specimens, and create specimen slides.
(3) *Transcriptions*: Transcribe results; correct records; manage pathologists' schedules.

The type of specimens in number, type, and how they are billed vary vastly, as do the number of slides that make up a billable order across cytology and histology, sub-specialties within the pathology lab.

One more example is in infusion therapy, a unit of work, giving a shot to a patient can equate to 1 min of work, versus chemotherapy equating to one and a half hours of work. This is a huge disparity in the time to actually perform the work.

Standard Setting Based on Benchmark Databases Not Based on Actual Work Performed

A common approach in healthcare to developing departmental standards is to first collect actual volume and number of hours worked by pay period (typically 2 weeks), across a year-long time period. The labor standard is calculated as:

$$\text{Labor standard} = \text{Productive hours worked/units produced}$$

Finance will review a national productivity benchmark standard database, such as Premier, and assess where this proposed standard fits against the average or best productive organizations. The standard is then moved based upon a targeted productive hour, based on the standard best or better practice range. For instance, if the proposed target based upon the historical worked hours and volume is at the 50th percentile, equating to the average hospital in the database, and the organization wants to set a standard at better practice of the 25th percentile, this equates to only 25% of the organizations are more productive than this level. The standard is then moved to this more stringent level, without measuring or observing the actual work, or understanding the unavoidable delays that the workers cannot control. Only management can improve processes and reduce systemic unavoidable delays. The workers may assist in improving processes, but management must implement the process improvement program, approve projects and resources, and remove barriers and challenges to successful change. Another fundamental problem with an industry standard database is that many times the work activities and characteristics of the staff, such as skill levels, patient acuity, work design, and unavoidable delays is not an apples-to-apples comparison between the organization setting the standard and the organizations in the database. It is extremely difficult to assess the critical factors that contribute to worker's efficiency of the organization that is at the 25th best percentile.

An example of an apples-to-oranges comparison of the actual work from an organization in the database, compared to one of the organization's setting a standard was in the transportation department. The Director was being compared to a productivity standard that allowed four full-time equivalent (FTE) patient escorts for the entire hospital. When the director inquired to the 25th percentile organization he found that they had only four FTEs in a centralized patient transportation organization, but every department had patient escorts, and this work was not included in the standard.

Fixed Departments Expected to Fluctuate with Patient Volume

The volume measure applied is selected based on whether the department's volume fluctuates based on the patient volume or is a fixed department whose work does not change based upon the patient's needs. The better practice organizations do not expect fixed departments to vary their work schedules or paid time off with patient volume when they are not a patient-facing department. For example, the process improvement group in our hospital has three FTEs. If one of the Six Sigma analysts does not work, the department is "more" productive, yet we were not actively helping the organization reduce the inefficiencies that would actually make the organization more productive, because we were taking the required paid time off. Additionally, if a patient care department is not meeting its productivity standard, the fixed departments must cover for these departments, that they cannot control the work of, penalizing fixed departments for the nature of their job activities.

Standard is Developed Based on Aggregate Volume, But Work is Managed Daily

The standard is typically based on peak (best) volume across the pay periods for the year, yet the productive working hours are expected to be managed and controlled on a daily basis. The volume fluctuates daily, and the work hours are expected to align with the fluctuating volume, yet the standard is based on either the average bi-weekly pay period average or a select peak volume activity based on three to four pay periods. Worse yet, there may be no correlation between the standard that was realigned to meet the benchmark database 25th percentile range and the daily worked hours and volume.

Summary of the Fatal Flaws of Healthcare Productivity Setting Practices

This section discusses the fatal flaws of healthcare productivity standard setting practices, shown in Fig. 12.1.

- Setting arbitrary standards that the processes are neither designed for nor are capable of meeting
  - The industry-wide productivity standard database assumedly includes unavoidable nonproductive time. If an arbitrary "better" 25th percentile target standard is selected, this should include less unavoidable delays or nonproductive time than the standard at the 50th percentile. Arbitrarily moving the standard requires that some work is not performed. This work could be value-added,

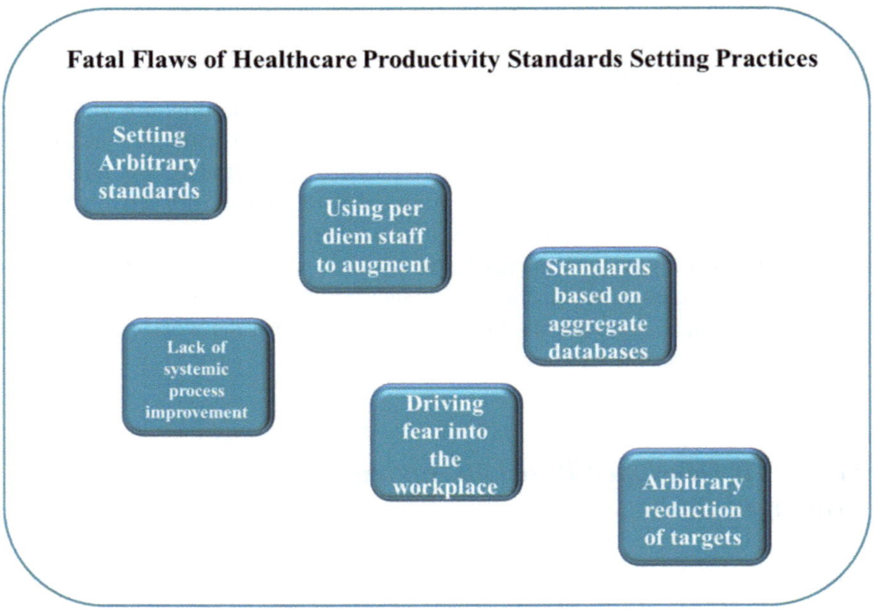

Fig. 12.1 Fatal flaws of healthcare productivity standards setting practices

such as bathing the patient prior to surgery, or non-value-added such as looking for a chart. Who decides what work is not performed? The nurse will probably spend the time to look for that chart because no orders can be entered without it, so does the value-added activity get missed? The root cause for a missing chart needs to be systematically investigated and removed, yet there won't be time for that with unrealistic productivity standards in place. Instead of using the time to improve processes when volume is low, like some world-class organizations do, staff is sent home and flexed off. We need to consciously either improve the work or decide what work is not done if we are going to set standards below what the process is capable of producing.

- Using per-diem staff to augment peak volumes when highly skilled staff is not available, which induces paycheck variability for our core staff
  - The assumption of using per-diem or part-time staff to augment peak volumes sounds reasonable in a low skill or resource plenty industry, but healthcare is far from this type of environment. Highly skilled clinical areas such as the surgery, emergency services, and critical care are examples of areas where skilled and available per-diem staff virtually do not exist. You are instead incorporating great variability in some of your most skilled and valued associates' pay checks week after week, season after season, and year after year. Additionally, hiring per-diem staff uses these resources for extended periods without giving them needed benefits, and causes increased hiring costs, lack of retention, and the need to constantly recruit and re-hire per diems. This costly cycle will continue.
- Not basing standards on the actual work performed and the existing design and capability of the processes
  - The standards are based on aggregate benchmark databases that are not an apples-to-apples comparison of the work and give little insight into the process factors that impact efficiency and productivity.
- Arbitrary reduction of productivity targets
  - Arbitrarily reducing productivity standards based on target benchmarked percentiles, not based on eliminating inefficiencies in the work, can lead to inaccurate standards.
- Not incorporating systemic process improvement
  - Not incorporating systemic process improvement to eliminate process inefficiencies and make the work truly more efficient causes workers to select activities to be eliminated that may be necessary for the quality and safety of the patient. Process improvement is the only sustainable way to be more productive and efficient.
- Driving fear into the workplace

– This type of labor management causes great fear, uncertainty, apathy, and disgruntled employees. Expecting associates to meet productivity standards that are unrealistic, not based upon the actual work and processes, and for unavoidable delays that they cannot control, creates fear and poor morale.

Recommended Standard Development Methodology

Following is a recommended standard development methodology, shown in Fig. 12.2.

- Base the productivity standards on the actual work across service lines. This includes creating a cross-functional productivity standard based on the functions that do the work across a service line. This will help eliminate cross-purpose goals, such as transportation or housekeeping flexing off to meet productivity standards, when patients are being discharged.
- Create productivity standards based on daily work and patient volume patterns.
- Create standards based on day of week, time of day, seasonality, and other appropriate factors.
- Perform work sampling to identify unavoidable delays.

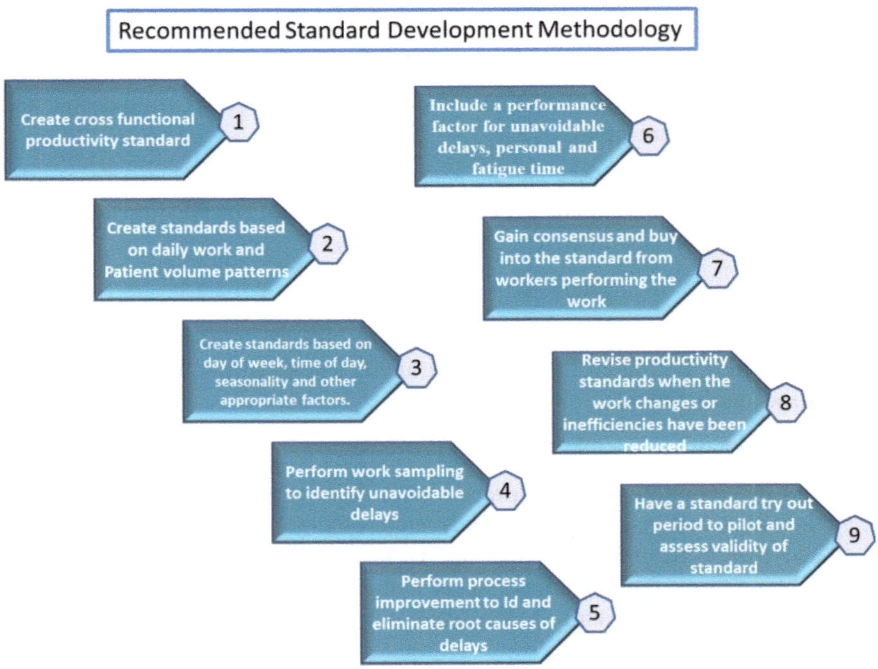

Fig. 12.2 Recommended standard development methodology

- Perform process improvement Kaizens to identify and eliminate root causes of unavoidable delays.
- Ensure that the standard includes a performance factor that accounts for unavoidable delays, personal and fatigue time.
- Gain consensus and buy-in to the standard from the workers performing the work.
- Revise productivity standards when the work changes or inefficiencies have been systematically reduced or eliminated.
- Have a standard try-out period to pilot and assess the validity and appropriateness of the standard.

If the FTEs must be managed and reduced due to low volumes:

(1) Grow the volume. Hire business development associates to develop business with physicians and surgeons to build volume.
(2) Proactive reduction of services. When the volume is low, we need to proactively determine which services need to be reduced to avoid burning out our associates and expecting them to meet unrealistic standards.

Conclusions

This chapter presents flaws in current labor standard setting in healthcare and proposes application of best practices for developing labor standards to meet the critical need of managing labor in healthcare, where labor costs can exceed 50% of the costs of doing business.

References

Smith, A. (1776). *An inquiry into the nature and causes of the wealth of nations.* Methuen & Co., Ltd.
Smith, G. L., Jr. (1978). *Work measurement, a systems approach.* Grid Publishing.

Chapter 13
Applying Modeling and Simulation to Reduce Time and Cost of Implementing Process Improvements

Abstract Modeling and simulation is an analytical, statistically based process modeling tool that can model a real-world system in its current state, model a future state with enhanced improvement recommendations, as well as assess the impact of incorporating the identified improvements. This tool can reduce the time, cost, and complexity of implementing improvements by "testing" the impact with modeling and simulation prior to the actual physical implementation.

Keywords Emergency department · Modeling and simulation · Healthcare

Introduction

In Chap. 3, we presented a case study that significantly improved the processes in an Emergency Department. The initial improvements were implemented within 3 months and reduced the overall patients' length of stay by 30%. The remaining improvements were more complex; focused on the diagnostic, discharge, and admission to the hospital processes; and took more than a year and a half. In this case study, we describe an approach to use modeling and simulation to understand the current process and revise the model to propose future recommendations and identify the impact of the improvements before physically implementing them. This modeling approach can reduce the time, energy, effort, and costs of implementing and test improvements within the organization's operational excellence programs. It also enables the ability to collect data on statistics that provide the estimated impact of changes to the system, before it is actually implemented. Modeling and simulation software has enabled non-computer programmers to embrace these tools. Industrial and systems engineers or management engineers will have the process modeling skills to leverage these tools for process improvement.

S. Furterer, *Lean Six Sigma for the Healthcare Enterprise*,
https://doi.org/10.1007/978-1-4471-7617-6_13

Modeling and Simulation Background

Modeling and simulation is a collection of methods and technology to understand a system consisting of people, processes and technology, and model and simulate this process to identify the impact of improvements prior to physically making changes to the system.

Modeling and simulation techniques started in the 1960s, with home-grown computer code required to be developed to simulate the systems. The computer coding usually required analysts with expertise in both process modeling and coding. Over the years, computer programs were developed that embedded scripts that allowed non-computer coders to embrace the technology and practice.

There are many different types of simulation programs. Discrete event simulation is typically used for process modeling and improvement. Discrete events are activities that occur within a system at points in time, where products are created within steps or events in a process, or when services are provided, such as in a hospital's emergency department. People, processes, and technologies combine to create a system that is modeled and improved. Queueing theory provides the underlying mathematical theories embedded in discrete event simulation to model the system. A group of entities waits within a queue for service to be provided by limited resources. For an example, a part needs to be manufactured on a milling machine that has a fixed capacity, depending upon the number of milling machines, and the operators trained and available to perform the milling activities. It is similar in a hospital system, where there are specialized pieces of equipment, such as a CT scanner, an MRI machine, or an emergency room where services are provided by technologists, nurses, physicians, and ancillary staff. Economically, there can only be so many specialized medical machines, nurses, physicians, and staff. There are set hours of the services that are provided. An important aspect of modeling and simulation is the importance of a process analyst having the knowledge and skills to understand the system and model it. In Chap. 3, Fig. 3.3, we provided a sample process map that described the patient arrival process to the emergency department. The LSS team mapped the entire set of processes within the emergency department, which provided excellent insight into how the emergency services were provided to the patients. The Voice of Process (VOP) data from the Define and Measure phases of the LSS project can be used as inputs to the simulation model.

In the next sections, we will describe the use of the Arena software®, and the modeling and simulation approach applied to model, simulate, and suggest improvements in an emergency department.

Helpful Terms and Definitions

Following are some terms and definitions that are used within the modeling and simulation methodology.

Arrival rates: The rates at which an entity arrives for service within the system.

Attributes: A common characteristic of all entities, where the specific value can differ from one entity to another (Kelton, Zupick, and Ivey, 2024).

Continuous simulation: The state of the system changes continuously over time, such as the levels in a filling system for cooking oil.

Discrete simulation: Change occurs in the system at points in time, such as manufacturing parts arriving to be assembled, machined, or painted, or patients arriving to be assessed and treated in an emergency department.

Entity: Dynamic objects or actors that move around the system.

Events: Occurrences that happen at an instant of simulated time that may change attributes, variables, or statistics accumulated.

Model and Modeling: A representation of a real-world system. Models can be physical, mathematical, a drawing or flow chart, and also simulated by a computer program.

Process: A sequence of activities, people, and technology that create work in the form of products or services.

Queue: A waiting area or line where entities wait for service within the simulation model.

Queuing theory: The mathematical theories that enable modeling of queues, arrivals of entities, and processing within the system.

Replications: Running the simulation model multiple times, replicating the simulation conditions to gain a measure of the variability in the system.

Resources: People, equipment, tools, etc. that are used to provide service.

Simulation: An approach that uses mathematical modeling to simulate a real-world system.

Statistics: Data collected during the simulation modeling runs.

System: A collection of people, processes, and technology that provides services or products.

Variables (Global): A piece of information that represents a characteristic of the entire system.

Modeling and Simulation Approach/Method

It is best to implement a structured approach to develop, test, validate, and apply simulation models, as shown in Fig. 13.1 and as follows:

(1) Define: Define the project goals. Identify the needs for improving the system, the stakeholders of the system, and the goals of the improvement effort.
(2) Measure: Understand and measure the existing system. Observe the existing system and develop detailed process maps of the flow. Collect data either manually or preferably from a computer system that provides times, arrivals, and entity information.

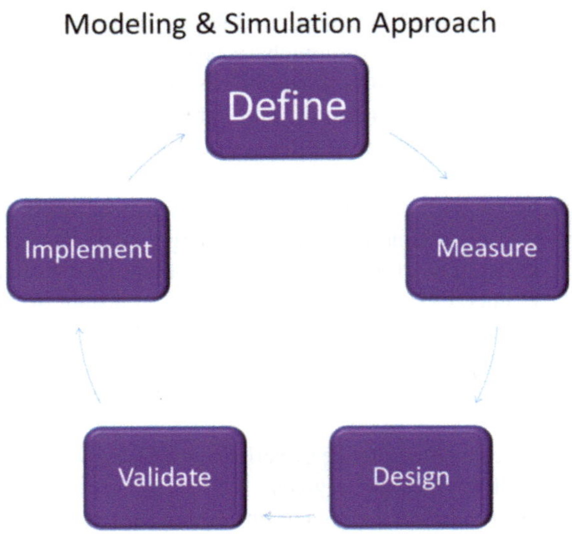

Fig. 13.1 Modeling and simulation approach

(3) Design: Design the model incorporating the goals, factors, variables, attributes, entities, process flow, experiments, and model to simulate the process.
(4) Validate: Test and validate the model that represents the existing system.
(5) Implement: Identify improvement recommendations, incorporate them into a future state model, and assess the potential impact of the improvements.

Gain approval to implement the proposed recommendations in the real-world system. Assess the impact of the programs and compare to the simulated estimated impacts. Share the successes with the process and system owners.

Common Modules for Modeling

Arena is a modeling and simulation software that can be used to develop discrete event simulations used for modeling processes. We will discuss some of the most common modules used to model the simulation. These modules include the script or codes that enable the simulation to operate. Originally when simulation started as a discipline, each process analyst had to write their own code within a simulation modeling language such as SIMAN, which Arena uses as the modeling language, but embeds scripts within the modules for easy use by the analyst. Software, such as Arena, have enabled the vast use of simulation without having to code to create most useful models. As shown in Fig. 13.2, the Basic Process modules used in most of the simulation modeling will be defined (Kelton et al., 2014). A module is the basic building block to create the flow in the simulation. Modules can be flowchart and data types. Flowchart modules provide process flow and activities within the simulation that the entities travel through to receive services or have activities performed. Data modules define and enable data collection on the entities, resources, and queues within the simulation. The most common modules are create, dispose, process, decide, and assign.

Create—Creates your entity to enter into the simulation to receive service. Examples of entities are a patient, part, and customer. Entities can be created based on a random exponential distribution, a schedule based on the number of entities per hour, a constant, or an expression.

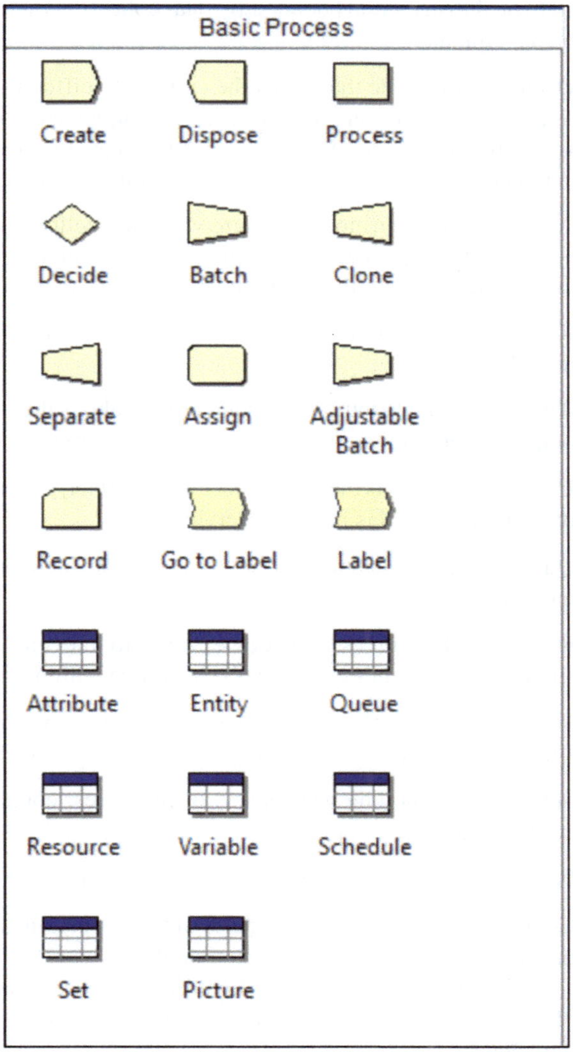

Fig. 13.2 Arena basic process modules

Dispose—This module disposes of the entity after it receives service in the simulation and its activities are complete, or the entity exits the simulation because it meets the exit criteria. Entity statistics can be recorded when the entity is disposed.

Process—A process is a set of activities performed for or on the entity. An example is a patient going through the triage process, where they collect patient information and vitals to assess the patient acuity for prioritizing them into the service queue. Resources that are used within the process module are defined. The logic action of the process is defined, such as "Seize-Delay-Release," where the resource is seized to be used, there is a delay during which time the activity is performed and then the resource is released from use. The delay type which can follow many different statistical distributions defined by data collection on the current state of the process being simulated.

Decide—A module that sends the entity into different paths in the simulation, such as a patient, can be an inpatient or an outpatient and goes through different activities within the simulation. The paths are determined by chance or condition. The decision can include a two-way path or n-way (multiple paths).

Assign—An assign module assigns variables, entity types and pictures, and attributes. These attributes can assign the current simulation type for calculating throughput or cycle times and tallying the number of entities that go through the model.

Record—A record module records statistics about the entity as it goes through the model.

Attribute—An attribute is a characteristic about the entity, such as type (inpatient, outpatient or admitted patient or discharged patient).

Entity—An entity is the object, person (customer, patient), part, etc. that goes through the simulation and is provided service.

Queue—A queue enables entities to wait in line for service to be provided at a process module.

Resource—Resources provide services to the entities as they go through the simulation. The entities will grab a resource within a process module.

Variable—A variable allows you to define a variable that can be used throughout the simulation instead of re-defining an attribute for each process module.

Set—The set module groups sets of resources, counters, tallies, entity types, and entity pictures.

Picture—Defines a picture to be used to animate the entity.

Connections—Flow chart modules are connected using lines to define the flow of the entity through the simulation model.

Animations using pictures of entities and queues can be used to provide the flow of entities, the use of resources and pictures of queues with the entities waiting at the queues.

Modeling and Simulation Applied to an Emergency Department Case Study

Define Phase

In this step, we define the project goals and identify the needs for improving the system, the stakeholders of the system, and the goals of the improvement effort.

The same project charter, goals, and stakeholders from Chap. 3 is used in this Define phase, as shown in Table 13.1.

The goal of the modeling and simulation project is to develop a model that will represent the current state of the processes. Once the current state model is validated, improvements are identified and incorporated into a future state model. Experiments can be designed varying different factors and levels of the factors that represent improvement recommendations. The future state model is run, and the impact of the improvements is assessed using operational metrics that are aligned to the Critical to Satisfaction (CTS) characteristics. The CTS' and metrics that were developed for the

Table 13.1 Project charter

Project Charter

Project Overview, Problem Statement, & Goals
Provides a description of the problem to be solved. Be as descriptive as possible. Include improvement goals that are SMART (Specific, Measurable, Attainable, Realistic, and Time-Based).
Project Overview: This project is focused on improving patient throughput in the ED. About 40% of the patients seen are admitted to the hospital, vs. 60% seen are discharged. Problem Statement: The Emergency Department is experiencing delays in moving the patient through the ED in a timely manner. For January through April, patients took an average of 5.8 hours for a patient to be seen, treated, tested and discharged and an average of 8.7 hours for a patient to be seen, treated, tested and admitted. There are excessive delays and an average of 6.5% of the patients left without being seen. The total bypass hours for the past year were 341 hours. Goal(s): Improve ED throughput time to 3 average hours for discharged patients and 5 average hours for admitted patients. There could be an improvement in patient satisfaction and quality of care due to the synergistic relationship between throughput, quality and satisfaction. The goal of the modeling and simulation project is to develop a model that will represent the current state of the process, and develop a model that can simulate and validate improvements that can reduce the implementation time of process improvements. Scope Statement: The scope includes the ED processes starting from patient entrance, to triage, treat, transport, test/diagnose, disposition and discharge/admit. Projected Financial Benefits: Improved revenue; increased volume (reduction of Left Without Being Seen (LWBS), increased volume through increased performance; reduced costs through improved efficiency (time per patient).

Table 13.2 Critical to satisfaction (CTS) criteria

Critical to satisfaction (CTS)	Metric (short title)	Operational definition (metric description)
Patient throughput time	Total LOS	Time from when the patient arrives to when they are discharged from the ED (physically leaves)
Patient throughput time	Triage time	Time from arrival until patient is triaged
Patient throughput time	EDP time	Time from when the patient is seen by the ED physician to when they complete their disposition
Patient wait time during visit	Wait time for ED	Time after triage until the ED physician sees the patient
Patient wait time during visit	Orders to leave ED	Time from when orders are received to when the patient physically leaves the ED
Patient satisfaction	% left without being seen	The number of patients left without being seen/Total number of patients signed into ED
Quality of care	Lab time	Time to perform labs: time from ordered by physician to results provided to physician
Quality of care	Diagnostic time	Time to perform diagnostic tests: Time from ordered by physician to results provided to physician
Quality of care	Register time	Time from arrival in ED to the patient is registered
Quality of care	Admit time	Time from diagnosis to when the patient is admitted to inpatient floor

original LSS project discussed in Chap. 3 are shown in Table 13.2. The simulation model must be validated to ensure that it represents the real process and then used to simulate improvement recommendations and their impact on the real process. This will reduce the time and expense of implementing improvements into the real future state process.

Measure Phase

In the Measure phase, we understand and measure the existing system. By observing the existing system, we then develop detailed process maps of the flow. We collect data either manually or preferably from a computer system that provides activities, times, arrivals, and entity information. Depending upon what part of the process we want to implement, dictates whether we use a value stream map, a SIPOC, or a more detailed process map.

The value stream map for the emergency services is shown in Fig. 13.3. The SIPOC (Table 3.3) and the first page of the emergency services process map (Fig. 3. 3) are shown in Chap. 3 and can be referred to for additional knowledge about the processes.

The Value Stream Map (VSM) has a value stream for the discharged patients and the patients who are admitted to the hospital. The reason to have a stream for each is that the process and wait times are different for each stream, and the Length of Stay (LOS) is longer for sicker patients who will have more diagnostics and be admitted to the hospital. The discharged patients spend a total average time in the system of 347 min or 5.8 h, about an hour is spent in value-added time. The admitted patients spend a total average time of 524 min or 8.7 h, with about 79 min being spent in value-added time. The VSM and knowledge of the processes are used to develop the process model within the simulation model, as shown in Fig. 13.4. The process flow that the patients go through starts with the patient entering the Emergency Department (ED). The patient gets triaged and gets assigned to an ED bed, when one is available. The triage activity is embedded in the time to ED bed activity. There is a wait time or delay before each activity, identified on the lower steps of the

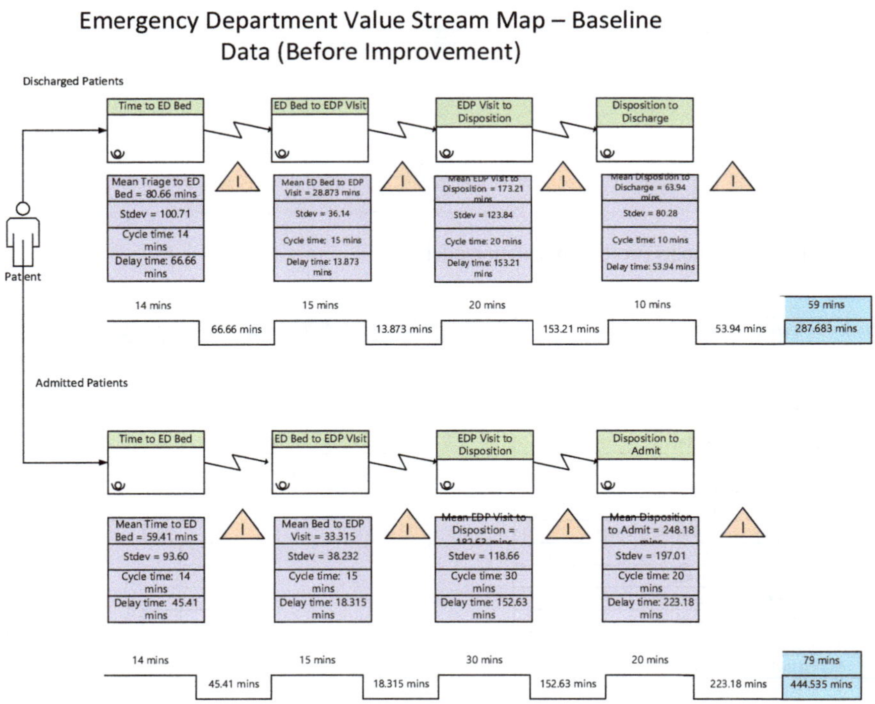

Fig. 13.3 Emergency services value stream map

timeline below the activities. The next activity includes the activities associated with seeing the nurse and waiting for the ED Physician (EDP). The next activity is seeing the EDP, performing labs, diagnostics, treatments, and the disposition of whether to discharge or admit the patient. The last step includes the discharge activities if the patient is to be discharged or the admission activities if they are to be admitted to the inpatient floor of the hospital. The goal is to keep the simulation as simple as possible, while still representing the real ED process flow. The ED Bed Board data for the current state was used to develop the distribution times, including the delay times within each process activity.

The high-level steps in the process and their relative relationships to each other based on timing spent in each step are shown in Fig. 13.5. The actual current state times from the ED Bed Board were used as inputs to the simulation model.

The four main processes are included in Fig. 13.5, the Time to ED Bed, Time from Bed to Physician, ED Visit time, and disposition to leave ED.

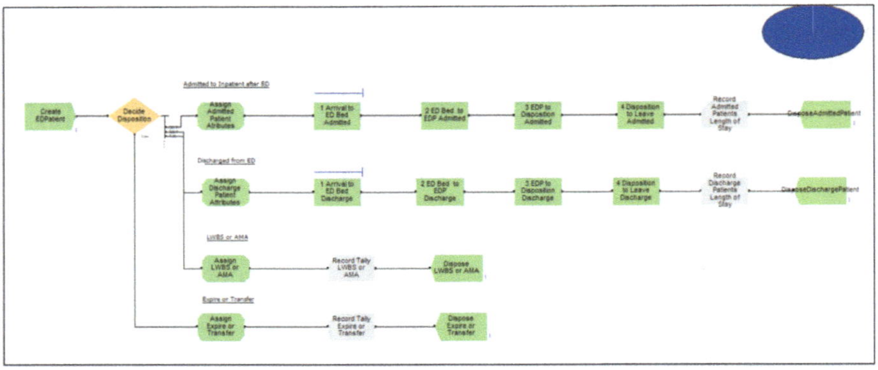

Fig. 13.4 Process model for the simulation model

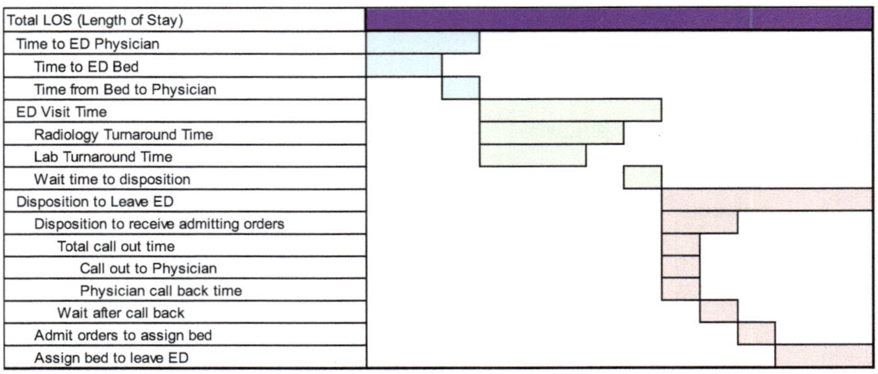

Fig. 13.5 Baseline ED process timeline

The times used in the simulation to assess the process times from Fig. 13.4 (first variable below) were used as the process activities in the simulation (second variable name below):

- Patient arrival to placed in ED Bed = Arrival to ED Bed.
- ED Bed to seen by EDP = ED Bed to EDP.
- Seen by EDP to disposition = EDP to disposition.
- Disposition to leave ED (discharged or admitted) = Disposition to leave ED.

The number of patients admitted and discharged during the simulation run was also tracked and were pulled from the ED Bed board for the current state time period.

Design Phase

We designed the factors, variables, attributes, entities, process flow, experiments, and model to simulate the process.

The factors incorporated into the model based on the current state data from the ED Bed Board system included the following:

(1) Patient arrival patterns each hour.
(2) Patient disposition, discharged or admitted. These patients would follow different flows, to align to the different process times in the real system.

Variables

A schedule was used within the simulation model developed from the actual patient arrival patterns per hour.

Entities

The model entities were ED patients.

Model Process Flow

The following model process flow was developed. A patient entity is created at the start of the simulation with a create module. The patient arrives based upon an arrival schedule (Table 13.3), as designated as patient arrivals per hour. A statistical distribution could also have been used, but the data didn't fit any of the existing statistical distributions well. The staffing is embedded within the ED bed schedule which identifies the number of ED beds (resources) available across 24 h of the day, as shown in Table 13.4. It was assumed that physicians covered six patients (rooms) at a time, nurses covered three patients (rooms) at a time, so staffing was aligned to the beds available.

The create module is shown in Fig. 13.6. The create module has a maximum capacity of 135 patients within a 24-h period, as does the ED.

After the patients arrived in the simulation, their eventual disposition was decided with a decide module. The percentages aligned to the actual percentage of patients

Table 13.3 Patient arrival schedule

Number patient arrivals	Hour of day
1.39	0
1.21	1
0.96	2
1.14	3
0.87	4
0.84	5
1.39	6
2.01	7
3.33	8
5.32	9
6.64	10
6.35	11
5.78	12
5.63	13
5.11	14
4.54	15
4.39	16
4.31	17
3.72	18
4.18	19
3.29	20
2.79	21
2.22	22
1.80	23

Table 13.4 ED bed board schedule

Number of available ED beds	Hours available
25	8 (hours 0:00 am 8:00 am)
40	10 (hours 8:00 am to 6:00 pm)
30	6 (hours 6:00 pm through midnight)

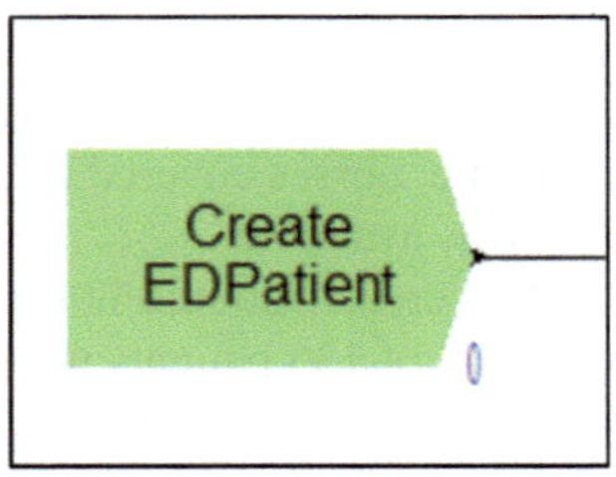

Fig. 13.6 Create module for arriving patients

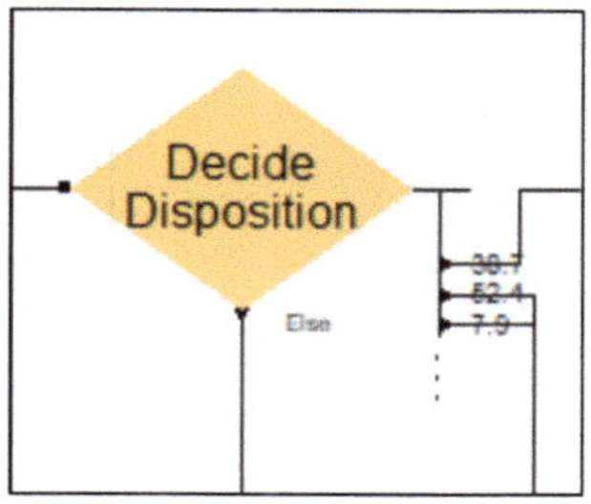

Fig. 13.7 Decision module for patient disposition

that were admitted (38.7%), discharged (52.4%), and left the ED without being seen or served (7.9%), due to them deciding not to wait for service Left Without Being Seen (LWBS), or leaving against medical advice (AMA), or being transferred to another facility or expiring (1%). The decision module is shown in Fig. 13.7.

Assign modules are used to assign the disposition attributes, admitted, discharged, LWBS or AMA, and Expire or Transfer, as shown in Fig. 13.8. The TNOW variable was used to assign the time that the patient arrived to the simulation and the tally of the number of patient arrivals by disposition category was also assigned.

For the LWBS or AMA patients, and the Transfer or Expire patients, a record module records the tallies, and a dispose module disposes of the entities, as shown in Fig. 13.9.

Four process modules are used to process the patients through the ED processes for the admitted and discharged patients. Separate streams and process modules are used for each disposition type. For the admitted patients, the process models are (Fig. 13.10) as follows:

(1) Arrival to ED bed admitted.
(2) ED bed to EDP admitted.
(3) EDP to disposition admitted.
(4) Disposition to leave admitted.

Table 13.5 shows the distributions applied for each process module. The mean times for each distribution were derived from the ED Bed board averages. The standard deviations were reduced due to the additive nature of the processes and not wanting to increase error in the model by having the standard deviations too high for each process.

A record module is used to record the statistics for counts of patients discharged or admitted, and the length of stay for each patient. The record and dispose modules are shown in Fig. 13.11. The dispose modules dispose of the entity, when the patient leaves the ED.

Animations were developed to show the patients moving through the ED system, which helps to validate that the process flow is working appropriately, and it helps process owners who are not versed in modeling and simulation to understand the model shown in Fig. 13.12.

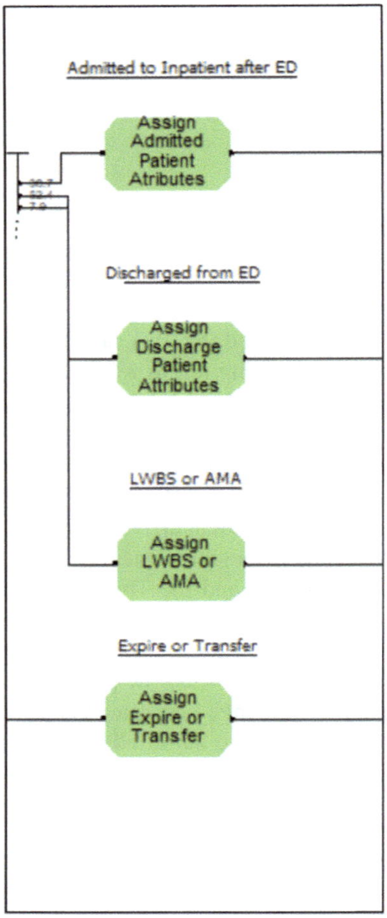

Fig. 13.8 Assign modules for disposition attributes

A warmup period of 1440 min was applied within the simulation setup, since the ED is never without patients. The simulation was run for 24 h a day, and for 7 replications (days). Statistics were collected on entities, queues, processes, and resources.

Simulation results for current state process:

The simulation process metrics for the current state process are shown in Table 13.6.

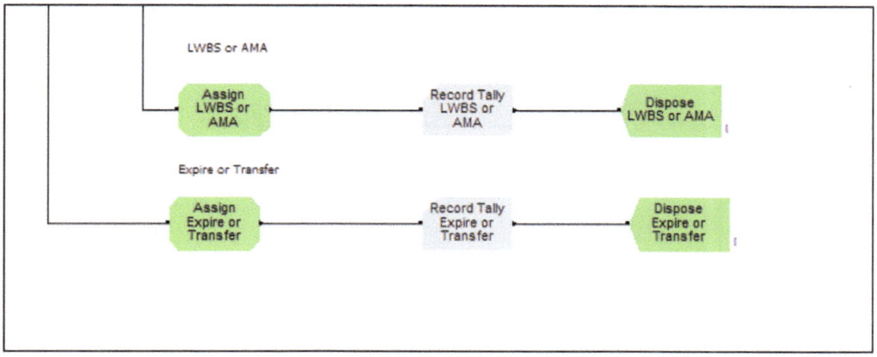

Fig. 13.9 Record and dispose modules

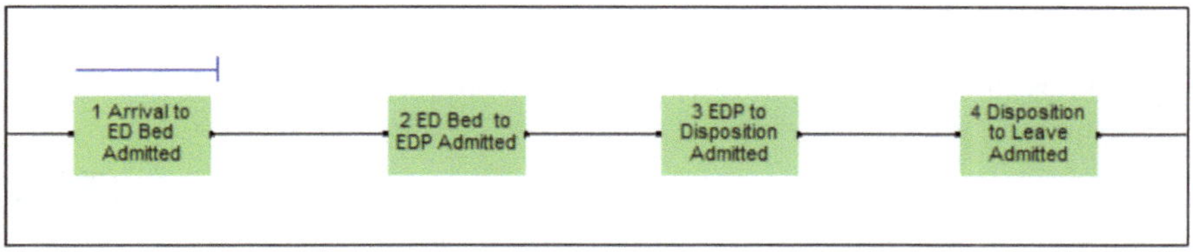

Fig. 13.10 Process modules for admitted patients

Table 13.5 Process, distribution, mean and standard deviations current state

Process	Distribution	Mean (minutes)	Standard deviation
Admitted disposition			
(1) Arrival to ED bed	Normal	48	8.5
(2) ED bed to EDP	Normal	31.82	5.8
(3) EDP to disposition	Normal	182.96	10.7
(4) Disposition to leave	Normal	222	11.1
Discharge disposition			
(1) Arrival to ED bed	Normal	68	8.9
(2) ED bed to EDP	Normal	27.09	5.5
(3) EDP to disposition	Normal	171.06	10.7
(4) Disposition to leave	Normal	58.53	7.3

Validate Phase

To validate the model there are four main types of validation:

(1) Face validity is about whether a system makes sense.
(2) Construct validity is assessing whether the system behaves as it is intended to.

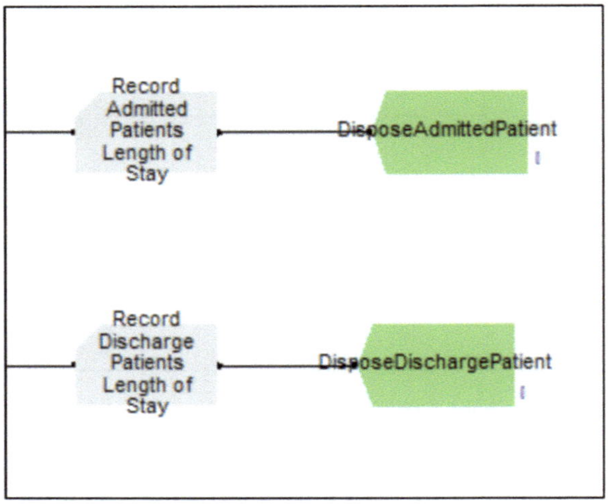

Fig. 13.11 Record and dispose modules

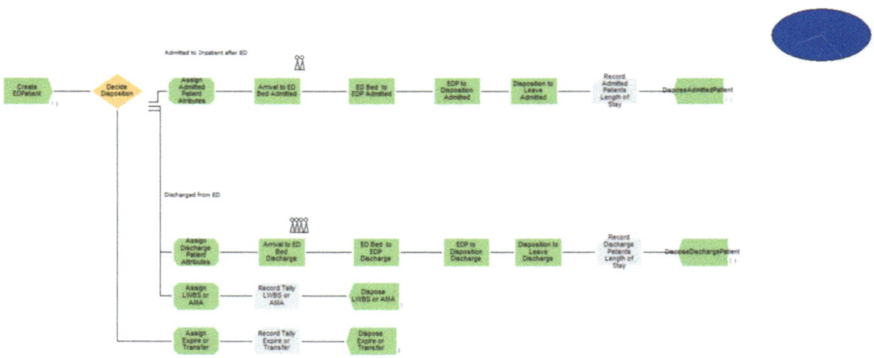

Fig. 13.12 Model run with animation

(3) Content validity is assessing whether the simulated system is similar to the real system.
(4) Criterion validity assesses whether the simulation model outcomes are similar to the real system.

To measure the face and construct validity, the analyst reviewed the simulation animation as it ran, and it behaved as the real system. To measure the content and criterion validity, the process analyst performed hypothesis tests on the process metrics, as shown in Table 13.7. The (S) signifies that these numbers are simulated results, and the (R) signifies that those numbers are real-world system results. We tested and validated the model that represents the existing system. The overall length of stay for admitted and discharged patients and each activity in the ED value stream were

Table 13.6 Simulation process metrics results

Metric	Average	Standard deviation
Admitted disposition		
Length of Stay (LOS)	487.25	6.90
(1) Arrival to ED bed	49.07	2.15
(2) ED bed to EDP	31.72	1.70
(3) EDP to disposition	183.41	2.32
(4) Disposition to leave ED	221.56	1.50
Discharge disposition (minutes)		
Length of stay (LOS)	325.67	2.76
(1) Arrival to ED bed	67.96	1.90
(2) ED bed to EDP	27.40	0.86
(3) EDP to disposition	171.02	1.52
(4) Disposition to leave	58.60	1.09

aligned to the current state of the real ED system. A two-sample t-test was performed to compare the actual mean to the simulated mean. The null hypothesis is that there is no difference between the means for the sampled distributions. The alternative hypothesis is that there is a difference in the means. If the p-value is below 0.05, then we reject the null hypothesis and conclude that there is a difference in means and practically that the simulation doesn't represent the real-world system well. For the total Length of Stay (LOS) and the four processes for both admitted and discharged patients, the null hypothesis is not rejected, so we can conclude that the simulation is validated and does represent the real-world system.

The total Length of Stay (LOS) histograms for admitted and discharged patients from the ED Bed Board systems for the current state of the ED processes are shown in Figs. 13.13, and 13.15, respectively. The two-sample t-tests for the simulated data compared to the real system data for the admitted and discharged patients, respectively, are shown following each of the histograms, in Figs. 13.14 and 13.16.

The average LOS for admitted patients is 486.9 min with a standard deviation of 164.7 min.

The average LOS for discharged patients is 326.7 min with a standard deviation of 147.4 min.

Implement Phase

We identified improvement recommendations, incorporated them into a future state model, and assessed the potential impact of the improvements. The following improvement recommendations (Table 13.10) were identified along with the proposed impact on the average activity times. The impacts on the average activity times were incorporated into the simulation system.

Table 13.7 Simulation process metrics results (S = simulation results; R = real results)

Metric	Average	Standard deviation	Sample size	P-value of hypothesis test	Simulation different from real system?
Admitted disposition					
Length of Stay (LOS)	S) 487.25 R) 486.87	S) 6.90 R) 164.68	S) 33 R) 3778	0.897	No
(1) Arrival to ED bed	S) 49.07 R) 49.51	S) 2.15 R) 71.23	S) 23 R) 3778	0.723	No
(2) ED bed to EDP	S) 31.72 R) 31.82	S) 1.70 R) 33.56	S) 24 R) 3778	0.877	No
(3) EDP to disposition	S) 183.41 R) 182.96	S) 2.32 R) 113.75	S) 27 R) 3778	0.813	No
(4) Disposition to leave ED	S) 221.56 R) 222.59	S) 1.50 R) 123.32	S) 33 R) 3778	0.611	No
Discharge disposition					
Length of Stay (LOS)	S) 325.67 R) 326.69	S) 2.76 R) 147.44	S)37 R) 5193	0.626	No
(1) Arrival to ED bed	S) 67.96 R) 70.02	S) 1.90 R) 79.26	S) 29 R) 5193	0.075	No
(2) ED bed to EDP	S) 27.4 R) 27.09	S) .86 R) 30.36	S) 30 R) 5193	0.491	No
(3) EDP to disposition	S) 171.02 R) 171.06	S) 1.52 R) 115.32	S) 35 R) 5193	0.980	No
(4) Disposition to leave	S) 58.60 R) 58.53	S) 1.09 R) 53.15	S) 37 R) 5193	0.927	No

The team developed an improvement plan based on eliminating the root causes, discussed in Chap. 3. Improvements were categorized into the following themes:

- Hospital bed assignment.
- Triage and waiting room.
- Admitting physician call out and call back.
- Patient transport.
- Patient flow.
- Minor care staffing.
- Diagnostic tests.
- ED and floor reporting process.
- Discharge process.
- Dashboards, metrics, and performance monitoring.

Hospital Bed Assignment

The first improvement category was focused on the bed board process and system. The electronic bed board was not being utilized to its full capacity. They needed to better

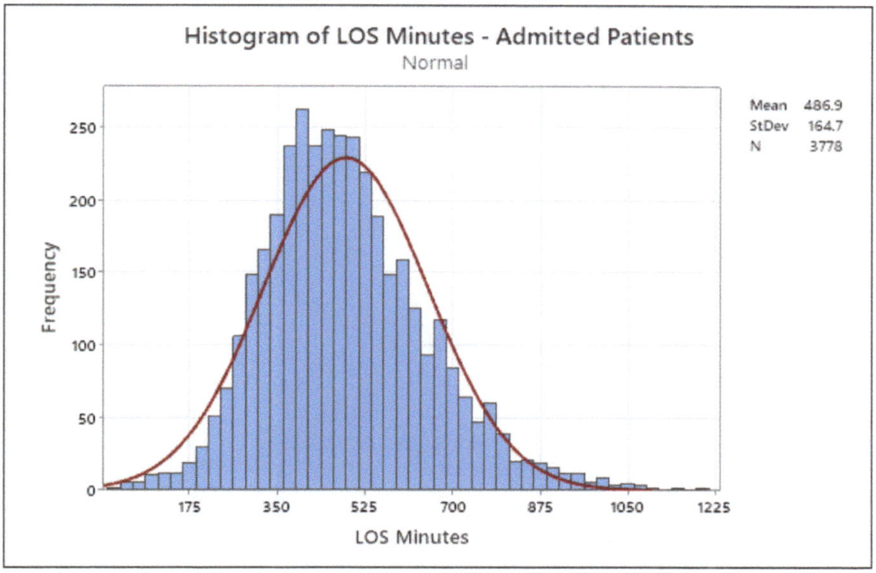

Fig. 13.13 Histogram of Length of Stay (LOS) minutes admitted patients—actual results

utilize the bed board system to provide hospital-wide bed availability and visibility of beds. This would improve the time to assign beds for admitted ED patients. This should positively impact the *disposition to Leave ED time f*or admitted patients.

The Assistant Nurse Managers would need to provide the information of the bed availability on a real-time basis. Prior to the pilot, the Admitting Department assigned beds. The team proposed that the Nurse Supervisors should assign the hospital beds, to reduce the amount of time and non-value-added activities involved in the Admitting Clerk calling to the floors to request open beds.

They also proposed implementing pre-emptive bed assignment for those patients most likely to be admitted. This would enable the earlier assignment of the beds by requesting beds when the admitting orders were initially received by the nurses.

Triage and Waiting Room

When a surge in patient volume occurs, they propose to add a triage nurse to more quickly triage the patients. When there are ED beds available, they will place the incoming patients into an ED bed immediately, and triage the patient in the ED bed, instead of triaging them in the triage room and then sending them back to wait in the waiting room. They also will do the medication reconciliation in the ED room, not in the triage room. This should positively impact the *arrival to ED bed time.*

Admitting Physician Call Out and Call Back

The team proposed to add a dedicated communication secretary during the peak ED census times of 2–10 pm. This dedicated person would focus on calling the admitting

WORKSHEET 1

Admit - LOS - Two-Sample T-Test and CI

Method

μ_1: population mean of Sample 1

μ_2: population mean of Sample 2

Difference: $\mu_1 - \mu_2$

Equal variances are not assumed for this analysis.

Descriptive Statistics

Sample	N	Mean	StDev	SE Mean
Sample 1	33	487.25	6.90	1.2
Sample 2	3778	487	165	2.7

Estimation for Difference

Difference	95% CI for Difference
0.38	(-5.38, 6.14)

Test

Null hypothesis	H_0: $\mu_1 - \mu_2 = 0$
Alternative hypothesis	H_1: $\mu_1 - \mu_2 \neq 0$

T-Value	DF	P-Value
0.13	944	0.897

Fig. 13.14 Hypothesis test for admitted patients' length of stay

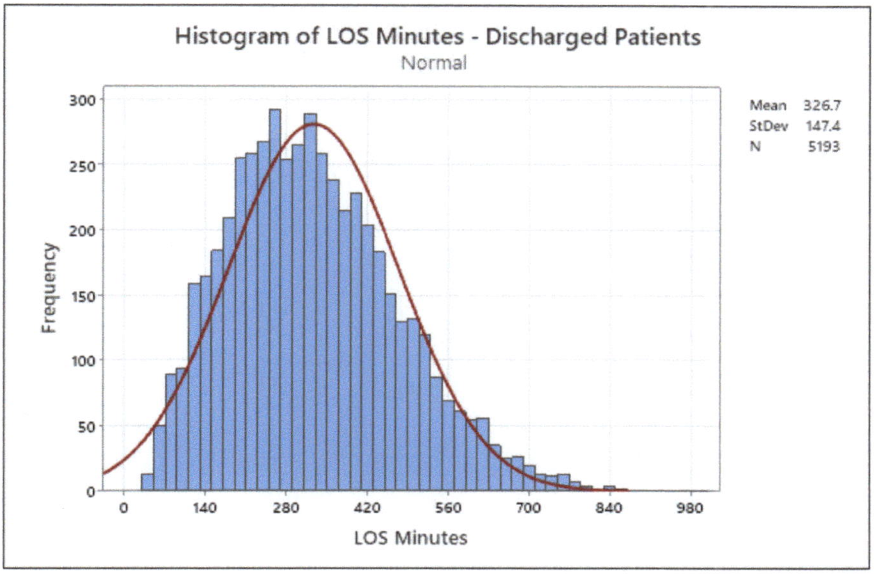

Fig. 13.15 Histogram of Length of Stay (LOS) minutes discharged patients—actual results

physicians for admitting orders. The medical policy requires the admitting physician to be called every 15 min for three times, and then the on-call physician can be called. Previously, the admitting physicians were not called consistently or quickly. The average time to call out to the admitting physicians was 13 min with a standard deviation of 20 min. The average time for the physician to call back was 29 min with a standard deviation of 34 min. There was also an average of 1.8 call attempts before the admitting physicians called back. This should positively impact the *disposition to Leave ED time f*or admitted patients.

Patient Transport

The team added an ED dedicated patient escort during the peak admitting times of 3–11 pm. We prepared a cost/benefit analysis estimating a rate of return based on proposed increased ED volumes from gaining additional patients and reducing the number of patients leaving the ED without treatment. The benefit-to-cost ratio was 13.4 to 1 (benefits to cost). The desired benefit-to-cost ratio is greater than 1. We showed that there was a positive return on investment when adding additional staff for the dedicated communication secretary, the patient escort, and the Nurse Supervisors for assigning beds. This should positively impact the entire LOS, but especially the *disposition to Leave ED time for both admitted and discharged patients.*

Patient Flow

The team proposed to add additional electronic alerts on the ED Bed Board to improve visibility of the status of each patient and which step in the process they are at in

WORKSHEET 1

Discharge - LOS - Two-Sample T-Test and CI

Method

μ_1: population mean of Sample 1
μ_2: population mean of Sample 2
Difference: $\mu_1 - \mu_2$

Equal variances are not assumed for this analysis.

Descriptive Statistics

Sample	N	Mean	StDev	SE Mean
Sample 1	37	325.67	2.76	0.45
Sample 2	5193	327	147	2.0

Estimation for Difference

Difference	95% CI for Difference
-1.02	(-5.13, 3.09)

Test

Null hypothesis H_0: $\mu_1 - \mu_2 = 0$
Alternative hypothesis H_1: $\mu_1 - \mu_2 \neq 0$

T-Value	DF	P-Value
-0.49	4237	0.626

Fig. 13.16 Hypothesis test for discharged patients' length of stay

the emergency services value chain. The team also developed an ED surge plan with volume and ED room availability triggers. This should positively impact the entire LOS.

Minor Care Staffing

The team realigned minor care staffing to add a physician for minor care when ED volume increases. This could positively impact the entire LOS for mostly discharged patients.

Diagnostic Tests

The team implemented both lab and imaging diagnostic testing improvements. Since many of the imaging exams that expose a patient to radiation (X-rays, CT scans) require a pregnancy test for women of childbearing age, we decided to fast-track pregnancy and other tests required for radiology exams. The team also created another Lean Six Sigma team to investigate and reduce the number of imaging exams and to provide appropriate CT scans. The amount of radiation exposure that CT exams give can be equivalent to 1000 X-rays, so it is critical to ensure that all tests are medically appropriate.

To decrease the time to draw the specimens for tests, we added an alert to the electronic ED Bed Board to track the time it takes for the phlebotomist to draw the specimens. This could positively impact the *EDP to disposition time.*

ED and Floor Reporting Process

We redesigned the reporting process to streamline the communication and hand off for admitted patients from the ED to the floors or units. We also automated the transfer form to reduce the time it took for the ED nurses to re-write information that was already electronically captured in the ED Electronic Medical Record. This would positively impact the *disposition to leave ED for the admitted patients.*

Discharge Process

We formed another Lean Six Sigma team to investigate and improve the discharge process for the floors, so that more beds are available when they are needed to admit patients from the ED. This could positively impact the *disposition to leave ED times* for admitted patients.

Dashboards, Metrics, and Performance Monitoring

We developed a dashboard, metrics, and performance monitoring processes to measure performance aligned to flow and throughput. We incorporated necessary performance measures to ensure metrics and performance is met. We also developed a flow dashboard that identified where each patient is in the ED value chain, and which process is holding the patient from moving through the value chain and being released. This could positively impact the entire LOS.

Table 13.8 illustrates which process is mainly impacted by the improvements in each category. An "X" signifies that there is a relationship between the improvement category and the process, so that reduced process times are anticipated.

Table 13.8 Improvement recommendations and proposed average

Improvement category	(1) Arrival to ED bed	(2) ED bed to EDP	(3) EDP to disposition	(4) Disposition to leave ED
Hospital bed assignment (admitted patients)				X
Triage and waiting room	X			
Admitting physician call out and call back (admitted patients)				X
Patient transport				X
Patient flow	X	X	X	X
Minor care staffing (for discharged patients)	X	X	X	X
Diagnostic tests			X	
ED and floor reporting process (admitted patients)				X
Discharge process (admitted patients)				X
Dashboards, metrics, and performance monitoring	X	X	X	X

We revised the simulation as follows to incorporate the potential improvements affecting the following activity times (shown in Table 13.9):

Arrival to ED Bed Processes

(1) If an ED room is available when the patient arrives to the ED, then the patient is placed immediately into the ED room and is triaged in the room. This should reduce the triage time, and subsequently the *arrival to ED Bed time* by 71–74%

ED Bed to EDP Processes

(2) Incorporating patient flow monitoring, and dashboards can reduce the *ED bed to EDP* by 37–38%%.

EDP to Disposition Processes

Table 13.9 Simulation process times for improvements

Process	Distribution	% improvement	Mean (minutes)	Standard deviation
Admitted disposition				
LOS		30	341	
(1) Arrival to ED bed	Normal	71	14	8.5
(2) ED bed to EDP	Normal	37	20	5.8
(3) EDP to disposition	Normal	13	160	10.7
(4) Disposition to leave	Normal	26	163	11.1
Discharge disposition				
LOS		30	228	
(1) Arrival to ED bed	Normal	74	18	8.9
(2) ED bed to EDP	Normal	38	17	5.5
(3) EDP to disposition	Normal	7	159	10.7
(4) Disposition to leave	Normal	20	47	7.3

Table 13.10 Simulation process metrics results improved (S = simulation results; R = real results)

Metric	Average	Standard deviation	Sample size	P-value of hypothesis test	Simulation different from real system?
Admitted disposition					
Length of Stay (LOS)	S) 356.67 R) 357.19	S) 3.15 R) 112.11	S) 30 R) 4789	0.771	No
(1) Arrival to ED bed	S) 14.49 R) 14.52	S) 1.20 R) 26.55	S) 22 R) 4776	0.948	No
(2) ED bed to EDP	S) 20.29 R) 19.82	S) 1.22 R) 23.3	S) 22 R) 4767	0.272	No
(3) EDP to disposition	S) 158.80 R) 159.96	S) 2.45 R) 84.77	S) 25 R) 4783	0.380	No
(4) Disposition to leave ED	S) 163.62 R) 163.28	S) 1.94 R) 78.71	S) 29 R) 4788	0.776	No
Discharge disposition					
Length of Stay (LOS)	S) 241.35 R) 240.38	S) 3.49 R) 106.45	S)34 R) 8348	0.507	No
(1) Arrival to ED bed	S) 17.56 R) 18.02	S) 1.48 R) 25.19	S) 26 R) 8337	0.254	No
(2) ED bed to EDP	S) 17.41 R) 16.52	S) 1.33 R) 20.35	S) 26 R) 8328	0.011	Yes
(3) EDP to disposition	S) 159.69 R) 158.73	S) 1.37 R) 89.55	S) 30 R) 8341	0.343	No
(4) Disposition to leave	S) 47.04 R) 47.20	S) 1.08 R) 47.56	S) 31 R) 8341	0.773	No

(3) Incorporating improvement in labs and diagnostic testing can improve the *EDP to disposition time* by 7–13%

Disposition to Leave ED Processes

(4) Incorporating improved hospital bed assignments, enhanced admitting physician call-out and call-back protocols, patient transport improvements, ED and floor reporting, and dashboards with alerts can reduce the *disposition to leave ED times* by 20–26%.

The actual results for 3 months after they started piloting improvements are shown in Figs. 13.17, 13.18, and 13.19.

Metrics	Baseline		Target		Improved		% Improvement	
	Admitted	Discharged	Admitted	Discharged	Admitted	Discharged	Admitted	Discharged
Total LOS:	8.7 hrs	5.8 hrs	5 hours (43%)	3 (48%)	6.1 hrs	4.1 hrs	30%	29%
% LWBS:	6.50%		3.50%		0.60%		91%	
Total Time to EDP:	93	109	35	35	32		66%	
Triage Time:	34	34	15	15	13		62%	
Wait time triage to EDP:	65	79	20	20	25		62%	
Time to ED Bed:	61	81	35	42	16		74%	
Bed to Physician:	34	55	20	20	16		53%	
Treat/Diagnose to Disposition Time:	192	175	109	91	166		14%	
Radiology Diagnostic Time:	138	270	154	154	170		-23%	
Lab Time:	119				57		109%	
EDP Time Visit:	192	175	109	91	175		9%	
Disposition to Leave ED	223	64	127	33	171	47	23%	27%
Disposition to get admit orders:	103		59		70		32%	
Total Call Out Time:	41		23		24			
Call Out to Physician	13		7		5			
Physician Call Back	29		17		19			
Wait After Call Back:	37		21		46			
Admit Orders to Assign Bed:	42		24		44		-5%	
Assign Bed to Leave ED:	112		64		67		40%	

Fig. 13.17 Real-world improvement results 3 months after start of pilot

Fig. 13.18 Real-world improvement results for admitted patients

Statistics

Variable	N	Mean	StDev
TOTAL_LOS	4789	357.193	112.107
1) Time to ED Bed	4776	14.5216	26.5503
2) Time from Bed to EDP	4767	19.8179	23.2994
3) Time from EDP to Disposition	4783	159.955	84.7686
4) Disposition to Leave ED	4788	163.276	78.7027

Fig. 13.19 Real-world improvement results for discharged patients

Statistics

Variable	N	Mean	StDev
TOTAL_LOS	8348	240.382	106.450
1) Time to ED Bed	8337	18.0200	25.1921
2) Time from Bed to EDP	8328	16.5197	20.3455
3) Time from EDP to Disposition	8341	158.731	89.5501
4) Disposition to Leave ED	8341	47.1994	47.5603

Once we implemented and improved the real-world system, we then assessed the impact of the improvements and compared them to the simulated estimated impacts. The comparisons of the simulation results after improvements compared to the real-world results are shown in Table 13.10. Only the ED bed to EDP simulated time for discharged patients varies from the real system. The simulation results align very well with the real-world results.

Conclusions

Modeling and simulation are powerful tool that enables developing a model of the real-world operations in a hospital's clinical and nonclinical processes. It provides a demonstration of the potential for improving the actual system prior to implementing the improvements. It can help gain buy-in from leadership and financial decision-makers who provide budgets for improvement programs. This chapter illustrated an example of implementing modeling and simulation in an Emergency Department. The model was built based upon the real-world ED Bed Board data distributions and validated to this data. The simulation model was then revised based on actual improvements that were implemented and again validated to demonstrate the strength of the model. This model could easily be revised incorporating new ideas for improvement, incorporating additional complexity to assess resource utilizations, prioritization of patients based upon acuity levels, and many other enhancements.

Reference

Kelton, W. D., Sadowski, R. P., & Zupick, N. B. (2014). *Simulation with Arena* (6th ed.). McGraw-Hill.

Chapter 14
Applications of Artificial Intelligence for Improving Processes in Healthcare

Abstract Artificial intelligence is a growing focus area in healthcare and other industries. There is great potential for applying AI in healthcare applications including healthcare research, pharmaceutical and clinical studies, and clinical care applications. At the same time, there are many ethical challenges when implementing AI in healthcare. In this chapter, an overview of the definition of AI is provided and the important ethical challenges that should be considered. We also provide two examples of applying AI to improving processes in an Emergency Department to reduce length of stay, and reducing the turnaround time and improving on-time starts in the operating room. We use a value stream map and a process architecture map with value analysis as the foundational data and tools provided to the AI generative tool to evaluate and identify improvements for the processes. We demonstrate where AI excels, the risks of applying AI and how the process analyst should engage with the AI tools to ensure positive operational outcomes.

Keywords Emergency department · Operating room turnaround · Operating room on-time starts · Artificial intelligence · Applications of AI in healthcare

Introduction

Artificial intelligence (AI) is an approach to develop algorithms that can learn, make decisions, understand and analyze data, and create and generate new content. Machine learning is a subset of AI that can learn from data and make predictions and decisions based upon the data. Deep learning is a further subset of machine learning that applies deep neutral networks to enable processing of images, sound, and text data. Generative artificial intelligence is rising in popularity and use. Generative AI understands and analyzes data, but can also create new content. Generative AI can enable machines to design images, write text, and even compose music. This content can be indistinguishable from human creations. Healthcare is an area where there is great potential and many different types of applications, including in healthcare research, pharmaceutical and clinical studies, and clinical care applications. There is

S. Furterer, *Lean Six Sigma for the Healthcare Enterprise*,
https://doi.org/10.1007/978-1-4471-7617-6_14

much research on the ethical challenges of implementing AI in healthcare that also need to be considered (Carnahan, 2025).

In this chapter, an overview of the definition of AI is provided, some AI applications in healthcare and some of the important developmental, application-based, and ethical challenges that should be considered. We also provide two examples of applying AI to improving processes in an Emergency Department to reduce length of stay, and reducing the turnaround time and improving on-time states in the operating room. We use a value stream map and a process architecture map with value analysis as the foundational data and tools provided to the AI generative tool to evaluate and identify improvements for the processes. We demonstrate where AI excels, the risks of applying AI, and how the process analyst should engage with the AI tools to ensure positive operational outcomes.

Applications of AI in Healthcare

The application of AI in healthcare has been growing substantially over the last 25–30 years. Although AI has its roots in the 1950s with a paper from Turing (1950), it has only been since the end of the twentieth century, with the resurgence of neural networks and deep learning, that the applications of AI have been able to excel. Some of the AI applications in healthcare include the areas of clinical and pharmaceutical research and studies and clinical patient care applications (Adirim, 2025). AI models can help to detect cancer faster and earlier than before, make more accurate medical diagnoses, and even care for the elderly using robots (Li et al., 2023).

Challenges of AI in Healthcare

There are many challenges of AI in healthcare. Since AI models begin with data, this is also the first challenge of AI, to ensure the appropriate and secure use of the data. Protecting patient confidentiality is critical and part of the Health Insurance and Portability and Accountability Act (HIPAA) enacted in 1996. Patient records are used to train AI models, and this data must be secured. Another challenge is when the AI statistics demonstrate that the model performs well, yet the model results don't make sense. Cooper et al. (1997) created an AI model to predict patient mortality that was found not to make clinical sense. The neural network model predicted that patients with asthma had a lower risk of mortality, but only because they were hospitalized more often than non-asthmatics, resulting in more aggressive treatment. This makes clinical sense, but the model didn't consider this, or provide an intelligible model that aligned to that practice (Carnahan, 2025). Another AI challenge is when a clinician follows the AI model decision, and a patient has a bad outcome, who is to blame, the clinician or the model? This is one of the ethical questions that arises with use of AI models.

Ethical Challenges with the Use of AI

AI models should have accuracy, explainability, transparency, and compatibility (Li et al., 2023). Some ethical challenges related to explainability include the need to understand how the model works and makes decisions. When the inner workings of the model are not apparent and transparent, there can be ethical issues with using the model. Li et al. (2023) performed a systematic literature review that identified 12 AI ethical challenges in healthcare, as follows:

Justice and fairness—relates to the fair distribution of medical goods and services without discrimination among individuals. This issue includes sub-issues of bias, fairness, discrimination, and equality.

Freedom and autonomy—includes sub-issues of control, respecting human autonomy, and informed consent.

Privacy—data privacy and confidentiality.

Transparency—refers to the understanding of an AI system's decision-making process, incorporating explainability of the model.

Patient safety and cyber security—patient safety and potential harm related to the use of AI.

Data security of devices for data collection and the potential for hacking and unauthorized surveillance.

Trust—the importance of trust and the interaction between patients and artificial intelligence care providers.

Beneficence—acting in the best interests of others.

Responsibility—who is responsible for errors and correct use of AI.

Solidarity—this challenge includes issues of justice and equality when AI models are deployed in healthcare.

Sustainability—sustainability related to developing, deploying, and implementing AI solutions in healthcare.

Dignity—includes respect for human rights and freedom.

Conflict—conflicting goals and priorities in the application of AI.

There is increasing research into the ethical challenges of AI and strategies for dealing with these ethical dilemmas. It is important to have an awareness of these ethical challenges as we build and deploy AI tools and models.

Examples of Application of AI Tools for Enhancing Process Improvement in Healthcare

In the following sections, we provide two examples of applying AI to improving processes in an Emergency Department to reduce length of stay, and reducing the turnaround time and improving on-time starts in the operating room. We use a Value Stream Map and a Process Architecture Map with Value Analysis as the foundational data and tools provided to the AI generative tool to evaluate and identify improvements for the processes. We demonstrate where AI excels, the risks of applying AI, and how the process analyst should engage with the AI tools to ensure positive operational outcomes.

We will first describe the process improvement tool, data, and prompts provided to the AI tool. The outputs from the Co-Pilot AI tool will be shown in italics font, and then discussed and evaluated.

Application of the Co-Pilot AI Tool with a Value Stream Map (VSM) for the Emergency Department Throughput

The Emergency Department VSM with the value-added activities and non-value-added times is shown in Fig. 14.1.

The mean time, standard deviation, cycle time, and delay time for activities within the Emergency Department are provided in the VSM for the following activities:

(a) Time to ED Bed: Time from patient arrival to when they are placed in an ED bed/room.
(b) ED Bed to EDP Visit: Time from patient's placement in the ED bed to when the EDP visits the patient.
(c) EDP Visit to Disposition: Time from when the EDP visits the patient to when they enter the patient's disposition in the Electronic Medical Record (EMR).
(d) Disposition to Admit or Discharge: Time from disposition entered into the EMR to when the patient leaves the ED to be admitted to the floor, unit, transferred to another facility, or discharged.

There are two value stream paths with different activity times for patients who are admitted to the hospital versus those who are treated and discharged from the ED. The timeline shows the value-added times at the top of the timeline and the delays and wait times between the activities at the bottom of the timeline. On the far right side of the timeline, the top number is the sum of the value-added times, while the bottom number is the sum of the delays between the value-added activities. The text in the blue boxes is the author's assessment of the Co-pilot results.

The initial Co-Pilot prompt is

"Evaluate this Value Stream Map."

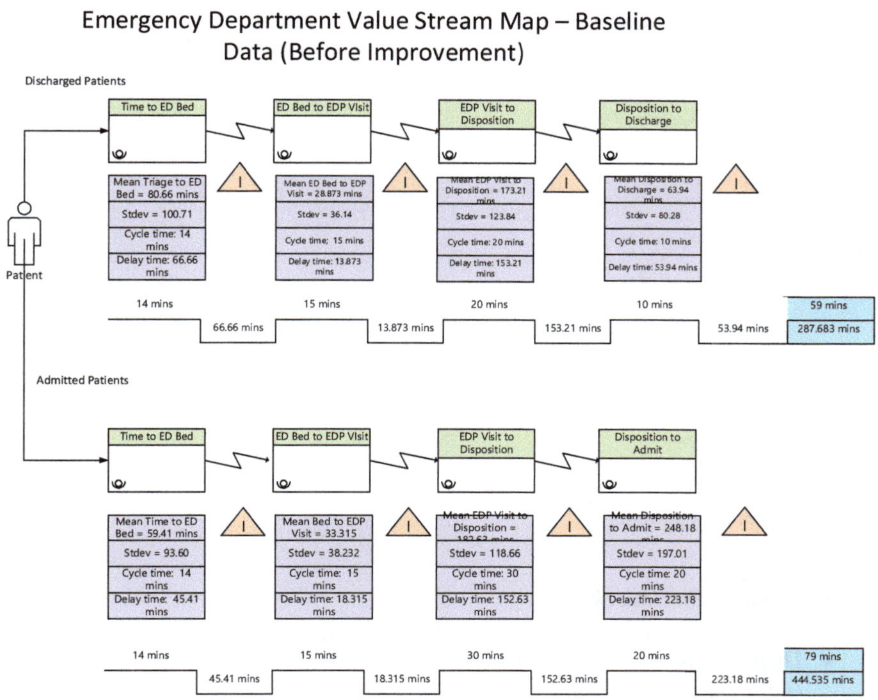

Fig. 14.1 Emergency department VSM

The Co-Pilot response:

Figure 14.2 shows the algorithm that Co-pilot performed in response to the initial prompt.

Thanks for sharing the baseline ED Value Stream Map. I read and summarized the key metrics from your PDF and then computed overall lead times, value-added (VA) versus non-value-added (NVA) proportions, and variability by step for both discharge and admit pathways. [Value Stream Map ED | PDF]

(1) ***Headline findings***

- ***Total Lead Time (mean)***
- ***Discharged**: ~**325.4 min** (5 h 25 m) with **59 min** VA and **266.4 min** NVA → **18.1% VA/81.9% NVA**.*
- ***Admitted**: ~**523.5 min** (8 h 44 m) with **79 min** VA and **439.5 min** NVA → **15.1% VA/84.9% NVA**.*
- *(Derived from the step means and cycle times listed in the uploaded map.) [Value Stream Map ED | PDF]*

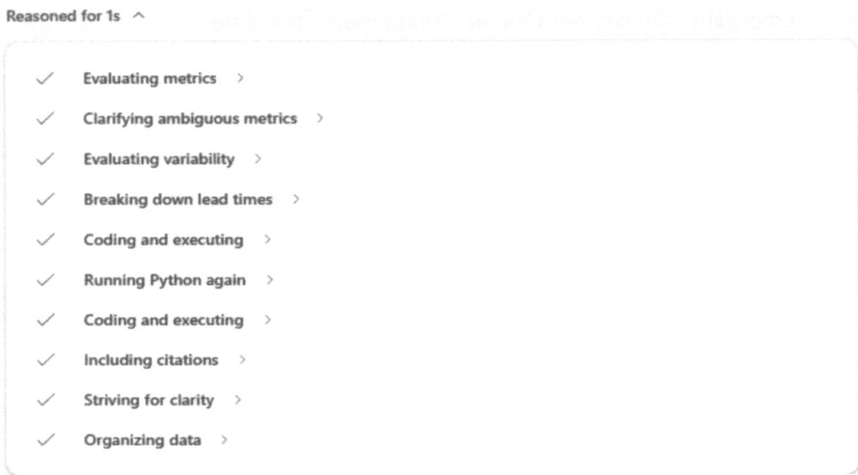

Fig. 14.2 Co-pilot algorithm

- ***Primary Bottlenecks***:
- ***EDP Visit → Disposition***: *Mean* ***173–183 min*** *with ~****153 min*** *delay.*
- ***Disposition → Admit (boarding)***: *Mean* ***248.2 min*** *with* ***223.2 min*** *delay (admit pathway). [Value Stream Map ED | PDF]*
- ***High Variability***: *Several steps show* ***CV ≥ 1.0*** *(stdev ≈ mean), notably* ***Arrival/Triage → ED Bed*** *and* ***Disposition → Discharge****, indicating unstable flow and inconsistent wait times. [Value Stream Map ED | PDF]*

(2) ***Step-Level View (Based on Your VSM Data)***

Tables 14.1 and 14.2 **show the summarized value stream data**

Discharged pathway *(assumption: using the "Time to ED Bed = 59.41 min" variant)*:

Values below originate from your file. *[Value Stream Map ED | PDF]*

Admitted pathway *(assumption: values specific to admitted track where present):*

Values below originate from your file. *[Value Stream Map ED | PDF]*

Table 14.1 Step-level view—discharged pathway

Step	Mean (min)	Std. dev. (min)	Cycle (min)	Delay (min)	CV
Arrival/triage → ED bed	59.41	93.60	14	45.41	1.58
ED bed → EDP visit	28.873	36.14	15	13.873	1.25
EDP visit → Disposition	173.21	123.84	20	153.21	0.71
Disposition → Discharge	63.94	80.28	10	53.94	1.26

Table 14.2 Step-level view—admitted pathway

Step	Mean (min.)	Std. dev (min.)	Cycle (min.)	Delay (min.)	CV
Arrival/Triage → ED bed	59.41	93.60	14	45.41	1.58
ED bed → EDP visit	33.315	38.232	15	18.315	1.15
EDP visit → Disposition	182.63	118.66	30	152.63	0.65
Disposition → Admit	248.18	197.01	20	223.18	0.79

Note on duplicated measures in the PDF: I see two variants for the first two steps and for "EDP Visit → Disposition" (e.g., "Mean triage → bed = 80.66" versus "Mean time → bed = 59.41"; "ED Bed → EDP Visit = 28.873" versus "Bed → EDP Visit = 33.315"; and "EDP Visit → Disposition = 173.21" versus 182.63). I treated **59.41/28.873/173.21** as the discharged track and **59.41/33.315/182.63** as admitted. If those pairings differ in your data definition, I can recompute instantly once you confirm which measure applies to which track. *[Value Stream Map ED | PDF]*

The headline findings and step-level view is a summary from the VSM.

(3) ***What the VSM Tells Us (Diagnosis)***

1. ***Front-End Access****: The* ***arrival/triage → bed*** *step averages* ***~59.4 min*** *with high variability (CV* ***1.58****), suggesting batching, bed assignment friction, or triage/registration rework. [Value Stream Map ED | PDF]*
2. ***Provider Contact: Bed → EDP visit*** *averages* ***~29–33 min****; variability remains high (CV* ***1.15–1.25****), often symptomatic of uneven provider load balancing or delays awaiting initial orders/imaging. [Value Stream Map ED | PDF]*
3. ***Diagnostic/Decision Phase: EDP visit → disposition*** *is the* ***largest single delay*** *(****~153 min*** *of non-value time), pointing to prolonged diagnostics, consult waits, or result turnaround gaps. [Value Stream Map ED | PDF]*
4. ***Discharge Processing*****:** ***Disposition → discharge*** *still consumes* ***~64 min*** *with* ***~54 min*** *delay; opportunities likely exist in transport, paperwork, meds-to-beds, and patient instructions. [Value Stream Map ED | PDF]*
5. ***Boarding (Admit Path)*****:** ***Disposition → admit*** *averages* ***~248 min*** *with* ***~223 min*** *delay—this dominates admit lead time and will mask improvements upstream unless addressed (classic capacity synchronization issue). [Value Stream Map ED | PDF]*

The diagnosis is similar to the assessment that the Lean Six Sigma team provided (Chap. 3).

(4) ***Targeted improvements (Lean + Flow)***

A. ***Front-end & provider access (arrival → EDP visit)***

- ***Provider-In-Triage/Split Flow for ESI 3–5:*** *Start orders at triage and create "results pending" areas to reduce bed blocking.*
- ***Bed Assignment Rules and Visual Bed Board:*** *Standardize pull, eliminate "bed shopping"; set a* ***takt*** *based on arrival rate; and staff to demand.*
- ***Bedside Registration and Triage Standard Work:*** *Remove serial handoffs; limit re-triage; and standardize triage order sets.*
- ***Load Balancing:*** *Dynamic assignment to providers based on real-time WIP and even out panel sizes.*

The triage targeted improvements were similar to what the Lean Six Sigma team implemented. They focused first on the triage process, where if an ED bed was available, they moved the patient to the ED room, and then the ED nurse triaged the patient in the ED room. This eliminated the delays and bottlenecks related to lack of triage room and nurse capacity. They also incorporated standard protocols/order sets when the patient was moved to an ED room. Additionally, lower acuity patients (less ill patients with acuity-level 4) were moved to a fast-track area to increase capacity for higher-acuity patients (acuity levels 1, 2, and 3).

B. ***Diagnostic/Decision Phase (EDP Visit → Disposition)***

- ***Parallel Processing of Labs/Imaging****: Auto-fire critical orders at triage for qualifying chief complaints; shorten the critical path.*
- ***Consult SLA and Escalation****: Time-box consult response; define escalation tiers.*
- ***Diagnostic Stewardship****: Reduce low-yield tests; leverage clinical pathways to cut cycle time from* ***20/30 min*** *VA and compress the* ***~153 min*** *wait. [Value Stream Map ED | PDF]*

A new Lean Six Sigma lab process improvement project was implemented to address the long lab diagnostics times. Additionally, a fast-track pregnancy test was implemented as appropriate to reduce the time to imaging diagnostics. Moving lower acuity patients to a fast-track area for care also helped reduce the long lead times in the diagnostic area. Another CT scan Lean Six Sigma project was performed to develop protocols for reducing CT use, which also reduced potential CT overuse and diagnostic times.

C. ***Discharge Throughput***

- ***Discharge Huddles Each Hour****: Pre-identify likely discharges early; prep instructions and meds; and schedule transport before disposition.*
- ***Meds-to-Beds and Discharge Lounge****: Move final steps off the ED bed; free capacity sooner.*

The ED Lean Six Sigma team implemented patient transportation improvements that helped to reduce the discharge times. They also incorporated a new field to better track when the patient physically left the ED versus when they were removed from the ED Bed Board system, this reduced the time by about 30 min by incorporating more accurate data collection for the discharge process.

C. ***Boarding Reduction (Admit Pathway)***

- ***Daily Bed-Management Cadence****: Hospital-wide admit target windows; early* ***bed requests at decision****, not after all orders.*
- ***Inpatient Surge Protocol****: Flex units, "admit relief" teams, and prioritize early inpatient discharges (pre-11 a.m.) to create bed openings.*
- ***ED Observation/Short-Stay Unit****: Offload qualifying admits to OBS protocols to avoid long inpatient bed waits.*

The ED Lean Six Sigma team implemented a separate holding area for patients that were in ED observation or who would be transferred to a separate facility. They also implemented more standard protocols for calling admitting physicians, to reduce the call-out times, and to get admitting orders more quickly. They incorporated ED surge plans for when the ED arrivals were surging, to notify the floors and units that they would have increased admissions. They streamlined the ED to floor nurse reporting and hospital bed assignment processes.

(5) ***Expected impact (first pass)***

If we conservatively cut ***EDP→Disposition*** *delay by* ***60 min*** *and* ***Disposition→Admit*** *delay by* ***120 min****, you would see*:

- ***Discharged lead time*** *drop from* ***~325*** *to* ***~265 min*** *(VA share rises from* ***18%*** *→* ***~22%****).*
- ***Admitted lead time*** *drop from* ***~524*** *to* ***~404 min*** *(VA share from* ***15%*** *→* ***~20%****).*
- These changes typically translate to meaningful reductions in LWBS, left-before-triage, and crowding pressure on arrival/triage.

The actual improvements were about 30% to 246 min for the discharged patient lead times, and 29% to 366 min for the admitted patients. The Left Without Being Seen (LWBS) percent decreased from 6.5 to 0.6%. Patient capacity increased from an average of about 85–100 patients.

(6) ***Measurement & control***

- ***CTQs to Track Weekly****: Time to bed, time to provider, decision time, discharge processing time, and boarding minutes (admit).*
- ***SPC/Run Charts****: Plot medians, 80th percentile, and special-cause detection; publish during ED flow huddles.*
- ***PDSA Cadence****: 2-week cycles focused on one bottleneck at a time (start with* ***EDP → Disposition*** *and* ***boarding****).*

The scorecard metrics that were tracked are shown in Fig. 14.3. It includes the total time to EDP, treat/diagnose to disposition time, and disposition to leave ED. The baseline, target, and implemented improvement times along with the percentage improvements are shown in the scorecard. Additionally dashboards with alerts for daily control were incorporated into the ED. Statistical process control charts were proposed as control mechanisms but were not implemented. Additional PDSA improvements were incorporated after the initial pilot implementation to continue improvements.

(7) ***A couple of clarifications to fine-tune the calculations***

1. *In your PDF there are duplicate metrics for the first and second steps (e.g.,* ***80.66 vs. 59.41*** *for time to bed;* ***28.873 vs. 33.315*** *for bed → provider) and*

Metrics	Baseline		Target		Improved		% Improvement	
	Admitted	Discharged	Admitted	Discharged	Admitted	Discharged	Admitted	Discharged
Total LOS:	8.7 hrs	5.8 hrs	5 hours (43%)	3 (48%)	6.1 hrs	4.1 hrs	30%	29%
% LWBS:	6.50%		3.50%		0.60%		91%	
Total Time to EDP:	93	109	35	35	32		66%	
Triage Time:	34	34	15	15	13		62%	
Wait time triage to EDP:	65	79	20	20	25		62%	
Time to ED Bed:	61	81	35	42	16		74%	
Bed to Physician:	34	55	20	20	16		53%	
Treat/Diagnose to Disposition Time:	192	175	109	91	166		14%	
Radiology Diagnostic Time:	138	270	154	154	170		-23%	
Lab Time:	119				57		109%	
EDP Time Visit:	192	175	109	91	175		9%	
Disposition to Leave ED	223	64	127	33	171	47	23%	27%
Disposition to get admit orders:	103		59		70		32%	
Total Call Out Time:	41		23		24			
Call Out to Physician	13		7		5			
Physician Call Back	29		17		19			
Wait After Call Back:	37		21		46			
Admit Orders to Assign Bed:	42		24		44		-5%	
Assign Bed to Leave ED:	112		64		67		40%	

Fig. 14.3 Actual ED scorecard

for ***EDP → Disposition (173.21 vs. 182.63)****. Are these* ***triage-to-bed versus arrival-to-bed****, and* ***discharged versus admitted*** *splits? If so, my pairing above should be correct; if not, please share the intended mapping. [Value Stream Map ED | PDF]*

2. *What time period and* ***sample size (n)*** *do these means/stdevs represent (weekday/weekend splits, seasonality)?*
3. *Do you segment by* ***ESI*** *and* ***arrival mode*** *(walk-in versus EMS), or by* ***chief complaint****? Variability suggests those subflows might behave very differently.*

If you can confirm those points (and, ideally, your ***daily arrivals*** *and* ***admit rate****), I'll produce a concise dashboard with updated VA/NVA %, CVs by segment, and a prioritized improvement ROI model tailored to your ED.*

> We did not request more detailed analysis based on arrival mode, daily arrivals, or admit rate from Co-Pilot, although the Lean Six Sigma team did perform detailed regression and Analysis of Variance (ANOVA) on the ED Bed Board data.

PROMPT

"What would be the best metrics to use?"

Co-Pilot

Great question! For an ED Value Stream Map, the best metrics are those that capture ***flow efficiency, variability, and patient impact****. Here are the key ones I recommend:*

1. ***Lead Time Metrics***
    - ***Total Lead Time (Arrival → Disposition)***
      Measures overall patient journey time.
    - ***Segment Lead Times***
      Break down by steps:
        - *Arrival/Triage → Bed*
        - *Bed → Provider*
        - *Provider → Disposition*
        - *Disposition → Discharge/Admit*
2. ***Value-Added vs Non-Value-Added***
    - ***Cycle Time Per Step (VA)***
      Actual work time (e.g., triage, provider exam).
    - ***Delay Time Per Step (NVA)***
      Waiting time between activities.

- ***% Value-Added***

$$\text{VA\%} = \frac{\text{Total Cycle Time}}{\text{Total Lead Time}} \times 100$$

3. ***Flow & Variability***

- ***Coefficient of Variation (CV)***

$$\text{CV} = \frac{\text{Stdev}}{\text{Mean}}$$

High CV indicates unstable processes.

- ***80th/90th Percentile Times***
  Better than averages for patient experience.

4. ***Throughput & Capacity***

- ***Patients per hour (arrival rate)***
- ***Bed turnover time***
- ***Provider-to-patient ratio***

5. ***Outcome & Quality***

- ***LWBS (left without being seen)***
- ***Door-to-Provider Time***
- ***Boarding Time (Disposition → Admit)***
- ***ED length of stay (LOS)***

6. ***Improvement Tracking***

- ***Takt Time versus Cycle Time***
  Align staffing to demand.
- ***First-Time Quality***
  % of patients with no rework (e.g., repeat triage).

*Would you like me to: ☑ Build a **dashboard template** with these metrics (including VA/NVA and CV calculations)?*

*☑ Or create a **prioritized KPI list** for your ED improvement project with targets and benchmarks?*

The proposed co-pilot dashboard is shown in Fig. 14.4. The actual ED dashboard used similar metrics.

Emergency Department Flow Dashboard Template

1. ***Executive Summary***

	Metric	Target	Current	Trend
1	Door-to-Triage (min)	≤ 10	[Insert]	↑/↓
2	Door-to-Bed (min)	≤ 30	[Insert]	↑/↓
3	Door-to-Provider (min)	≤ 30	[Insert]	↑/↓
4	ED Length of Stay – Discharge	≤ 240	[Insert]	↑/↓
5	ED Length of Stay – Admit	≤ 360	[Insert]	↑/↓
6	Boarding Time (Disposition→Admit)	≤ 120	[Insert]	↑/↓
7	LWBS %	< 2%	[Insert]	↑/↓

Fig. 14.4 Co-pilot proposed ED dashboard

- ***Date Range***: *[Insert]*
- ***Total ED Visits***: *[Insert]*
- ***Admit Rate***: *[Insert]*
- ***LWBS Rate***: *[Insert].*

Tables 14.3, 14.4, 14.5 and 14.6 show the Co-Pilot generated scorecard metrics.

2. ***Key Performance Indicators (KPIs)***
3. ***Flow Efficiency Metrics***
4. ***Variability and Reliability***
5. ***Percentile Performance***
6. ***Outcome and Quality Metrics***
   - ***LWBS Count and %***: *[Insert]*

Table 14.3 Key performance indicators

Metric	Target	Current	Trend
Door-to-triage (min)	≤10	[Insert]	↑/↓
Door-to-bed (min)	≤30	[Insert]	↑/↓
Door-to-provider (min)	≤30	[Insert]	↑/↓
ED length of stay—discharge	≤240	[Insert]	↑/↓
ED length of stay—admit	≤360	[Insert]	↑/↓
Boarding time (disposition → admit)	≤120	[Insert]	↑/↓
LWBS %	<2%	[Insert]	↑/↓

Table 14.4 Flow efficiency metrics

Step	Mean (min)	Cycle (VA)	Delay (NVA)	% VA
Arrival/Triage → Bed	[Insert]	[Insert]	[Insert]	[Insert]
Bed → Provider	[Insert]	[Insert]	[Insert]	[Insert]
Provider → Disposition	[Insert]	[Insert]	[Insert]	[Insert]
Disposition → Discharge/Admit	[Insert]	[Insert]	[Insert]	[Insert]

Total Lead Time: [Insert] min
% **Value-Added**: [Insert]%
% **Non-Value-Added**: [Insert]%

Table 14.5 Variability and reliability

Step	Mean	Stdev	CV
Arrival/Triage → Bed	[Insert]	[Insert]	[Insert]
Bed → Provider	[Insert]	[Insert]	[Insert]
Provider → Disposition	[Insert]	[Insert]	[Insert]
Disposition → Discharge/Admit	[Insert]	[Insert]	[Insert]

Interpretation: CV > 1 indicates high variability

Table 14.6 Percentile performance

Metric	Median	80th %	90th %
Door-to-bed	[Insert]	[Insert]	[Insert]
Door-to-provider	[Insert]	[Insert]	[Insert]
ED LOS—discharge	[Insert]	[Insert]	[Insert]
ED LOS—admit	[Insert]	[Insert]	[Insert]

- ***Return Visits Within 72 h***: *[Insert]*
- ***Patient Satisfaction Score***: *[Insert].*

7. ***Visuals***

    - ***Run Charts***: *Door-to-provider, LOS trends.*
    - ***Pareto Chart***: *Delay contributors.*
    - ***Heat Map***: *Hourly arrival versus staffing.*

Usage Notes

- *Update weekly or monthly.*
- *Highlight steps with highest delay and CV for targeted improvement.*
- *Include color coding (green = on target, yellow = watch, red = off target).*

Would you like me to ***add sample data and charts*** *to make this dashboard ready for presentation?*

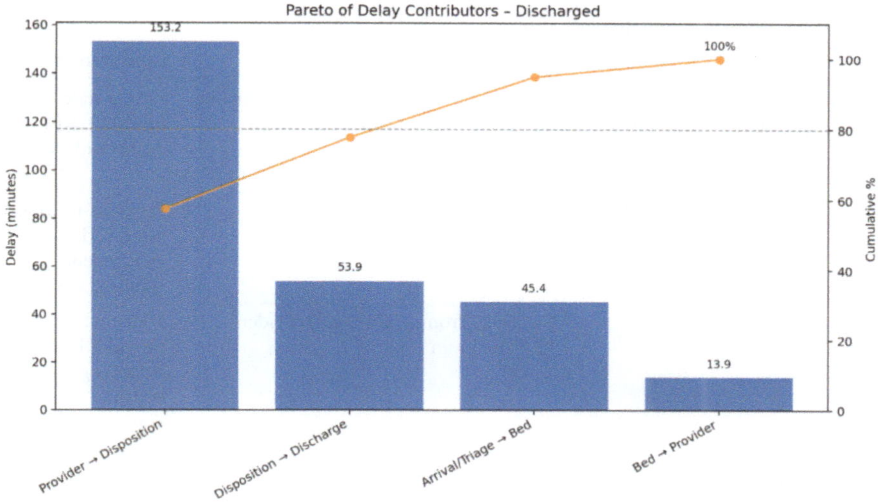

Fig. 14.5 Pareto chart of delays

Figure 14.5 shows the Co-Pilot generated Pareto Chart of the delay times by activity.

PROMPT: Over the next 12 months, create a schedule for achieving a future state for monthly Kaizen events.

12-Month Kaizen Event Schedule for Achieving Future State VSM

Goal: Transition from baseline ED performance to the modeled future state (LOS Discharge ≤ 240 min, LOS Admit ≤ 360 min, Door-to-Provider ≤ 30 min, Boarding ≤ 120 min) through monthly Kaizen events.

Schedule Overview

- ***Duration***: *12 months.*
- ***Cadence***: *1 Kaizen event per month (3–5 days each).*
- ***Focus***: *Sequentially address major bottlenecks and supporting processes.*

Month-by-Month Plan

The month-by-month plan is shown in Table 14.7.

Event Structure

Table 14.7 Month-by-month plan

Month	Focus area	Objectives	Key metrics	Deliverables
1	Kickoff and baseline review	Validate current VSM, set targets, form Kaizen teams	LOS, door-to-provider, boarding	Future state VSM finalized, KPI dashboard live
2	Arrival/triage → bed	Reduce time from 59.4 → 30 min	Door-to-bed	Provider-in-triage pilot, bed assignment standard work
3	Bed → provider	Cut delay from 13.8 → 5 min	Door-to-provider	Dynamic provider load balancing, bedside registration
4	Provider → disposition (diagnostics)	Streamline labs/ imaging	Provider → disposition	Parallel processing workflow, consult SLA
5	Provider → disposition (consults)	Reduce consult delays	Consult turnaround	Escalation protocol, consult response timer
6	Disposition → discharge	Target ≤ 30 min	Discharge cycle	Discharge lounge, meds-to-beds process
7	Disposition → admit (boarding)	Reduce boarding from 223 → 100 min	Boarding time	Bed management cadence, early bed request process
8	Observation unit launch	Offload admits	OBS utilization	OBS unit SOP, staffing plan
9	Staffing alignment	Match takt to demand	Takt versus staffing	Heat map analysis, flex staffing plan
10	IT and automation	Optimize EHR workflows	Order-to-result time	Auto-fire triage orders, alert system
11	Sustainment and control	SPC charts, audits	KPI stability	Control plan, audit checklist
12	Review and scale	Evaluate results, plan next cycle	LOS, VA%	Executive report, next-year Kaizen roadmap

- ***Pre-work***: *Data collection, stakeholder alignment.*
- ***Kaizen Week***: *Map current process, root cause analysis, implement quick wins.*
- ***Post-work***: *Monitor metrics, adjust SOPs, prep. for next event.*

Next Steps

1. *Confirm Kaizen team composition (ED leadership, nurses, providers, ancillary services).*
2. *Provide hourly arrival and staffing data for Month 9 heat map analysis.*
3. *Schedule leadership report-outs after Months 6 and 12.*

Would you like me to ***create a Gantt chart*** *for this 12-month Kaizen roadmap or* ***prepare a PPT summary*** *for leadership presentation?*

The 12-month Gantt chart is shown in Fig. 14.6.

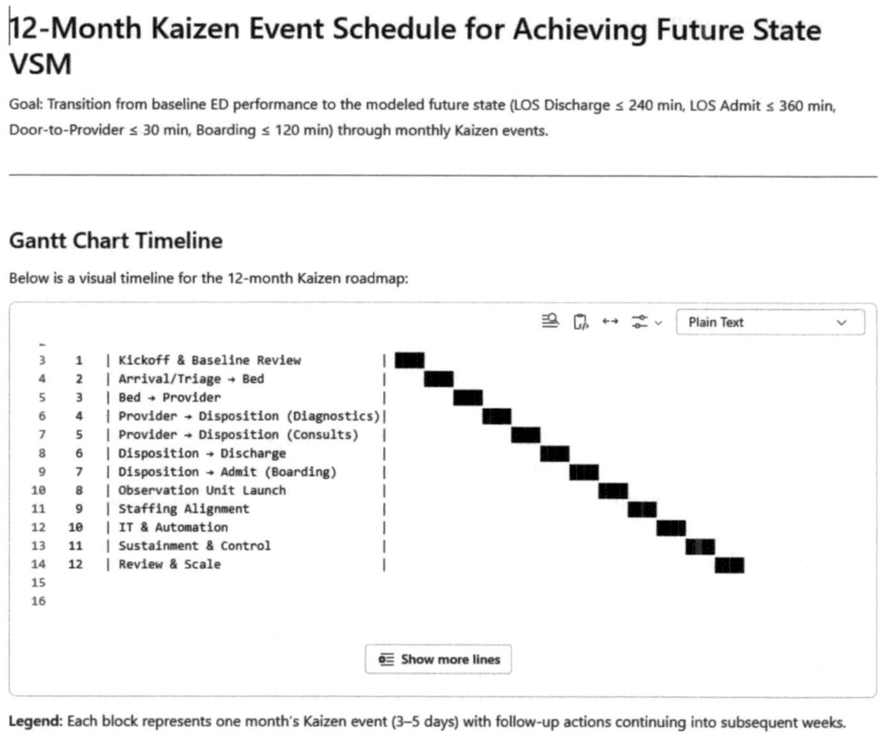

12-Month Kaizen Event Schedule for Achieving Future State VSM

Goal: Transition from baseline ED performance to the modeled future state (LOS Discharge ≤ 240 min, LOS Admit ≤ 360 min, Door-to-Provider ≤ 30 min, Boarding ≤ 120 min) through monthly Kaizen events.

Gantt Chart Timeline

Below is a visual timeline for the 12-month Kaizen roadmap:

Plain Text

```
1  | Kickoff & Baseline Review                |
2  | Arrival/Triage → Bed                     |
3  | Bed → Provider                           |
4  | Provider → Disposition (Diagnostics)     |
5  | Provider → Disposition (Consults)        |
6  | Disposition → Discharge                  |
7  | Disposition → Admit (Boarding)           |
8  | Observation Unit Launch                  |
9  | Staffing Alignment                       |
10 | IT & Automation                          |
11 | Sustainment & Control                    |
12 | Review & Scale                           |
```

Show more lines

Legend: Each block represents one month's Kaizen event (3–5 days) with follow-up actions continuing into subsequent weeks.

Fig. 14.6 Kaizen Gantt chart for future state VSM. *Notes* • Events are sequential but allow overlap for sustainment activities; • Use color coding in PPT or Excel for clarity (green = completed, yellow = in progress, red = upcoming)

Assessment of Co-Pilot Use for the Emergency Department

The graphics in Co-Pilot are very limited. We asked for a future state value stream map to be developed, but it didn't create a usable one.

Overall the Co-Pilot gave helpful improvement recommendations. However, the Lean Six Sigma team is still going to need to understand the Lean and Six Sigma concepts, tools, and recommendations to be able to successfully train others and implement these changes. Additionally, the project leader and manager will need to know how to make change happen successfully, and get over resistance to change from process owners.

Application of the Co-Pilot AI Tool with a Process Architecture Map (PAM) for the Operating Room (OR) Turnaround and On-Time Starts

The Process Architecture Map (PAM) was developed by the author to integrate a process map with the information and process architecture elements, including the roles of who performs each activity, the technology and information systems, and the information used in each activity. Each activity is identified as either value-added (a green check mark), limited value-added (yellow triangle) or non-value-added (red X). Additionally, the PAM incorporates a lean analysis, combining a value analysis with a waste analysis. The OR Turnaround and On-Time Starts process maps are shown in Figs. 14.7, 14.8, 14.9 and 14.10. Figure 14.11 illustrates the value analysis information provided to Co-Pilot. Figure 14.7 shows the process architecture map with the lists of stakeholders, information, and technology used across the process maps.

The lean analysis reports identify the lean wastes for each limited value-added and non-value-added activities, shown in Figs. 14.12, 14.13 and 14.14.

Following is the prompt provided along with the PAM and the value analysis Information (Fig. 14.11):

PROMPT: Evaluate the process map for the operating turnaround and on-time start cases.

Following is the algorithm performed by Co-Pilot, in Fig. 14.15.

Here's a structured evaluation of the ***Operating Room Turnaround*** *and* ***On-time Starts for OR Cases*** *process maps you shared, focusing on waste, bottlenecks, role*

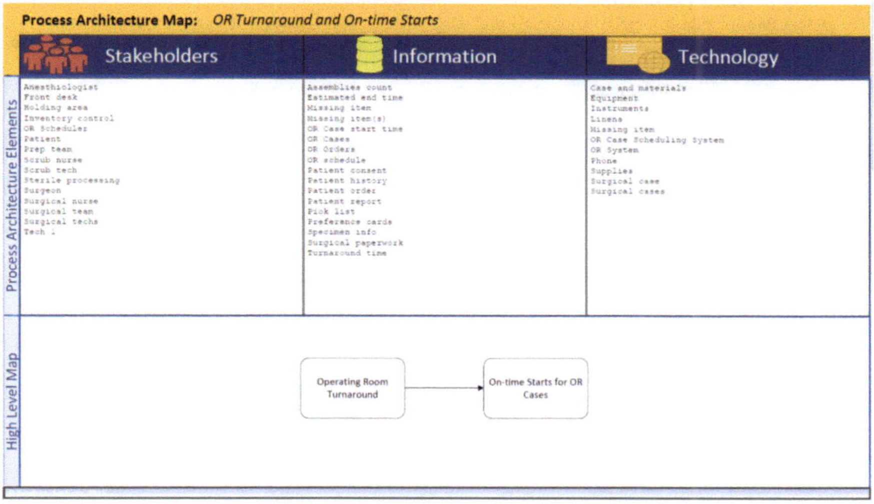

Fig. 14.7 Process architecture map—OR turnaround and on-time starts

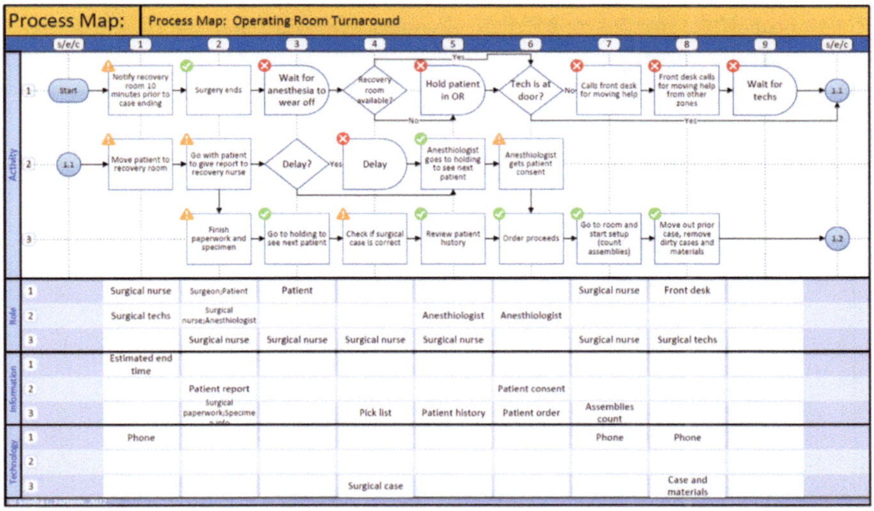

Fig. 14.8 Process map—operating room turnaround—Page 1

clarity, and actionable improvements. I also overlay the ***value analysis*** *lens (value-added, limited value-added, non-value-added) to highlight where effort should shift. [Process Ma…e Analysis | PDF], [Value Analysis | Word]*

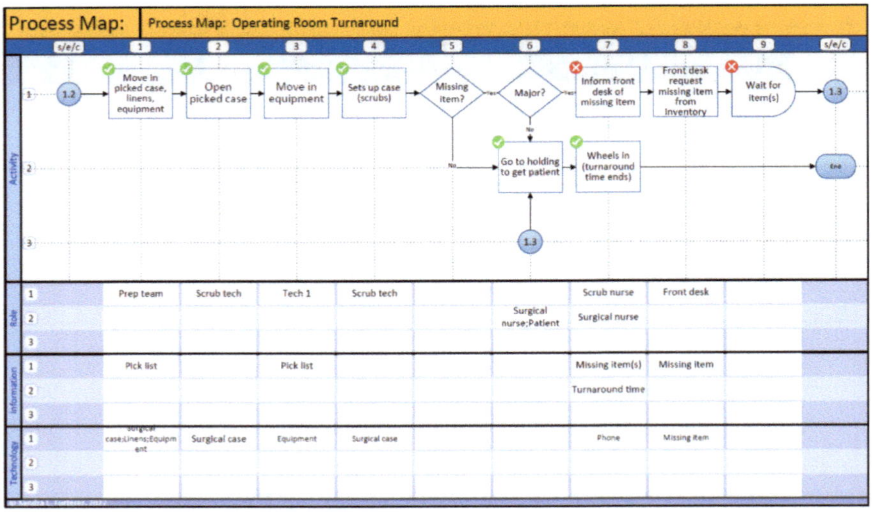

Fig. 14.9 Process map—operating room turnaround—Page 2

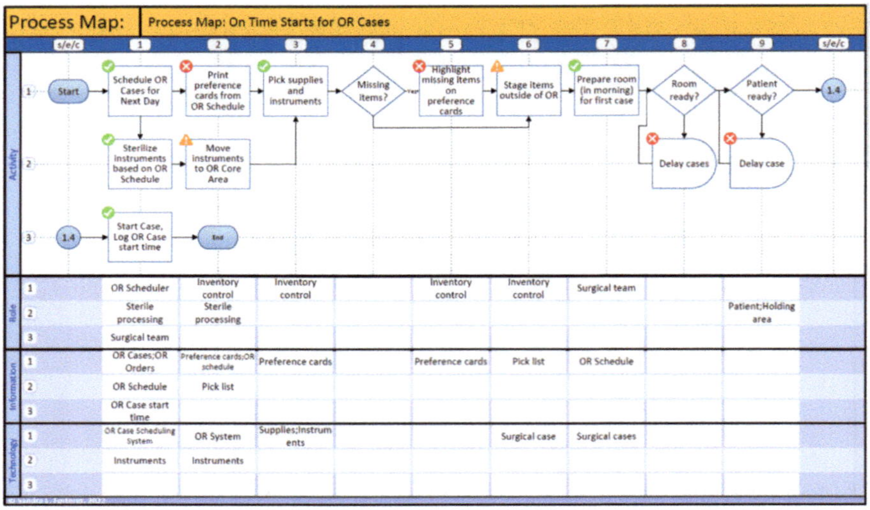

Fig. 14.10 Process map—on-time starts

Executive summary (what's working vs. where to focus)

Strengths

- *Clear swimlanes and role handoffs across anesthesia, nursing, techs, front desk, inventory control, sterile processing, and scheduling; decision points are explicit ("Room ready?", "Patient ready?"). [Process Ma…e Analysis | PDF]*

Value Analysis

- Value added activities: Activities that the customer would pay-for, that add value for the customer
- Limited value added activities: Those activities that are necessary, such as regulatory activities, that don't add value from a customer perspective.
- Non-value added activities: Activities that the customer would not want to pay-for, or don't add value for the customer. Some are necessary, such as for legal, financial reporting, documentation reasons, others are unnecessary and should be reduced or eliminated.

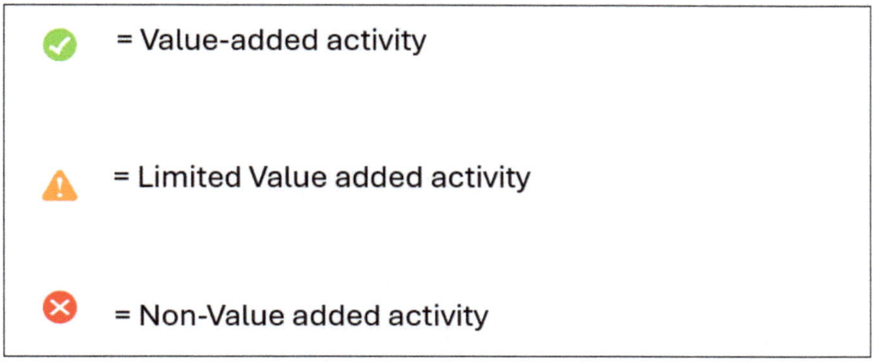

Fig. 14.11 Value analysis information provided to Co-Pilot

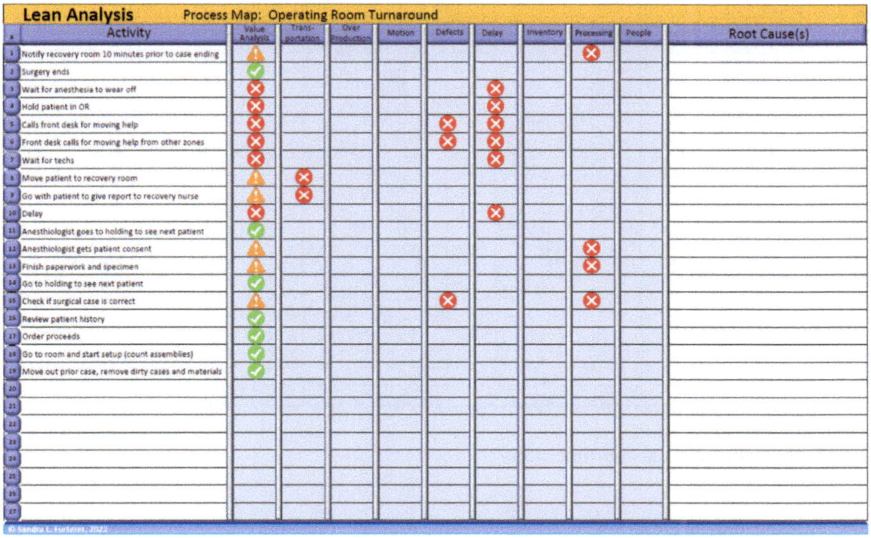

Lean Analysis — Process Map: Operating Room Turnaround

#	Activity	Value Analysis	Trans-portation	Over Production	Motion	Defects	Delay	Inventory	Processing	People	Root Cause(s)
1	Notify recovery room 10 minutes prior to case ending	⚠							✗		
2	Surgery ends	✓									
3	Wait for anesthesia to wear off	✗					✗				
4	Hold patient in OR	✗					✗				
5	Calls front desk for moving help	✗				✗	✗				
6	Front desk calls for moving help from other zones	✗				✗	✗				
7	Wait for techs	✗					✗				
8	Move patient to recovery room	⚠	✗								
9	Go with patient to give report to recovery nurse	⚠	✗								
10	Delay	✗					✗				
11	Anesthiologist goes to holding to see next patient	✓									
12	Anesthiologist gets patient consent	⚠							✗		
13	Finish paperwork and specimen	⚠							✗		
14	Go to holding to see next patient	✓									
15	Check if surgical case is correct	⚠				✗			✗		
16	Review patient history	✓									
17	Order proceeds	✓									
18	Go to room and start setup (count assemblies)	✓									
19	Move out prior case, remove dirty cases and materials	✓									
20											
21											
22											
23											
24											
25											
26											
27											

Fig. 14.12 Lean analysis—operating room turnaround Page 1

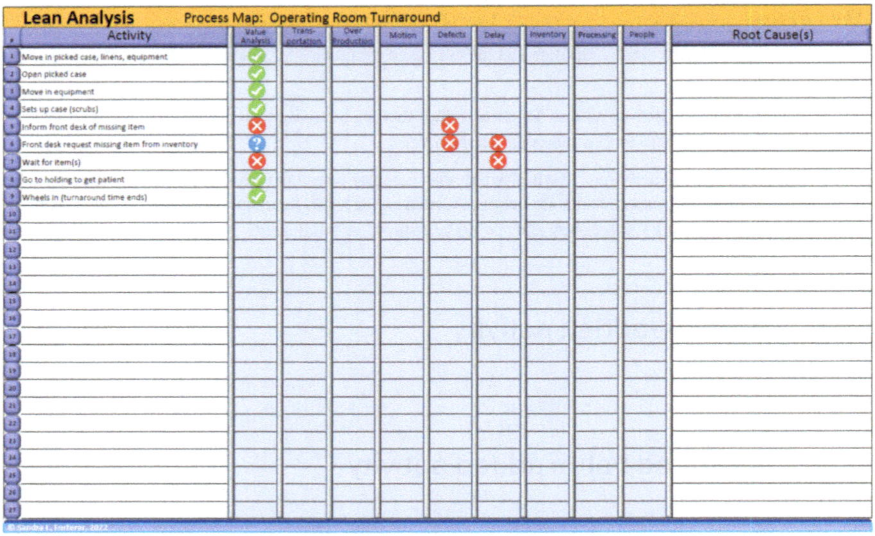

Fig. 14.13 Lean analysis—operating room turnaround—Page 2

Fig. 14.14 Lean analysis—on-time starts

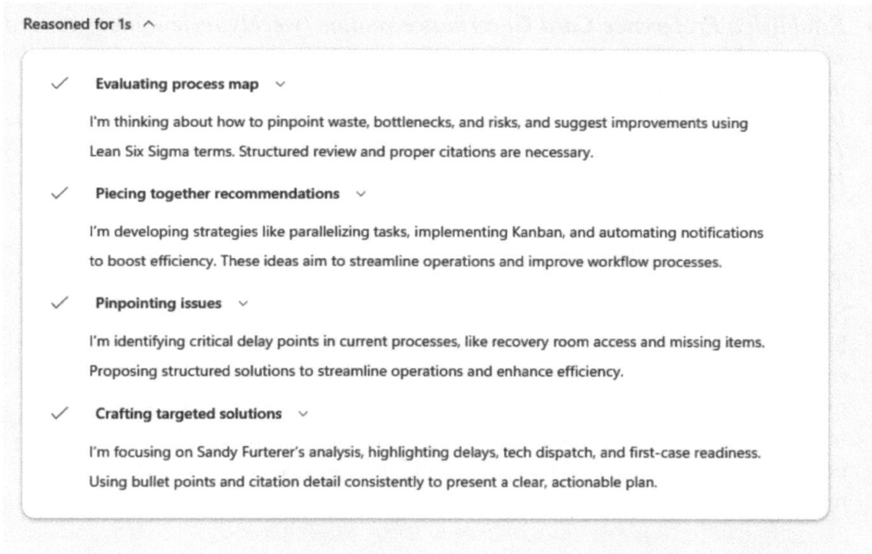

Fig. 14.15 OR turnaround Co-Pilot algorithm

- *Upstream activities for on-time starts (preference cards, picking, staging, sterilization, and moving instruments to the OR core) are spelled out—this enables proactive readiness management. [Process Ma…e Analysis | PDF]*

Top Constraints/Delay Drivers

1. ***Transport and recovery availability*** *cause waiting after case end (front desk calls for help, techs not at door, recovery room may not be available). [Process Ma…e Analysis | PDF]*
2. ***Missing items*** *discovered during setup (inform front desk → request from inventory → wait), which can push "wheels in" and extend turnaround. [Process Ma…e Analysis | PDF]*
3. ***First-case readiness gates*** *("Room ready?" and "Patient ready?") that frequently trigger* ***delay cases*** *or* ***delay case****, suggesting upstream variability in room prep. or patient flow. [Process Ma…e Analysis | PDF]*

This is a good assessment of where the main inefficiencies in the OR turnaround process exist.

Three Highest Impact Fixes

- *Create a* ***dedicated, pre-assigned transport dispatch*** *with an automated "10-min" alert, SLAs, and escalation visual management. [Process Ma…e Analysis | PDF]*

- *Establish a* ***Preference Card Governance*** *routine (weekly review, data stewardship) and* ***kit completeness checks*** *the afternoon before surgery to surface missing items* ***before*** *room setup. [Process Ma…e Analysis | PDF]*
- *Implement a* ***First-Case Readiness Checklist*** *(room + patient) with a 24-h and 1-h pre-op gate, supported by huddles and on signaling to resolve blockers early. [Process Ma…e Analysis | PDF]*

The Co-Pilot suggested improvements are sound. The actual OR Lean Six Sigma team did incorporate the 10-min alerts, dashboard alerts to identify bottlenecks with visual management throughout the OR. They also incorporated preference card updates by surgeons on a more continual basis. Improvements to reduce missing items related to surgical case items and sterilized instruments were also incorporated. They also incorporated a first-case readiness checklist attached to the pick lists and the surgical cases. These checklists made it easy to verify missing items the day prior and the day of the case.

Current State Observations (By Flow) and Waste Analysis

A. ***Operating Room Turnaround (post-case to "wheels in")***

   - ***Notification + Handoff:*** *"Notify recovery room 10 min prior to case ending"; if recovery not available or techs aren't at the door, nurse calls the front desk; front desk calls other zones; team waits for techs. Primary wastes: Delay, Transportation, Motion. [Process Ma…e Analysis | PDF]*
   - ***Patient Transfer:*** *Move patient to recovery; surgical nurse and anesthesia give report; then anesthesia heads to holding for next patient; nurse completes paperwork/specimen and goes to holding. Risk: parallelization not fully exploited; paperwork may be internal setup time. [Process Ma…e Analysis | PDF]*
   - ***Room Turnover and Next-Case Prep.:*** *Nurse checks case correctness and patient history; order proceeds; start setup (count assemblies); techs move out dirty materials. Wastes: Motion/Defects if assemblies are miscounted or case data mismatches. [Process Ma…e Analysis | PDF]*
   - ***Setup and Missing Items Loop:*** *Move in picked case/linens/equipment, open case, setup; if items are missing, inform front desk → inventory control → wait; nurse goes to holding to get patient;* ***turnaround ends at "wheels in."*** *This late discovery is the dominant driver of extended TAT. [Process Ma…e Analysis | PDF]*

The current state observations were sound.

Value Analysis Overlay (Examples)

- ***Value-Added***: *"Move patient to recovery and give report," "Start setup (count assemblies)," "Wheels in" (directly advances the case for the patient). [Process Ma...e Analysis | PDF], [Value Analysis | Word]*
- ***Limited Value-Added***: *Required documentation (paperwork/specimen), regulatory checks (patient history verification)—necessary, but not patient-perceived value. [Process Ma...e Analysis | PDF], [Value Analysis | Word]*
- ***Non-Value-Added***: ***Waiting for techs*** *and* ***waiting for missing items****; aim to eliminate or minimize through design. [Process Ma...e Analysis | PDF], [Value Analysis | Word]*

The value analysis summary is sound.

B ***On-time Starts for OR Cases (Day Before to Case Start)***

- ***Schedule and Prepare***: *OR Scheduler sets next day's cases; Inventory Control prints preference cards, picks supplies/instruments, highlights missing items, stages items outside OR; Sterile Processing sterilizes per schedule and moves instruments to OR core. [Process Ma...e Analysis | PDF]*
- ***Morning Prep. and Gates***: *Surgical team prepares room for first case;* ***room ready?*** *If not,* ***delay cases***; ***patient ready?*** *If not,* ***delay case***; *start case and log start time when both are "go." [Process Ma...e Analysis | PDF]*

Waste hot-spots

- ***Information Timing***: *Highlighting missing items on preference cards is good, but if fulfillment isn't closed before morning, the downstream setup still waits. Delay/Defects in the info → action loop. [Process Ma...e Analysis | PDF]*
- ***SPD Capacity and Timing***: *Sterilization aligned to schedule is critical; any lateness cascades into room-ready delays. Delay/Inventory (work-in-process). [Process Ma...e Analysis | PDF]*
- ***Patient Readiness***: *Late consent/clearance or holding-area delays push start times. Delay/Defects in pre-op process. [Process Ma...e Analysis | PDF]*

The waste hot-spots summary is sound.

Role Clarity and Handoffs (Where RACI Can Tighten)

- ***Front Desk*** *acts as dispatch for transport and missing-item requests; establishing response SLAs and escalation criteria will reduce "Wait for techs" and "Wait for item(s)" loops. [Process Ma...e Analysis | PDF]*

- ***Inventory Control** owns picking and staging; add **kit completeness verification** and a **time-bound replenishment window** the day before (e.g., by 16:00) to prevent morning discovery. [Process Ma…e Analysis | PDF]*
- ***Sterile Processing** synchronizes instrument sterilization with schedule; publish a **readiness board** to the OR core by case so room teams can see instrument status visually. [Process Ma…e Analysis | PDF]*
- ***Anesthesia + Surgical Nurse**: Move consent and patient readiness checks **earlier** (pre-op clinic or holding) to minimize morning gate failures. [Process Ma…e Analysis | PDF]*

Actionable Improvements (Lean/Six Sigma + OR operations)

(1) ***Turnaround Time (TAT) Reduction—Apply SMED and Parallelization***

- *Convert Internal Tasks to External: Complete paperwork/specimen and verify next case details **outside** the room while the prior patient is in recovery; start staging equipment and supplies **before** the room is free. [Process Ma…e Analysis | PDF]*
- ***Pre-assigned Transport Tech** with an automated **"T-10 min" alert** (from nurse/anesthesia) and visual timers; set a **response SLA*** (e.g., *$\leq$3 min) and track. [Process Ma…e Analysis | PDF]*
- ***Standard Work** for room turnover: clear sequence—remove dirty materials → bring in picked case → equipment positioning → open case—validated* via *a checklist to catch defects early. [Process Ma…e Analysis | PDF]*

The OR Lean Six Sigma incorporated standard work for room turnover, and developed a video, and verification within the performance reviews of the OR techs.

(2) ***Missing-Items Prevention—Preference Card Governance***

- *Weekly cross-functional review (surgeon/nurse/SPD/inventory) of preference cards and **pick lists**; commit to **one source of truth** and sunset duplicates. [Process Ma…e Analysis | PDF]*
- ***Kit Completeness** check by late afternoon (e.g., 16:00) with an **andon** trigger to resolve shortages same day; no case goes green unless the kit is complete. [Process Ma…e Analysis | PDF]*
- *Use **barcode/Kanban** for standard consumables; pre-assemble frequent procedure kits to reduce* ad hoc *picking variability. [Process Ma…e Analysis | PDF]*

> The OR Lean Six Sigma team incorporated a 5S project to remove unneeded tables, equipment, and supplies, as well as standardizing supply carts, OR room cabinets, and sub-sterile rooms. They assigned responsibilities for re-stocking of the carts, cabinets, and rooms.

(3) ***First-Case On-Time Starts—Robust Readiness Gates***

- ***T-24 h:*** *Case scheduled, instruments queued, supplies staged, consent verified;* ***T-1 h****: room prepped, SPD "ready," patient in holding, anesthesia assessment done; any red → visible escalation. [Process Ma...e Analysis | PDF]*
- *Conduct a* ***daily huddle*** *(charge nurse + anesthesia + SPD + inventory + front desk) to clear reds before start times. [Process Ma...e Analysis | PDF]*

The OR Lean Six Sigma team did incorporate a daily huddle to clear reds before start times.

(4) ***Visual Management and Flow***

- ***OR Core Readiness Board*** *listing each case with green/yellow/red for* ***room, patient, instruments, supplies, transport****; owners and SLA clocks visible. [Process Ma...e Analysis | PDF]*
- ***Transport Dispatch Board*** *with zones and techs; eliminate "call other zones" delays through capacity balancing. [Process Ma...e Analysis | PDF]*

The OR Lean Six Sigma team did incorporate tech zones and responsibilities.

Measurement: What to Track (Baseline → Control)

- ***Turnaround Time (TAT)****: Median and 90th percentile from "surgery ends" to "wheels in"; stratify by service, room, and time of day. [Process Ma...e Analysis | PDF]*
- ***On-Time Start %*** *(first case and all cases) with* ***reason codes*** *(room not ready, patient not ready, instrument late, missing items, transport delay). [Process Ma...e Analysis | PDF]*
- ***Missing Item Rate*** *per case and* ***preference card error rate; transport response time*** *from alert to arrival;* ***SPD readiness by case*** *(planned versus actual). [Process Ma...e Analysis | PDF]*
- *Use* ***Pareto charts*** *for delay reasons and* ***control charts*** *for TAT; review weekly in an OR operations huddle. [Process Ma...e Analysis | PDF]*

Turnaround Time (TAT), on-time start %, and OR delays were tracked.

Future State (High Level)

1. ***Pre-day Readiness***: *Preference card verified, kit complete, instruments sterilized and at core, patient consent confirmed—green status by 16:00. [Process Ma…e Analysis | PDF]*
2. ***Post-case***: *Automatic T-10 transport alert; nurse/anesthesia complete report while transport is en route; paperwork/specimen handled externally. [Process Ma…e Analysis | PDF]*
3. ***Room Turnover***: *Dirty materials out; staged supplies/equipment in; quick verification → open case; any shortage triggers immediate andon to inventory control. [Process Ma…e Analysis | PDF]*
4. ***First-Case Gate***: *Room and patient both green; start on time and log. [Process Ma…e Analysis | PDF]*

Co-Pilot developed a 30–60–90-day improvement plan.

Suggested 30–60–90-Day Plan

- ***Days 0–30 (Assess and Stabilize)***
  - *Time studies for TAT; instrument and transport response baselines. [Process Ma…e Analysis | PDF]*
  - *Stand up* ***daily readiness huddle*** *and visual boards; pilot transport SLAs in two rooms. [Process Ma…e Analysis | PDF]*
- ***Days 31–60 (Fix Flow and Pre-empt Shortages)***
  - *Launch* ***Preference Card Governance****; implement* ***kit completeness*** *check by 16:00; start andon for shortages. [Process Ma…e Analysis | PDF]*
  - *SMED-style standard work for turnover; externalize paperwork/specimen tasks. [Process Ma…e Analysis | PDF]*
- ***Days 61–90 (Lock In and Scale)***
  - *Expand transport dispatch model and readiness boards to all ORs; weekly Pareto and control chart review to drive continuous improvement. [Process Ma…e Analysis | PDF]*

Notes on Value Analysis Conventions Used

- ***Value-Added***: *Activities patients would pay for and that directly advance their care.*
- ***Limited Value-Added***: *Necessary (e.g., regulatory) but not patient-perceived.*
- ***Non-Value-Added***: *Waiting, rework, or unnecessary motion—reduce or eliminate. [Value Analysis | Word]*

A Few Clarifying Questions to Tailor the Redesign

1. *What's your current **median and 90th percentile TAT** by service line and room? [Process Ma...e Analysis | PDF]*
2. *First-case **on-time start %** last quarter, and the top three delay reasons? [Process Ma...e Analysis | PDF]*
3. *How many transport techs per shift, and do you use a **central dispatch** or **zone model**? [Process Ma...e Analysis | PDF]*
4. *Who owns **preference card data quality** today, and how often are cards audited? [Process Ma...e Analysis | PDF]*
5. *What are **SPD cut-off times** for sterilization batches, and do you publish instrument readiness by case? [Process Ma...e Analysis | PDF]*

*If you'd like, I can turn this into a concise **future-state checklist** for the team or a **one-page visual** you can use in huddles.*

*Your visual checklist is ready! It includes **Pre-Day Readiness**, **Morning Readiness**, **Turnaround Process**, and **Visual Management Tools**, plus key metrics for tracking performance.*

*Would you like me to **add color-coded icons** ☑, ⚠, ✖ for GREEN/YELLOW/RED statuses, or create a **flowchart version** for quick reference during huddles?*

Conclusions

The analysis and improvements for the OR turnaround and on-time starts generated by Co-pilot related to the PAM were not quite as detailed as that for the ED using a VSM. This is most likely due to the more detailed activity times and statistics available on the VSM versus the process maps.

The Co-Pilot evaluation, analysis, and proposed improvements gave sound and thoughtful considerations and potential improvement ideas. Most of these ideas were generated and implemented by the ED and OR Lean Six Sigma teams prior to this analysis. Co-Pilot can generate improvement plans that can help the project managers more quickly develop the workplans for piloting and implementing improvements. The AI analysis can also help to ask thoughtful questions that the team can use in brainstorming discussions, and as validation that the team is going in the right direction with their potential improvements. The AI tools do not eliminate or reduce the need for the Lean Six Sigma teams and process analysts or industrial engineers being trained in Lean Six Sigma and process improvement tools. It will be important for the teams to remain engaged to ensure that the improvements are embraced by the process owners. These AI tools are a good sanity check, and easy way to summarize improvement ideas, metrics, and Lean best practice areas for the teams to consider. Additionally, it is important to avoid the thinking that process experts are not needed, and that process owners can just blindly attempt to implement what AI gives as solutions. Even benchmarking other processes and industries can provide insights, but it doesn't mean that it is appropriate for every environment and process.

Acknowledgements The author would like to thank Dr. Elizabeth Cudney, Ph.D., for her expertise and insights on leveraging AI for this chapter. The Co-Pilot AI tool was used for this analysis that is shown in italics throughout this chapter.

References

Adirim, T. (Ed.). (2025). *Digital health, AI and generative AI in healthcare.* https://doi.org/10.1007/978-3-031-83526-1_1

Carnahan, D. (2025). Artificial intelligence fundamentals, In T. Adirim (Ed.), *Digital health, AI and generative AI in healthcare.* Springer Nature Switzerland. https://doi.org/10.1007/978-3-031-83526-1_4

Cooper, G. F., Aliferis, C. F., Ambrosino, R., et al. (1997). An evaluation of machine-learning methods for predicting pneumonia mortality. *Artificial Intelligence in Medicine, 9*(2), 107–138.

Li, F., Ruijs, N., Lu, Y. (2023). Ethics & AI: A systematic review on ethical concerns and related strategies for designing with AI in healthcare. *AI, 4,* 28–53. https://doi.org/10.3390/ai4010003

Turing, A. M. (1950). Computing machinery and intelligence. *Mind, LIX*(236), 433–460.

Chapter 15
Critical Success Factors and Future of Enterprise Performance Excellence

Abstract There are a wide variety of ways to structure an organization's Enterprise Performance Excellence program. This chapter describes practical guidance based on the literature and the author's experience for designing your Enterprise Performance Excellence program to help enhance the success of the program. We then describe a view into the future with the attempt at projecting where Lean Six Sigma for healthcare will evolve over the next decade.

Keywords Sigma program · Black belt · Sigma efforts · Sigma project · Project champion · Customer relationship management · Quality function deployment · Total quality management · Quality management system · Enterprise resource planning

Critical Success Factors for Enterprise Performance Excellence

Introduction

A critical success factor with respect to a Six Sigma program or projects is defined as "… the essential ingredients without which a project stands little chance of success" (Coronado & Antony, 2002). The author had developed framework components for successful implementation of Six Sigma in service-based organizations (Furterer, 2004) including local government that she adapted to the components for a successful enterprise-wide Lean Six Sigma program in healthcare. The key components are shown in Fig. 15.1, and will be described in the next section. The author reviewed the literature for the critical success factors identified in quality management and process improvement programs. These success factors mapped to each key component are shown in Table 15.1. These critical success factors provide support for the key components of the Enterprise Performance Excellence program.

S. Furterer, *Lean Six Sigma for the Healthcare Enterprise*,
https://doi.org/10.1007/978-1-4471-7617-6_15

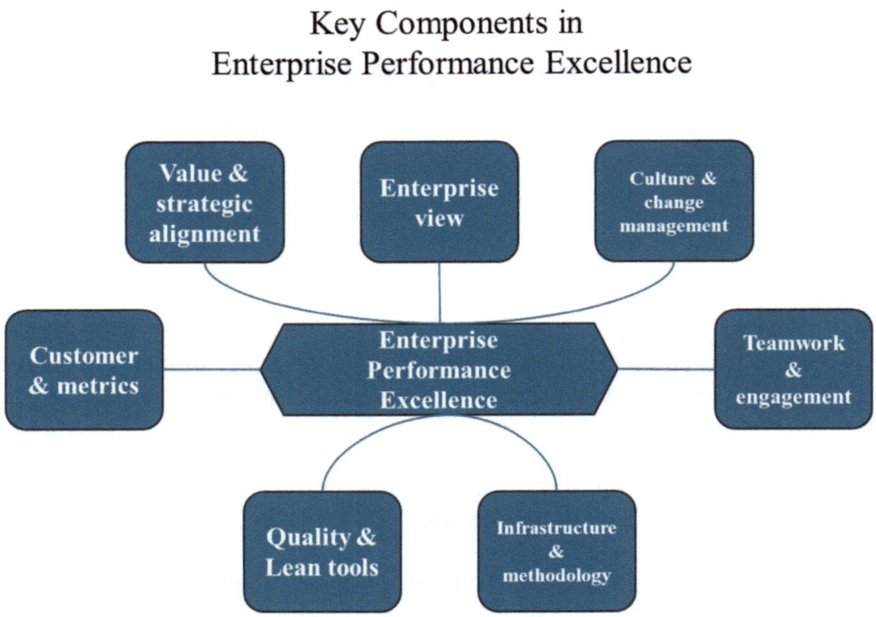

Fig. 15.1 Key components in enterprise performance excellence

Key Components

Value and Strategic Alignment

The value proposition is used to "sell" or convince management of the value of implementing a Lean Six Sigma program. There can be many facets related to the Lean Six Sigma value proposition, depending on who is defining the value. The value can be defined from a customer perspective first and foremost. We should focus on understanding the value and requirements from a voice of customer perspective. Another perspective can be from the perspective of reducing or avoiding costs as budgets are becoming increasingly stagnant or even decreasing. Yet another perspective could be to focus on the value of applying Design for Six Sigma to innovate the processes.

Lean Six Sigma can provide value by:

(1) Reducing or avoiding costs of providing healthcare services. One example of avoiding costs is the ability to not hire additional staff or resources due to streamlined processes and the release of additional staff capacity.
(2) Helping to better understand and satisfy customers' needs, including both internal and external customers, as well as partners, suppliers, and collaborators.
(3) Improving processes to innovate processes, eliminate waste, improve quality and productivity, and reduce variation.

Table 15.1 Critical success factors

Key component	Critical success factors
Value and strategic alignment	• Basing the Lean Six Sigma (LSS) project selection on key business priorities (Furterer, 2008) • Focusing on the key business areas (Furterer, 2008) • Top management support through leadership. This factor includes management's setting of goals and providing leadership and direction (Hoffman and Mehra, 1999) • Strategy that is complementary and integrated with existing policies and strategies (Dale, 1994) • Clear links between strategic goals and the change strategies (Newman, 1994) • Convince management of the value of LSS to the customer (Dale, 1994)
Enterprise view	• Systems thinking (Deming, 2000) • Value chain (Porter, 1980)
Culture and change management	• Top management commitment (Sureshchander et al., 2001) • Clear vision and mission (Sureshchander et al., 2001) • Values that support the changing culture (Newman, 1994) • Change strategy that builds flexibility to respond to customers, and focuses on the technical, political, individual, and organizational elements of change (Newman, 1994) • Social responsibility of the organization (Sureshchander et al., 2001)
Teamwork and engagement	• Training and education (Sureshchander et al., 2001) • Employee empowerment (Hoffman and Mehra, 1999) • Employee involvement and participation (Hoffman and Mehra, 1999) • Recruitment and selection systems that support the culture and program (Hoffman and Mehra, 1999) • Encourage innovation (Robertson and Gill, 2000) • Upward mobility and well-defined career path (Polster and Harris, 2000) • Teamwork (Robertson and Gill, 2000)
Infrastructure and methodology	• Visionary leadership (Sureshchander et al., 2001) • Project management with projects focused on critical to quality characteristics and financial benefits • Dedicated resources with black belt structure • Action plan and awareness • Top management support through financials, personnel, greater organizational autonomy, and effective oversight (Poister and Harris, 2000) • Design of services • Management of key business processes • Highly disciplined problem-solving approach • Program methodology • Continuous process improvement • Communication throughout the program • Benchmarking of processes, products, financials, greater organizational autonomy and effective oversight (Sureshchander et al., 2001) • Incorporate suppliers into the value chain

(continued)

Table 15.1 (continued)

Key component	Critical success factors
Quality and Lean tools	• Quality tools • Lean tools
Customer and metrics	• Outcome-based, results oriented performance measurement system (Hoffman and Mehra, 1999) • Measurement and feedback (Dale, 1994) • Measure customer satisfaction (Dale, 1994) • Link projects to financial benefits and business strategy (Dale, 1994) • Customer focused on defining needs and requirements (Poster and Harris, 2000) • Customer satisfaction as the main goal (Robertson and Gill, 2000)

Symptoms of environments where extensive waste and non-value-added activities exist can be used to convince management that improvement is needed. Some of the symptoms that are common are as follows:

- My employees complain that they need more staff or that morale is bad.
- Customers (internal and external) complain that communication is poor, there are many errors or rework, or there are long wait times.
- Expediting and fixing problems is rampant or the norm.
- You trust the system and processes, only if someone is there to guide the customer through the process (patient advocate).
- Audits (financial, etc.) show noncompliance.

The following function is a relational qualitative function, not a strict mathematical function, but can be used to represent the Lean Six Sigma overall value proposition:

Lean Six Sigma Value Proposition = Quality of Service Delivery × Trust × Financials.

where.

- *Quality of Service Delivery* is measured by defects in the service delivery process; the ability to meet customers' expectations; the ability to be responsive to customer's service delivery needs, which incorporates process measures related to productivity and cycle time; and variability of processes which can lead to errors and responsiveness issues related to service delivery.
- *Trust* is measured by the relationships built between internal and external customers, suppliers, partners, and collaborators and can be evaluated through customer and stakeholder surveys.
- *Financials* include the costs and revenues of providing services and can be measured by the reduction in costs related to the Lean Six Sigma program, and by the revenues that can be attained by providing excellent services.

A critical component of a Lean Six Sigma program and the DMAIC problem-solving approach encourages defining performance measures that can be used to assess the Lean Six Sigma Value Proposition, which will be discussed in the measurement component.

The second element of this component is to provide value by aligning to the strategic objectives and goals of the organization. Use of the Balanced Scorecard (Norton and Kaplan 1996) can provide a balanced performance measurement system, by measuring not only the financials, but the internal processes, the workforce, and the customer and growth opportunities. The Malcolm Baldrige National Quality Award Healthcare Criteria encourage tracking results in the following categories: healthcare and process outcomes; customer-focused outcomes; workforce-focused outcomes; leadership and governance outcomes; and financial and market outcomes. The key performance indicators can be identified for each result area, which can then enable identifying the strategic Process Improvement (PI) projects and priorities. Another great way to develop your enterprise-wide PI projects and priorities is to extract the Critical to Satisfaction (CTS) criteria from the patient satisfaction survey data. Many hospitals use nationally compared patient satisfaction benchmark data to assess where their hospital stands with regard to patient satisfaction and can even use correlation analysis to identify the CTS'. The correlations are derived for the customer survey questions that most highly correlate to the "would recommend your hospital" question. If noise or nurse communication is highly correlated to the "would recommend your hospital" ratings, then this provides a key focus to the Critical to Satisfaction and the Voice of the Customer.

Customer Focus

Defining the customer of the processes and services in a healthcare organization can be quite complex. There are potentially many customers of a healthcare organization including: (1) the patient, (2) internal employees or associates, (3) physician partners, and (4) suppliers and vendors. Whether these people are defined as customers or stakeholders is dependent upon how the scope of the Lean Six Sigma project is defined and which processes are part of the improvement effort. It is important that the project team comes to agreement as to who the customer is, for the different projects and components of the processes. It is important to engage the different customer and stakeholder groups to understand their "voice" and their requirements with respect to the services that they receive that are provided by the processes that need to be improved. Again, the patient satisfaction, associate, and physician surveys can be valuable tools in capturing the voice of the customer. Some hospitals are creating patient advisory boards that are used as an on-going tool to derive the customers' requirements. Customer grievance and complaint data are also tools to capture what needs to be improved. It is important to collect the complaints in a way (database) that we can trend, identify improvement areas, and assess that improvement has taken place.

Culture and Change Management

This framework component incorporates the cultural and change management aspects of the organization related to the Enterprise Performance Excellence Program. It includes developing a change strategy that addresses how to change the culture to focus on continuous improvement and measurement, focusing on the customer, the technical, political individual, and organizational elements of change. Mapping the political terrain (Bolman & Deal, 1997) can help to identify the stakeholders and their acceptance or resistance to change. A communication plan and strategy can be developed to address how to help the stakeholders become less resistant to change. The inherent structure of the DMAIC problem-solving approach, and the focus on data and measurement, helps with the culture and change management component. The project management components of Lean Six Sigma and clearly defining the mission, vision, and goals of the Lean Six Sigma projects also helps with changing the culture to embrace quality, productivity, and continuous improvement.

Also important to help change the culture is to thoroughly and continuously communicate the status of the Lean Six Sigma effort and projects. This communication can include formal project kick-off meetings where the mission, vision, and goals of the Lean Six Sigma program and projects are discussed. At the beginning of the program and projects, it is critical to identify and communicate the mission, vision, values, goals, roles, and responsibilities of the projects and project teams. Throughout the projects, it is important to provide continuous status of the projects through periodic status meetings and reports so that the organization is aware of the successes and learnings from the projects.

It is also important to show that the senior management team is committed to leading and owning the Lean Six Sigma effort. The CEO, senior management team, and the project champions should show their constant and continued commitment and support to the Lean Six Sigma effort. The project champions should be trained with an understanding of the Lean Six Sigma methodology and tools and should spearhead the project effort and remove obstacles and barriers to successful implementation. Project champions should be at a Director level or above, be well respected in the organization, and be willing to advocate for change.

We used the Dr. John Kotter change model (Kotter 2011) to implement improvements and manage change within our Enterprise Performance Excellence program.

(1) Acting with Urgency

    a. Gaining top management support and visibly communicating this support for the projects, emphasizing the importance of the project to the organization.

(2) Develop the Guiding Coalition

    a. Form the project team and properly scope the projects. Throughout the project we engage appropriate subject matter experts who help to develop the new process and generate improvement ideas.

(3) Developing a Change Vision

a. Develop the new processes, creating innovative ways to transfer this knowledge (social media, intranet, videos, online training, etc.).
b. Develop the implementation plan including a detailed work plan to stage the improvements.

(4) Communicate the Vision Buy-In

a. Develop and deploy a communication plan to communicate the changes and status of the improvement efforts.
b. Management must model the behavior of the new process, assess understanding and compliance to the new process.

(5) Empowering Broad-Based Action

a. Identify and remove obstacles to change. Implement suggestion systems to get ideas from any and all of the staff.
b. Incorporate accountability to adhere to the new processes.

(6) Generating Short-Term Wins

a. Implement a reward system to encourage change and improvement.
b. Communicate the wins and significant improvement to all staff and hospital administration.
c. Celebrate success and share best practices and lessons learned.

(7) Do Not Let Up

a. Incorporate the new metrics dashboards into the daily metrics monitoring and control.
b. Incorporate the process competencies into the performance evaluation system.

(8) Make Change Stick

a. Provide change management and Lean Six Sigma training to staff and management to ensure the succession of the change in thinking to a fact-based decision process and one of continuous improvement.

Teamwork and Engagement

Most of the quality and process improvement programs, such as Total Quality Management (TQM), Business Process Reengineering (BPR), Six Sigma, Lean, and Lean Six Sigma, incorporate elements to address managing the people in the organization. They advocate empowering employees to make and be accountable for decisions about their processes. They include training, education, and providing the right skills for people to do their jobs. The Lean Six Sigma program is based upon a team structure where team members solve problems together. This component also includes rewarding and recognizing employees for their involvement on project teams, their teamwork and efforts, and their successful implementation of process and financial improvements, and meeting the goals and objectives of the project. In a healthcare organization, it is important to engage the physicians to be advocates for change and improvement. Many physicians are partners, but not employees of the hospital, so clearly defining how they are to be engaged in the improvement program and projects is important. Since their involvement is vital to the organization's ability to implement process change and innovate new processes, this should be well planned and well managed. Appropriate resources should be engaged and committed to the problem-solving teams, where it is part of their regular duties and activities to improve the processes, not based on a voluntary after thought, or "if they get the time" to improve processes. Successful organizations do not send their clinical or nonclinical staff home when the patient census is low, and they engage them in problem-solving teams, projects, and activities.

Infrastructure and Methodology

The infrastructure and methodology component provides the structure to support the Lean Six Sigma effort. It includes defining a methodology of how to perform the problem-solving and improvement effort (DMAIC). It includes having visionary and committed leadership as a critical element of success. It includes incorporating a project management structure to ensure that the projects are successful and continually move toward a successful implementation. This component also includes developing a communication plan to communicate the mission, goals, objectives, and progress of the Lean Six Sigma program and projects. A common complaint of organizations new to Six Sigma is that it takes too long. This can be true when the following mistakes are made:

- Lean Six Sigma team members are not properly trained to understand when to stop collecting and analyzing data. They must define and understand the Critical to Satisfaction Criteria, measure the CTS, identify the root causes, and implement improvements to improve the CTS. The CTS is the common thread that helps you to understand when to stop, when you have collected enough data, and when you have improved.
- Defining targets for your metrics is important so that you know if you have improved enough, and to assess when you can turn your efforts to other projects.
- When resources are not committed to work on the projects.
- When project champions are not assigned who remove barriers, encourage the teams, and hold the teams accountable to complete their projects within the planned milestones.
- When teams do not create project plans, estimate phase milestones, and are not held accountable to complete their projects.

Selecting the appropriate problem-solving methodology is also important. Not every project needs extensive statistical analysis. We can apply four different problem-solving approaches:

(1) The typical DMAIC Six Sigma problem-solving methodology, for strategically important, unstructured, complex, and cross-functional problems, which should take from 4 to 6 months to complete.
(2) Rapid process improvement projects that are shorter in duration, usually within one department, are not as complex, and can be completed in 2 to 3 months.
(3) Kaizen events is another approach that is small, focused improvement efforts, where little data collection is needed, or it has already been done within a larger Six Sigma project, where 2 to 4 weeks can be spent designing and piloting a new or improved process.
(4) Design for Six Sigma which can be used to design and innovate new processes for new services.

The infrastructure component also includes how to organize the Enterprise Performance Excellence program and define the reporting structure. We will describe the infrastructure that is used at an acute care hospital for their Enterprise Performance Excellence program.

EPE Mission

To provide guidance, direction, mentoring, training, and improvement services to the hospital, to enable Enterprise Performance Excellence in all operations and processes.

The EPE Center provides the following services.

EPE Center Services

- Enterprise Performance Excellence program *methodology* and *structure* to enable performance improvement.
- *Guidance* and *mentoring* to EPE teams, Black Belts, Green Belts, White Belts, Champions, and Executive Steering Committee.
- *Training* and *Education* as appropriate to further the EPE program methodologies and tools.
- Performance improvement *consulting* expertise and *services.*

The Enterprise Performance Excellence Center Director reports directly to the CEO. The Executive Team prioritizes projects and provides guidance for the EPE projects, who should be trained, and are trained as, and act as project champions. They provide the typical Executive Steering Committee function of an enterprise-wide process improvement program. The EPE Center's Director is a Master Black Belt and Industrial Engineer, develops and deploys the governance infrastructure, facilitates complex, enterprise-wide improvement and innovation projects, and mentors and trains Green and Black Belts. Two Black Belts that are also Industrial Engineers are part of the EPE Center. They facilitate projects, engage in internal consulting activities on EPE projects, mentor Green Belt and Black Belt candidates, and train Green and Black Belts.

There are currently 20 Lean Six Sigma certified associates (1 Master Black Belt, 3 Black Belts, and 16 Green Belts). We have trained 43 Green Belt candidates, where some of the projects are still in progress. We plan to train another 30 Green Belt candidates and 15 Black Belt candidates in the next year. Our goal is to train all of the Directors within 2 years. We will have the majority (over 90%) of the Directors trained at the Lean Six Sigma Green Belt level by the end of next year. Most of the Green Belts and Black Belts are full-time associates that work on improvement projects in addition to their full-time duties. The Enterprise Performance Excellence Governance Structure is shown in Fig. 15.2. The EPE Center reports directly to the CEO. The EPE teams are formed and worked on projects based on the strategic goals and objectives of the organization. The EPE Center Master Black Belt and Black Belts mentor the Green Belt and Black Belt candidates. The Green Belt and Black Belts who have gone through the training work on additional improvement projects within their departments and some work on the enterprise-wide improvement efforts. We also have an EPE network that consists of anyone who is trained in Lean Six Sigma.

The Lean Six Sigma Network Mission is to

- Provide and share best practice processes.
- Provide and share Lean Six Sigma best practices, projects, etc.
- Mentor Green and Black Belts through their projects.
- Help on projects, data analysis, tools training, etc.
- Supports teams and the program.

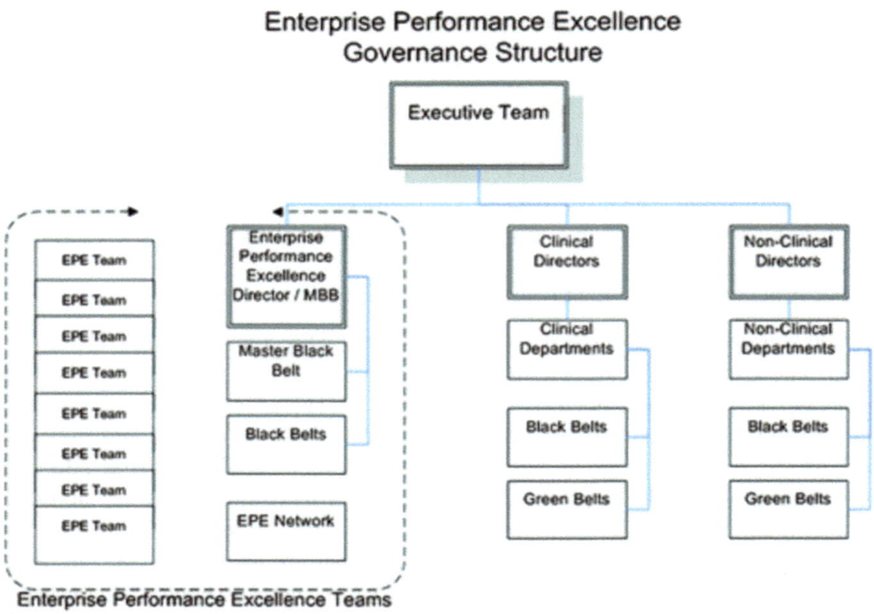

Fig. 15.2 Enterprise performance excellence governance structure

- Advocates for Enterprise Lean Six Sigma.
- Meets with teams to remove roadblocks.

Quality and Lean Tools

Quality and lean tools are valuable techniques to solve problems and improve the quality and productivity of the processes. Table 15.2 shows a mapping of the DMAIC phases to the most common tools that can be applied in a healthcare setting. Many of the tools are similar to other service environments, but somewhat different from a manufacturing setting, especially in the Lean area. The major Lean tools applicable in a healthcare setting are the Lean Wastes, the 5S, Why–Why diagrams, standard work, Kaizen, and visual control. Many of the commonly used Six Sigma tools in a healthcare setting are: process mapping, cause-and-effect analysis, stakeholder analysis, project charters/project management, SIPOC, statistical analysis, Quality Function Deployment, dashboards and scorecards, mistake proofing, failure mode and effects analysis, improvement plans, and control plans.

Table 15.2 Summary of Lean Six Sigma DMAIC most common tools

Define	Measure	Analyze	Improve	Control
• Project charter • Stakeholder analysis • SIPOC • Process map (high level) • Communication plan • Project plan • Responsibilities matrix • Items for Resolution (IFR) • Ground rules	• Process map • Value Stream Map (VSM) • Critical to Satisfaction (CTS) • Data collection plan • Quality Function Deployment (QFD) • Pareto chart • VOP matrix • Gage R&R • Cost of poor quality	• Cause-and-effect diagram, and matrices • Why–why diagram • Histograms and graphical analysis • Correlation analysis • Basic statistics • Sampling • Process analysis • Failure Mode and Effects Analysis (FMEA) • Gap analysis • Hypothesis tests • Summary of problems • Waste analysis • Lean analysis • Value analysis • 5S • Kaizen	• Recommendations • Improvement plans • Action plan • Cost–benefit analysis • Cost of poor quality • Future state process map • Design of experiments • Dashboards • Scorecards • Weighted cause-and-effect diagram or matrices	• Hypothesis testing • Design of experiments • Basic statistics • Graphical analysis • Sampling • Mistake proofing • Standard work • FMEA • Control plan • Process capability • DPPM/ DPMO • Statistical Process Control (SPC) • Standard work • Kaizen • Dashboards • Scorecards

Measurement (Metrics)

The measurement component includes defining and applying performance measurements and metrics that can be used to measure improvements in quality, productivity, variability, service delivery, and financials. The performance measurements should be defined within the problem-solving methodology at a process level. The metrics measure the Critical to Satisfaction characteristics to show improvement and ensure that we are meeting the customers' requirements.

The Executive Steering Committee should review the status of the projects and meet with the project leaders on a monthly or bi-weekly basis, using a dashboard that assesses the following status levels:

- *Green.* The project is progressing well and meeting the estimated DMAIC phase milestones.
- *Yellow.* The project is in jeopardy of missing an estimated DMAIC phase milestone, or there are significant barriers that are impeding progress.
- *Red.* The project has missed a DMAIC phase milestone.

Last, but of course not least, the teams and the organization should take the time to celebrate their success in improving processes and making change happen. Change is not easy, so we should also pause, be thankful, and celebrate with our team members. It is important to also track, measure, and monitor the status of the Six Sigma projects to ensure that they are completed.

The Future and Challenge of Lean Six Sigma in Healthcare

One of the exciting elements of the Lean Six Sigma evolution, has been how many somewhat diverse and at first glance, disparate methods and tools have come together to provide a more holistic and integrated toolkit for solving extremely complex problems.

The world is getting more complex each day. The problems are getting bigger and more complex, so the tools to solve these problems need to evolve as well.

We can go back to the Evolution of Quality graphic, in Fig. 2.1 of Chap. 2, adding the information stream and the business process management elements to set the stage for our discussion on the future evolution of Lean Six Sigma, shown in Fig. 15.3. The economy in the United States of America and in many other countries in the world is becoming an information, service, and knowledge age. Our economy has evolved from tangible, craft, agricultural, manufacturing economies, to an intangible information, service, and knowledge-based one. We discussed the progression from Statistical Process Control, which provided control of discrete manufacturing processes, broadening to Business Process Reengineering (BPR), and Total Quality Management (TQM). BPR and TQM incorporated a broader view of the quality management principles and philosophies that needed to be in place to effect change within the cultures that were applying these methods. Six Sigma brought a more structured problem-solving approach, the mentoring and training focus of the belt structure, and a broader toolkit of tools. The Lean side evolved from the Ford Production System that provided an assembly process to manufacture and assemble discrete products. Lean and Just-in-Time broadened the spectrum to include more of the supply chain elements. At this same time, information technology was advancing from non-integrated Material Requirements Planning (MRP) and Material Resource Planning (MRP) II applications focusing on managing the shop floor and purchasing processes to integrated Enterprise Resource Planning (ERP) and Customer Relationship Management (CRM) information systems that evolved into managing the entire supply chain. The supply chain focus provides an end-to-end view from sourcing of suppliers through the logistics and distribution networks, to the enterprise and back to the distribution channels to get products to the market. The three streams of quality, productivity, and information technology have integrated over the last few years into Lean Six Sigma Supply Chain, with many variations on the combinations of the names. The other elements that are evolving to enhance quality in organizations is Business Process Management and Business Process Management Systems which provide a holistic approach to managing, documenting, improving, and controlling our processes to enhance operational excellence. Business architecture, as discussed in Chap. 1, can help us to solidify and organize business process management.

For the future, the author sees this integration evolution approach continuing to more tightly couple and integrate the entire enterprise and supply chain, and the philosophies and tools within Lean, Six Sigma, Business Process Management, and the Supply Chain areas. As we look at the underlying elements of the three major methodologies, there are many challenges in healthcare that can be addressed by

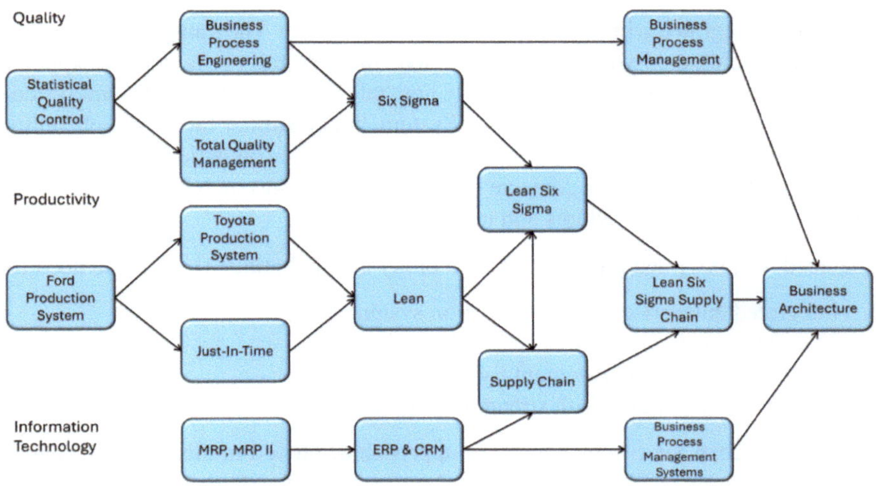

Fig. 15.3 Evolution of quality

integrating Lean, Six Sigma, Business Process Management, and Supply Chain. We will address the following challenges and how Lean Six Sigma and an Enterprise Performance Excellence view can help address these challenges:

- Information technology that enables processes, services, and measurement.
- New and varied problem-solving approaches.
- Process innovation with design for six sigma and TRIZ http://www.triz-journal.com/archives/what_is_triz/, 2013.
- Customer focus and engagement–voice of the customer and quality function deployment.
- Teamwork and engagement.
- Quality management system, incorporating business process management.

Information Technology that Enables Processes, Services, and Measurement

Enterprise Resource Planning (ERP) applications started in the manufacturing industry and have evolved to where they are very integrated, extremely customizable from a process perspective, and enable the business to leverage processes and information to meet customers' requirements. Healthcare has not been so lucky. Compared to ERP in manufacturing, "enterprise" systems in healthcare are far from meeting the needs of an enterprise in an integrated fashion. Electronic medical information systems may not be integrated with Supply Chain Management applications from a supply inventory control, charging, and billing perspective. The Materials Management side of Supply Chain Management may not be integrated with the Inventory Control or Point of Use functionality. The Emergency Department Bed Board System may not be integrated with the Hospital Bed Board System. The transportation system may not be integrated with the Hospital Bed Board or the room cleaning application. The Cardiovascular Lab system may not "talk" to the Operating Room System. There may be an Outpatient Scheduling System, but no similar Inpatient Scheduling System. The patient may be able to schedule their labs or outpatient diagnostic test and procedures, but if you become an inpatient, you have no such luxury of knowing when your inpatient procedures and tests are scheduled.

The need for integrated process and technology is not abating. There is the need for

1. Integrated enterprise-wide information systems.
2. Electronic medical record.
3. Business Process Management Systems.
4. User-friendly systems with streamlined processes.

How can Lean Six Sigma help? The approach discussed in this book in Chap. 1 to build the Strategic Business Process Architecture models can enable Lean Six Sigma by methodically documenting, analyzing, and prioritizing processes and process improvement efforts. Lean Six Sigma must be elevated to the enterprise-wide process view aligned to the organization's strategic goals and objectives. Metrics must continue to evolve to align to business drivers and focus on reducing the cost of doing business, enable improving revenues by tapping new markets, and ensuring customer satisfaction through measurement and improvement. Data and information must support the business, not exist for its own goals. The place for technology must be identified based on the business strategies and the processes that they must support. Leadership will play an even more critical part, and be necessary throughout the organization, not just at the top, but in the middle and at the grassroots of our organizations. Enterprise Business Architecture is an emerging framework that fundamentally begins with the needs of the customers and the business to align the strategies, organizations, information, global locations, processes, and timing of all of these elements to enable rapid and controlled change.

New and Varied Problem-Solving Approaches

With the integration of Lean, Six Sigma, Systems Engineering, and Supply Chain methodologies new and varied problem-solving approaches are warranted. There is no longer one-size-fits-all methodology that can solve the ever-increasing complex problems that are evolving in today's healthcare environment. We recommend incorporating the flexibility to apply various problem-solving methodologies that can support varying levels of complexity of variables and processes, need for cross-functional integration, and the need for different levels and types of analytical tools.

Different levels could include

- The typical DMAIC Six Sigma problem-solving methodology, for strategically important, unstructured, complex, and cross-functional problems, which should take from 4 to 6 months to complete.
- Rapid Process Improvement projects that are shorter in duration, usually within one department, are not as complex, and can be completed in 2 to 3 months.
- Kaizen Events are another approach which are small, focused improvement efforts, where little data collection is needed, or it has already been done within a larger Six Sigma project, where the 2 to 4 weeks can be spent designing and piloting a new or improved process.
- Lean projects focusing on 5S, Lean Wastes, and other Lean tools.
- Design for Six Sigma which can be used to design and innovate new processes for new services.

Process improvement in healthcare should continue to focus on mistake proofing, identifying, and preventing potential failure modes before they occur, and performing quick and effective root cause analysis when errors do occur.

Process Innovation Through Design for Six Sigma and TRIZ

Our organizations, processes, and information systems must be designed in the most agile ways to support innovation and speed to market. Functional silos must be eliminated to provide a cross-functional view of the businesses that we support. Design for Six Sigma and Theory of Inventive Problem-Solving (TRIZ) can be used to design innovative processes that better integrate patient care across the healthcare continuum (TRIZ Journal). We need to better integrate patient and healthcare information across provider networks, while protecting patient data. We need to balance an innovation focus while still ensuring compliance with accreditation, regulatory, and ethical challenges.

Customer Focus and Engagement: Voice of the Customer and Quality Function Deployment

The focus on the customer, not just competitors will need to grow each day where we more fully engage and align with our customers' needs. Customization must become a way of life, where we can reach out to each customer to understand their needs and customize our offerings to meet each of their needs in a reachable and cost-effective manner. Quality Function Deployment can be used to align the customer requirements to the process requirements. We should leverage Customer Relationship Management (CRM) systems to collect, track, and trend customer requirements and satisfaction data.

Teamwork and Engagement

Ultimately, we cannot do any of the above without first considering the needs of our people; empowering them; and enabling them with the training, skills, and tools to effectively do their jobs. We need to engage the entire supply chain, our physician partners, suppliers, vendors, as well as our associates to innovate, solve problems, and streamline the delivery of our processes and services.

Quality Management System Incorporating Business Process Management

To further integrate and manage our enterprise, we need to design or enhance our quality management systems to include the following elements:

- Leadership.
- Strategic planning.
- Customer focus.
- Workforce management.
- Process management.
- Performance measurement, information, and knowledge management.

Conclusions

We started this chapter with the caveat that there are many ways to set up an Enterprise Performance Excellence program with the goal of this chapter to provide some sound ideas for either setting up or enhancing your program. Lean Six Sigma and the Enterprise Performance Excellence methodologies and tools can leverage our ability to manage our enterprises.

We challenge the reader and student of Lean Six Sigma to understand the underlying principles embedded in Lean Six Sigma to be part of the revolution of the tools and philosophies that will continue the evolution of these amazing bodies of knowledge. The best of luck in your Lean Six Sigma and Enterprise Performance Excellence endeavors.

References

Bolman, L., & Deal, T. (1997). *Reframing organizations artistry, choice, and leadership* (2nd ed.). Wiley.

Coronado, R. B., & Antony, J. (2002). Critical success factors for the successful implementation of six sigma projects in organisations. *TQM Magazine., 14*(2), 92–99.

Furterer, S. (2004). *A framework roadmap for implementing lean six sigma in local governmental entities*. Ph.D. Thesis, University of Central Florida.

http://www.triz-journal.com/archives/what_is_triz/, 2013.

Kaplan, R. S., & Norton, D. P. (1996). *The balanced scorecard: Translating strategy into action*. Harvard Business School.

Kotter International. (2011). 8 steps for leading change. Kotter International—Home. 09 Jan 2011. http://www.kotterinternational.com/KotterPrinciples/ChangeSteps.aspx.

Index

A
Abdominal CT scans, 244
Abdominal imaging, 244
Action plan, 18, 58, 59, 455, 463
Analyse, 14, 23, 41, 42, 84, 164, 170, 173,, 201, 237, 240, 241, 255, 256, 258, 264, 268, 301, 302, 319, 463
Analysis of Variance (ANOVA), 53, 271, 433
Antibiotic protocols, 277
Appropriate diagnosis, 229, 231–233, 236, 237, 346, 347
Appropriate use of CT scans, 231, 236, 238, 241
Appropriate use of linen, 148
Artificial Intelligence (AI), 423–426, 451
Associate satisfaction, 19, 120, 128, 129
Audit control plan, 176

B
Benchmarking, 27, 36, 37, 256–258, 271, 451, 455
Business architecture, 2–5, 7, 9, 10, 344, 465, 467
Business Process Management (BPM), 2, 3
Business Process Management Systems, 3, 381, 465, 466
Business Process Reengineering (BPR), 2, 12, 459, 464
Business scenarios, 56

C
Cause-and-effect analysis, 58, 92, 256, 302, 357, 371–377, 463
Cause-and-effect diagram, 42, 43, 174, 201
Cause-and-effect matrix, 42, 98, 100, 137–139, 141
Cause-and-effect relationships, 14, 23, 41, 42, 84, 137, 201, 241, 287, 357
Cervical spine imaging, 245
Change management, 59, 76, 110, 113, 124, 158, 161, 225, 252, 455, 458, 459
Check sheet, 27, 36, 38, 39, 146, 199, 208, 210, 214, 215, 220
Coding, 167, 170, 173, 175, 328, 396, 436, 439
Computed Tomography (CT), The, 227
Conceptual model, 4, 383
Confidence intervals, 51
Control, 2, 12–14, 20, 23, 32, 49, 50, 55, 56, 62–65, 70, 77, 92, 110–112, 116, 117, 118, 121–123, 146–148, 151, 152, 158, 159–161, 163–165, 167–171, 176, 177, 181, 189, 195, 197, 204–208, 210, 215, 216, 222, 224, 242, 250, 256–259, 268–271, 281, 284, 285, 294, 297, 299, 302, 338–340, 383, 384, 386, 387, 389, 392, 425, 432, 438, 442, 446–450, 459, 463, 464, 466
Control charting, 258
Control plan, 14, 23, 62–64, 77, 110–112, 152, 158–160, 176, 177, 222, 224, 250, 256, 269, 281, 294, 299, 302, 338, 340, 383, 384, 438, 463
Correlation analysis, 50, 85, 87, 457, 463
Cost/benefit analysis, 57–59, 99, 102, 103, 144, 415
Cost of Poor Quality (COPQ), 15, 27, 40, 41, 59

S. Furterer, *Lean Six Sigma for the Healthcare Enterprise*,
https://doi.org/10.1007/978-1-4471-7617-6

Critical success factors, 1, 2, 113, 161, 225, 252, 299, 302, 453, 455
Critical to Cost (CTC), 20
Critical to Delivery (CTD), 20
Critical to Quality (CTQ), 256, 257, 455
Critical to Satisfaction (CTS), 14, 15, 18, 20, 21, 23, 31–33, 35, 36, 43, 51, 55, 57, 58, 61, 63, 71, 73, 78–80, 96–101, 116, 117, 120, 125, 127, 128, 166, 167, 190, 191, 194, 198, 199, 228, 229, 231, 232, 236, 237, 242, 248, 250, 251, 278, 281, 283–285, 297, 302–304, 307, 309, 325, 344, 346, 349, 370–377, 401, 402, 457, 460, 463, 464
CTS measures, 20, 21, 31, 32, 36, 63, 125, 198, 281, 283, 303, 460
Culture and change management, 455, 458
Culture change, 105
Customer focus, 456, 457, 469
Customer Relationship Management (CRM), 465, 468
Customer survey analysis, 53
Customer surveys, 30, 34, 53, 130, 457

D

Dashboards/scorecards, 58, 61, 66, 463
Data collection plan, 27, 31–33, 78–80, 125, 127, 198, 199, 234, 236, 283, 284, 302, 307, 309, 463
D-dimer lab, 240
Defects PerMillion Opportunities (DPMO), 42, 54, 55, 62, 63, 463
Define, 3, 6–9, 12–15, 17, 18, 20–24, 26–30, 32, 35, 36, 40, 42, 47–49, 51, 55, 70, 71, 75, 77, 78, 100, 110, 116, 123–125, 157, 164, 168–170, 173, 190, 195, 196, 198, 212, 227, 229, 233, 236, 237, 256, 277, 278, 281, 283, 302–306, 327, 344, 350, 351, 355, 371, 373, 380, 386, 388, 396–401, 430, 453–455, 457, 460, 461, 464
Define, Measure, Analyze, Improve, Control (DMAIC), 12–14, 18, 20, 36, 48, 57, 66, 70, 71, 116, 164, 168, 170, 189, 233, 255–257, 270, 271, 301, 302, 340, 456, 458, 460, 461, 463, 464, 467
Design for Six Sigma, 454, 461, 466–468
Design of Experiments (DOE), 60, 63, 256, 270, 271, 463
Design Phase, 405
Develop process capability, 14, 23, 41, 42, 84, 137, 138, 201, 205, 287, 288
Discharge planning, 215, 370, 381
DMAIC Six Sigma problem-solving methodology, 461, 467

E

8 wastes analysis, 205, 206
Electronic medical record, 103, 190, 284, 286, 289, 381, 417, 426, 466
Emergency departments, 69, 111, 252
Emergency services value chain, 70, 71, 77, 102, 417
Engagement, 5–9, 75, 76, 113, 123, 124, 147, 225, 299, 339, 374, 455, 466, 468
Enterprise performance excellence, 1, 147, 148, 176, 453, 454, 458, 461, 462, 466, 469
Enterprise planning, 5, 6, 8, 9
Enterprise Resource Planning (ERP), 465, 466

F

Failure Mode and Effect Analysis (FMEA), 49, 50, 62, 63, 463
Failure modes, 49, 50, 467
5S, 42, 48, 58, 145, 146, 148, 150, 152, 158, 159, 163, 164, 166–171, 173–179, 187, 188, 212, 219–221, 448, 463, 467
5S analysis, 48
5S audit, 159, 170, 171, 177, 178, 187, 188
Flow control board, 147, 148, 151, 159
Focus groups, 15, 20, 21, 27, 30, 31, 34, 36, 37
Ford production system, 12, 465
Full Factorial Experiment, 60
Full Time Equivalent (FTE), 346, 348, 389, 393
Functional decomposition, 3, 350, 351, 355
Future plans, 62, 66, 110, 112, 152, 160, 222, 225, 250, 251, 294, 299, 383
Future state, 7–9, 14, 23, 57–60, 62, 99, 104, 144, 145, 205, 221, 224, 225, 242, 289, 370, 376, 378, 395, 398, 401, 402, 411, 437–440, 463

G

Gage R&R (Repeatability & Reproducibility) study, 40, 463

Generative AI, 423
Graphical analysis, 34, 42, 55, 62, 87, 463

H
Histogram, 27, 36, 38, 42, 53, 55, 61, 83–85, 106, 131, 134, 286, 295, 312, 314, 316–318, 331, 333–336, 379, 380, 411, 413, 415, 463
House of quality, 35, 57, 58, 352–354
Hypothesis test, 51–53, 108, 156, 294, 337, 410, 412, 414, 416, 419, 463
Hypothesis testing, 51–53, 62, 63, 302, 463

I
Improve, 3, 7, 10, 12–14, 23, 29, 35, 37, 40, 43, 48–50, 57, 58, 63, 65, 69–72, 88, 91, 99, 102, 103, 115, 116, 123, 144, 145, 161, 163–165, 170, 172, 174, 175, 181, 189, 195, 202, 205, 224, 225, 242, 255, 256, 265, 266, 277, 279, 281, 289, 297, 301–303, 319, 322, 324, 325, 331, 338, 339, 354, 370, 381, 382, 389, 391, 401, 413, 415, 417, 420, 454, 460, 463
Improvement opportunities, 37, 44, 99, 144, 205, 242, 289, 370
Improvement recommendations, 14, 15, 20, 23, 35, 43, 49, 50, 57–62, 72, 190, 191, 209–212, 222, 270, 302, 395, 398, 401, 402, 411, 418, 440
Includes severe sepsis, 277
Information technology, 2, 20, 26, 70, 75, 113, 151, 327, 352, 370, 380, 465, 466
Infrastructure and methodology, 455, 460
Inpatient scheduling, 354, 370, 372, 374, 376, 380, 466
Interventional radiology, 301, 305, 306, 309, 315, 340, 374
Item For Resolution (IFR), 24, 25, 463

J
Just-in-Time (JIT), 12, 465

K
Kaizen, 13, 29, 48, 49, 58, 196, 393, 437–439, 461, 463, 467
Kaizen blitz, 196
Kaizen events, 13, 48, 49, 437, 461, 467
Kanban pull system, 13

L
Labeling, 163, 167, 175
Labor management, 374, 385, 392
Labor standards, 385, 387, 388, 393
Lean, 1–4, 9, 11–13, 47, 48, 60, 69, 71, 102, 103, 105, 111, 147, 148, 158, 161–164, 189, 205, 219, 222, 223, 225, 251, 271, 277, 281, 301, 322–324, 339, 340, 344, 345, 417, 429–431, 433, 440, 443, 444, 446, 448, 449, 451, 453–460, 462–467, 469
Lean analysis, 440, 443, 444, 463
Lean manufacturing, 13
Lean production cells, 13
Lean Six Sigma, 1–5, 9, 10, 12, 13, 69, 71, 102, 103, 105, 111, 116, 147, 148, 158, 161, 162, 164, 189, 222, 223, 225, 251, 271, 277, 281, 301, 339, 340, 417, 429–431, 433, 440, 446, 448, 449, 451, 453–460, 462, 464–466, 469
Lean Six Sigma Supply Chain, 465
Lean thinking, 11, 13
Lean wastes, 324, 440, 463, 467
Left without being seen, 33, 69, 71, 72, 79, 86, 89, 90, 98, 100, 101, 402, 407, 432, 434
Length of stay, 22, 69, 80, 83, 87, 90, 92, 98, 100, 101, 106, 109, 201, 287, 345–349, 352, 354, 370, 371, 373, 375, 377, 383, 395, 403, 407, 410–416, 419, 423, 424, 426, 434, 435
Light levels, 261, 265, 269
Linear regression analysis, 51
LS spine imaging, 246
Lux, 257, 258, 263, 265, 266, 268, 269

M
Machine learning, 423
Manage change, 14, 23, 62, 110, 152, 222, 250, 294, 458
Material Requirements Planning (MRP), 465
Material Resource Planning (MRP) II, 465
Measure, 13, 14, 18, 20, 21, 23, 27–29, 32, 39, 40, 42, 44, 48, 50, 54, 55, 60–62, 64, 77, 78, 84, 103, 110, 112, 123–125, 129, 149, 152, 168, 169, 195, 196, 222, 227, 233, 235, 241, 250, 256–258, 281, 283, 285, 286,

288, 294, 307, 309, 350, 354, 374, 382, 383, 389, 396, 397, 402, 410, 417, 429, 456, 464
Measurement (metrics), 464
Measurement systems analysis, 39
Measure results, 62, 110, 152, 222, 250, 294, 383
Metrics, 3, 15, 24, 27, 28, 31, 32, 34, 39, 55, 60, 61, 63, 78, 80–82, 84, 99, 103, 104, 106, 108, 111, 113, 136, 140, 146, 158, 159, 161, 174, 192, 208, 210, 212, 215, 216, 219, 221, 222, 224, 235, 250, 257, 270, 283, 286, 288, 294, 297, 299, 307, 308, 310, 328, 338, 340, 341, 367, 370, 374–376, 381, 382, 384, 401, 408, 410–412, 417–419, 427, 432–436, 438, 439, 451, 456, 459, 460, 464, 467
Mistake proofing, 13, 62–64, 208, 210, 215, 463, 467
Modeling and simulation, 395–398, 401, 407, 421
Modeling and simulation approach, 158, 396–398

N
Neonatal Intensive Care Unit (NICU), 255

O
On-time starts (of cases), 116, 126, 127, 129, 132, 134–136, 139–142, 145, 151–154, 159, 160, 168, 193, 224, 310, 314, 315, 318, 329, 337, 338
Operating model, 343–345, 384
Operational definition, 27, 32, 33, 36, 37, 61, 78–80, 127, 128, 198, 199, 235, 236, 283, 284, 307, 309, 402
Optimize design, 60
Orbital CT/facial bones sinus imaging, 246
Order entry and diagnostics, 370, 379
OR turnaround, 116–118, 125, 126, 137, 143–150, 152, 157, 158, 160, 163, 176, 440, 441, 445, 451
OR turnaround time, 117, 119, 121, 122, 126, 127, 140, 145, 165–167

P
Pareto chart, 36–38, 53, 127, 130, 132, 134, 200, 237, 248, 249, 251, 256, 264–266, 271, 324, 325, 436, 437, 449, 463
Patient satisfaction, 15, 19, 22, 33, 37, 69, 71–75, 78, 79, 109, 113, 117, 119, 120, 212, 214, 220, 306, 309, 320, 321, 325–327, 402, 436, 457
Patient transport, 78, 80, 99, 102, 173, 176, 346, 347, 370, 380, 381, 389, 412, 415, 418, 420, 431
Patient volume metrics, 81, 82
P-D-C-A, 62, 65
Percent of CT scans, 236, 246, 247, 252
Percent of CT scans by modality, 236, 237, 249–251
Performance factor, 386, 387, 393
Performance targets, 14, 23, 57, 58, 60, 78, 99, 104, 144, 145, 205, 221, 289, 293, 299, 370
Personal and fatigue time, 393
Pilot improvements, 77, 99, 105, 106, 113, 144, 146, 205, 221, 225, 242, 246, 289, 293, 328
Predictive bed demand, 370, 376
Process analysis, 42, 44, 47, 58, 463
Process Architecture Map (PAM), 302, 310, 355, 423, 424, 426, 440, 441
Process capability, 14, 23, 41, 42, 54–56, 62, 63, 84, 92, 93, 137, 138, 140, 143, 144, 201, 205, 209, 256, 261, 262, 270, 271, 287–289, 297, 298, 463
Process capability study, 55
Process control plan, 14, 23, 62, 63, 110, 152, 222, 224, 250, 294, 299, 383, 384
Process map, 27–29, 44, 58, 60, 62–64, 77, 78, 104, 105, 125, 126, 145, 197, 221, 224, 225, 234, 235, 282, 283, 307, 310–314, 322, 350, 355, 356, 376, 378, 396, 397, 402, 403, 440–442, 451, 463
Process requirements, 32, 468
Process standardization, 146
Process throughput and time metrics, 80
Productivity, 12, 13, 19, 20, 28, 117, 118, 120, 125, 128, 130, 165, 166, 192, 193, 198, 346, 348, 357, 368, 370, 371, 373, 374, 377, 379, 386, 389–393, 454, 456, 458, 463–465
Productivity measurement, 370, 374
Project charter, 14–16, 22, 23, 26, 71, 72, 116–118, 166, 190, 191, 229, 230,

233, 278, 279, 302, 303, 344, 345, 401, 463
Project plan, 23, 59, 71, 77, 116, 123, 190, 195, 229, 233, 278, 281, 460, 463
Project scorecard, 14, 23, 57, 58, 60, 61, 99, 104, 144, 145, 205, 221, 289, 293, 370
Pull work based on patient demand, 208, 212, 219

Q
Quality and lean tools, 463
Quality Function Deployment (QFD), 30, 35, 58, 352, 463, 466, 468
Quality management, 1, 2, 11, 12, 283, 453, 459, 464, 466, 468
Quality management system, 466, 468

R
Rapid process improvement, 461, 467
Receive linen and return soiled linen, 176
Recommendations for improvement, 58
Refractory septic shock, 277
Renal colic, 245
Replenish linen, 167, 219
Replication opportunities, 62, 65, 110, 111, 152, 158, 222, 225, 250, 251, 294, 299, 383
Root cause analysis, 205, 270, 439, 467
Root causes, 14, 20, 23, 36, 38, 41–44, 49, 57, 72, 84, 92, 93, 96, 98–101, 129, 137, 138, 144, 145, 161, 170, 173, 201, 203, 205, 207–209, 220, 224, 227, 241, 252, 264, 265, 287, 301, 319–321, 325–327, 357, 370, 371, 373, 375, 377, 379, 381, 393, 412, 460

S
Scorecard, 14, 23, 57, 58, 60–63, 66, 99, 104, 110, 111, 144, 145, 152, 158, 160, 179, 205, 220–222, 224, 250, 289, 293, 294, 299, 339, 340, 370, 382–384, 432, 435, 457, 463
Scorecard data, 62, 63, 110, 111, 152, 158, 222, 224, 250, 294, 299, 383, 384
Sepsis, 277–290, 293–295, 297, 299
Septic shock, 277–280, 282, 289, 290
Shadow boards, 175, 220
Shine, 164, 167, 169, 170, 173, 220, 221
Signals for work, 208, 212, 219
Simulation, 380, 395–405, 407, 408, 410–412, 418, 419, 421
Six Sigma, 1–3, 5, 9–14, 20, 22, 31, 37, 39, 40, 44, 48, 54, 57, 61, 63, 69, 71, 102, 103, 105, 111, 115, 116, 147–150, 158, 161, 162, 164, 189, 222, 223, 225, 227, 228, 233, 251, 252, 256, 270, 271, 277, 278, 281, 299, 301, 302, 339, 340, 344, 389, 417, 429–431, 433, 440, 446, 448, 449, 451, 453–469
Sort, 21, 163, 165, 167, 169, 170, 219
Stakeholder analysis, 7, 14, 15, 17–20, 23, 71, 73, 74, 110, 116, 117, 119–122, 167, 168, 190, 192, 193, 229, 231, 232, 278–280, 302–306, 344, 346–348, 463
Standardize, 49, 65, 147, 164, 167, 169, 170, 173, 175, 208, 212, 217, 220, 430
Statistical analysis, 22, 30, 85, 376, 379, 461, 463
Statistical Process Control (SPC), 2, 12, 62, 256, 302, 432, 463, 464
Statistical Process Control (SPC) charts, 64, 269–271, 432
Statistics, 33, 39, 42, 52, 62, 63, 79, 85, 127, 153–155, 189, 236, 256, 284, 337, 382, 395, 397, 400, 407, 408, 424, 451, 463
Straighten, 48, 164, 167, 169, 170, 173, 219
Straighten (set in order), 164
Strategic Business Process Architecture (SBPA), 4–10
Strategic business process architecture modeling, 5, 7, 8
Strengths, Weaknesses, Opportunities, and Threats (SWOT), 6
Suppliers-Input-Process-Output-Customer (SIPOC), 7, 15, 17, 27, 28, 73, 76, 117, 118, 122, 123, 167, 169, 192, 195, 231, 232, 279, 280, 290, 302, 304, 308, 348, 349, 403, 463
Supply chain, 115, 465–468
Supply chain management, 115, 466
Sustain, 48, 49, 164, 167, 169, 170, 176, 177, 220
Systems engineering, 1, 15, 71, 256, 270, 344, 384, 467

T
Takt time, 91, 202, 434
Takt time analysis, 91

Takt-time-driven one-piece-flow product cells, 13
Teamwork, 455, 459, 460, 466, 468
Theory of constraints, 71, 115
Theory of Inventive Problem-Solving (TRIZ), 466, 468
Total Quality Management (TQM), 2, 12, 459, 464
Toyota Production System (TPS), 11, 12
Train, 14, 23, 57, 58, 61, 99, 107, 108, 144, 146, 152, 205, 221, 225, 242, 250, 289, 293, 340, 424, 440, 461, 462
Translate VOC into technical requirements, 35

U
Unavoidable delays, 386, 387, 389, 390, 392, 393

V
Validate causes, 287
Validate measurement system, 23, 77, 83, 124, 136, 196, 200, 281, 286
Validate phase, 409
Value analysis, 42, 302, 322–324, 423, 424, 426, 440, 441, 447, 450, 463
Value and strategic alignment, 454, 455
Value chain, 3, 70, 71, 77, 91, 102, 103, 111, 113, 115, 116, 211, 350, 352, 354–357, 367, 368, 370–375, 377, 380–384, 417, 455
Value chain integration, 370, 372
Value chain process metrics, 370, 382
Value proposition, 454, 456, 457
Value stream map, 27, 29, 58, 196, 201, 302, 307, 309, 350, 352, 403, 423, 424, 426, 427, 433, 440, 463
Value Stream Mapping (VSM), 350
Visual controls, 169, 197, 208, 210, 215
Visual management, 170, 175, 220, 445, 446, 449, 451
Voice of Customer (VOC), 23, 71, 73, 77, 78, 124, 130, 190, 194, 196, 198, 233, 237, 256, 278, 281, 285, 315, 344, 349, 351, 454
Voice of Customer (VOC) Survey, 256, 263, 270, 315
Voice of Process (VOP), 23, 77, 78, 124, 130, 196, 198, 233, 238, 256, 281, 286, 308, 350, 355, 396
Voice of the customer, 15, 27, 29, 30, 35, 41, 130, 198, 237, 256, 257, 302, 344, 355, 457, 468
VOP matrix, 27, 36, 58, 61, 62, 463
VQ scan, 244

W
Waste analysis, 42, 47, 58, 440, 446, 463
Wells clinical model, 243, 244
Why–Why analysis, 138, 203
Why–Why diagram, 43–46, 58, 92, 96, 97, 463
Work layout, 206, 208, 212, 219

MIX
Papier aus verantwortungsvollen Quellen
Paper from responsible sources
FSC® C105338

If you have any concerns about our products,
you can contact us on
ProductSafety@springernature.com

In case Publisher is established outside the EU,
the EU authorized representative is:
Springer Nature Customer Service Center GmbH
Europaplatz 3, 69115 Heidelberg, Germany

Printed by Libri Plureos GmbH
in Hamburg, Germany